P9-DCX-569

DISCARDED
FROM
UNH LIBRARY

Risk Management of Chemicals

Risk Management of Chemicals

Edited by

Mervyn L. Richardson, BSc., C. Biol., M.I. Biol., C. Chem., F.R.S.C. Principal, Birch Assessment Services for Information on Chemicals (BASIC)

The Proceedings of an International Conference on 'Risk Management of Chemicals – Can Chemicals Be Used Safely?', organized by The Royal Society of Chemistry and held at the University of Surrey, Guildford, UK, 13–16 July 1992

FECS
Event 159

ISBN 0-85186-467-8

A catalogue record for this book is available from the British Library

Published by The Royal Society of Chemistry,
Thomas Graham House, Science Park, Cambridge
CB4 4WF

Printed in Great Britain by Redwood Press Ltd., Melksham, Wiltshire

BUCKINGHAM PALACE.

The publication of this review of Risk Management of Chemicals by The Royal Society of Chemistry in its 151st year is a valuable contribution to the understanding of this complex issue.

The hazards associated with chemicals have been well understood for a great many years, but the effective management of the risks in the use of chemicals is a more recent development. There is no dispute about the value of chemicals in health care, agriculture, food and many other industries, but experience has shown that these advantages are seldom gained without penalties. Techniques that seemed sensible and safe at the time have all too frequently turned out to have unexpected and long term hazards. The assessment and management of the risks are therefore becoming increasingly important in the larger process of managing the global environment.

Philip

1991

Editor's Preface

The aim of 'Risk Management of Chemicals' is to highlight the essential role of chemistry and related scientific disciplines in the multidisciplinary approach to risk management as applied to chemicals and the underlying chemistry.

The use of chemicals and chemical processes is an essential element in the promotion of human well-being. Chemicals are used extensively by all societies, irrespective of their stage of development. The benefits of chemicals are inestimable. However, chemicals can result in adverse effects on human health, and can have harmful consequences for the environment.

This book is a sequel to The Royal Society of Chemistry's publications 'Toxic Hazard Assessment of Chemicals' and 'Risk Assessment of Chemicals in the Environment', and the specialized area covered in 'Chemistry, Agriculture and the Environment'.

This book, which is also the proceedings of the Conference entitled *Risk Management of Chemicals – Can Chemicals be Used Safely*?, held at the University of Surrey, UK, 13–16 July 1992, covers five major areas: setting the scene, introduction to the management of risk, managing risk in manufacture, risk management from waste, and managing risk during chemical use.

The book reviews the current status of the management of the risks entailed in the synthesis, handling, use, and particularly disposal of the chemicals on which we all depend.

In today's society, note has also to be taken of risk perception and societal risk. These aspects involve moral, economic, and political judgements, and the total harm which may be suffered by a (human) population, now and in the future. Societal risk can be measured by the probability of a large accident or incident causing a defined number of deaths or injuries. Additionally, consideration has to be given to local (or distant) disruption to plant and services, such as electricity or water, or to waste emissions to the environment on a global basis.

For disposal of waste, risk management is of growing concern as the air/land/water environments, in addition to workplace environment, have to be protected. Any gross mismanagement could lead to a regression in our way of life.

The Editorial Board has attempted to minimize overlap between chapters. However, in dealing with such important topics, some overlap is inevitable. Such repetition should enhance the value of the contents of the book in view of the various and diverse experiences expressed by the authors, from European countries, India, China, and the United States of America.

As Editor, and Chairman of The Royal Society of Chemistry's Toxicology Subject Group, I am particularly fortunate in being able to draw on the advice and experience of members of this committee.

In particular, I have been well supported by members of my Editorial Board, J. Deschamps, J.H. Duffus, H.P.A. Illing, M. Mercier, P. Peterson, B. Samimi-Eidenbenz, G. Skholenok, and D. Taylor, and my friends and colleagues who have acted as referees and include: R. Atkins, J. Deschamps, J.H. Duffus, R. Fielder, S. Hubbard, H.P.A. Illing, G.V.McHattie, P. Peterson, H.I. Shalgosky, D. Stevenson, and D. Taylor.

The production of this book was made possible by the unstinting secretarial support of Pauline A. Sim of Gascoigne Secretarial Services, who acted as secretary to the Editorial Board. Liaison with board members in the UK and Switzerland, and authors from nine countries, was no mean task. Pauline undertook the retyping of all the manuscripts in camera-ready format under almost impossible time constraints, bearing in mind that the time was restricted because of the necessity for having copies available in time for the Conference on which this book is based.

Furthermore, I am indebted to members of The Royal Society of Chemistry's Books and Reviews Department, for their continuing support: R.H. Andrews, P.G. Gardam, and C.N. Lyall, and in particular A.G. Cubitt, for his unfailing assistance and attention to detail in the desk editing. Finally, my most sincere gratitude to my long-suffering wife Beryl, who so patiently accepted my working on this volume, sometimes seven days a week, and for tolerating the mountains of paper and post that editing such books produces.

MERVYN L RICHARDSON, EDITOR
PRINCIPAL, BIRCH ASSESSMENT SERVICES FOR INFORMATION ON CHEMICALS (B.A.S.I.C.)

Contents

Risk Management from Waste

Managing Risk During Chemical Use

A Preview

Mervyn L. Richardson

PRINCIPAL, BIRCH ASSESSMENT SERVICES FOR INFORMATION ON CHEMICALS (BASIC), 6 BIRCH DRIVE, MAPLE CROSS, RICKMANSWORTH, HERTS. WD3 2UL, UK

1 PRELIMINARY CONCEPT

It is vital prior to undertaking the management of any risk associated with a chemical substance or indeed any physical or biological agent that procedures for retrieving sufficient and suitable information are undertaken. This requires a clear understanding of the terminology involved.

Effective risk management involves the following stages:

Information retrieval (either from the public domain or new information from commissioned studies), validation, and interpretation;
Hazard assessment;
Exposure assessment;
Risk assessment; in turn leading to
Management of relevant activities;
Good risk management will result in the safe use of chemicals.

2 INFORMATION RETRIEVAL, VALIDATION, AND INTERPRETATION

Techniques for information retrieval vary from manual techniques to those involving computers. Anyone attempting any type of assessment first needs to obtain relevant, suitable, sufficient, and detailed information on the substance of concern, together with the details of the species, micro-organisms, plants, fish, birds, animals, or man, which may be exposed to that substance and through which exposure may occur, and the environments, *e.g.* sewage treatment works, water bodies, soil, air, the workplace, or domestic environments.

2.1 Manual Data Retrieval

The basic procedures are indicated by Sanderson[1] in The Royal Society of Chemistry's sister publication 'Toxic Hazard Assessment of Chemicals'. Manual sources will include a selection of handbooks, textbooks, containing for example physicochemical data, toxicological and environmental data, company material safety data sheets, government reports, monographs, and booklets produced by companies, particularly those produced by the International Programmes, *e.g.* The International Agency for Research on Cancer, the Environmental Health Criteria produced by the World Health Organization's International Programme on Chemical Safety (a joint WHO/UNEP/ILO programme), and various publications by the United Nations Environmental Programme and the Commission of the European Communities.

In addition there are of course many data in a wide variety of journals and abstracts, of the latter, perhaps the most important are Chemical Abstracts and Toxicology Abstracts.

Particular stress needs to be given to the older literature and whilst such data can in certain circumstances be flawed by 1990's standards, particularly in the case of ecotoxicology, they may be the only available data. Such data in many cases can only be obtained by manual searches by following restrospectively the references given in either standard text books or journals and again looking up the references quoted in such sources and even the references in such reference articles. Few, if any, such references can be retrieved by computer techniques.

2.2 Computer Sources

For the past 20 or so years, it has been possible to search a number of large remote computer bibliographic databanks and databases. It is important to understand that a databank contains preselected factual information in summary form, with a sophisticated search system to enable the relevant information to be located, whereas a database will provide references and in some cases abstracts of papers in the more recent literature.

The technique and some of the most relevant facilities are again outlined by Sanderson.[2] Possibly the best known of such facilities is Chemical Abstracts On Line, with both the abstracts themselves, and very sophisticated means of searching by structure or indeed by structure fragments to enable the toxicological and ecotoxicological data to be found in the on-line facilities accessable via the National Library of Medicine.

The 1990s have seen a more readily accessible form of these data in the form of CD-ROMs where the data can be accessed from a personal computer located on one's desk instead of having to reach the remote computer by telephone links. In the field of toxicology/ecotoxicology one such example is the pesticide data available on CD-ROM from data published by The Royal Society of Chemistry;[3] another is the Silver Platter CD-ROM NIOSHTIC which contains HSE line.

It has also been emphasized that there are suspect data in the literature. When there are gaps in information, it may be necessary to close such gaps using appropriate studies. This can be effected by commissioning studies. Test procedures relevant to human health may need to be conducted in animals if it would be unethical or inappropriate to carry out studies in man. When tests are commissioned it is sensible to conduct them to internationally accepted protocols (*e.g.* the OECD) and under Good Laboratory Practice conditions as this aids international acceptability. In addition to literature data, obviously data generated in house or by a contract laboratory have to be taken into consideration and the stages outlined below need to be no less rigorously applied to these data as well as literature derived data. In retrieving experimental observations, the test reports now invariably undertaken to the requirements of Good Laboratory Practice need to be thoroughly assessed.

2.3 Validation

Having retrieved the information, the next stage is to ascertain whether it is good, sound, and authentic. Are the data of any value? There are substantial differences between pipeline seals and the animals found in the sea. Validation also is necessary to accommodate misprints; at some time or other, have we all not encountered mg being printed instead of ng resulting in a difference of 1 million? Secondly, there is advancing knowledge; a compound, *e.g.* DDT, considered safe in 1970 may not be considered to be safe in 1990.[4] Lastly, there is the case of toxicological studies producing incorrect data. Further examples of these are described by Taylor.[5]

2.4 Interpretation of Data

Here, the problems can be more difficult and diverse, and will depend largely on whether one is to interpret data for toxicological or for ecotoxicological purposes. There are many excellent texts which deal with these problems. However, to quote the obvious, data retrieval for an ecotoxicological purpose, *e.g.* bird eggshell thinning, will be of no value whatsoever if the problem is one of liver carcinogenicity.

There are many uncertainties inherent in laboratory methods and in order to make extrapolations there is a need to proceed with great caution.

Thus, having assured oneself that the retrieved data are sound and reliable or as sound and reliable as one can ascertain within one's expert judgement, the next stage is to consider hazard assessment.

3 HAZARD

Hazard can be defined as the set of inherent properties of a chemical, mixture of chemicals, or a process involving chemicals which, under production, usage, or disposal conditions, make it capable of causing adverse effects to organisms or the environment, depending on the particular degree of exposure; in other words, it is a source of danger.

Briefly, hazards can be considered in principle as either individual hazards or combined hazards.

3.1 Individual Hazards

Individual hazards require consideration of specific information or specific substances usually related to their toxicology, exposure, and effects. The most usual outcome for a substance shown to be of significant hazard would be either to replace it with one clearly shown to be less hazardous or to use it in such a manner that any risk (see below) is minimized, *e.g.* by use in a containment area, or by use of appropriate protective clothing. In every case, it is important to stress that adequate training must be provided.

3.2 Combined Hazards

Combined hazards usually require consideration of far more complex principles as the hazards arise from a number of sources and may include many different substances, some or all of which may be unknown, each with perhaps different effects and requiring different methods of measurement and assessment. Typical examples of combined hazards would be highway run-off, agricultural run-off, or percolation into groundwater, or urban air pollution from motor vehicles. In the case of combined hazards by-standers can also be exposed.

3.3 Hazard Identification (Prediction)

This is the identification of the environmental agent, either in the workplace, or external environment of concern, its adverse effects, target populations, and the conditions of exposure.

4 RISK

It is important that risk is not confused with hazard as is often the case. Perhaps the most important consideration is that risk should always contain an element of quantification.

In non-technical terms, risk means that there is a probability of a generally unfavourable outcome. In technical terms, risk is the:

i) Possibility that a harmful event (death, injury, or loss) arising from exposure to a biological, chemical or physical agent may occur under specific conditions;

ii) Expected frequency of occurrence of a harmful event (death, injury, or loss) arising from exposure to a biological, chemical, or physical agent under specific conditions. Hence, risk can be considered as $\sum$ hazard x exposure.

4.1 Risk Assessment

This is the combination of four aspects: hazard identification (see above), risk characterization, exposure assessment, *i.e.* measurement, monitoring, and risk estimation. (See chapters by P. Koundajkian and H.P.A. Illing, and B. Broecker.)

It is the identification and quantification of the risk resulting from a specific use or occurrence (or disposal) of a chemical taking into account possible harmful effects on individual people or society of using the chemical in the amount and manner proposed and all the possible routes of exposure. Quantification ideally requires the establishment of dose–effect and dose–response relationships in likely target individuals and populations.

It is important to remember that a risk assessment really only commences to be examined when exposure is combined with hazard. It is also necessary to consider potential interactions of substances to produce toxic materials, either spontaneously in the environment, or as a consequence of metabolic transformation. These toxicities may be observed as ecological changes or as effects on human health.

This then leads to the subject of this book – Risk Management.

4.2 Risk Management

This is the management, decision–making, and active hazard control process involving consideration of political, social, economic, and engineering factors with relevant risk assessments relating to a potential hazard so as to develop, analyse, and compare regulatory options and to select the optimal regulatory response for safety from that hazard. It is a combination of risk evaluation, emission and exposure control, and risk monitoring. It is paramount that process control is emphasized. These procedures are shown in Figure 1.[6] (Reproduced with permission of the World Health Organization, Geneva.)

The question of risk perception also has to be addressed.

4.3 Risk Perception

This is the subjective perception of the gravity or importance of the risk based on the person's knowledge (and thus training) of different risks and the moral, economic, and political judgement of their implications.

In understanding these procedures it is vital to remember that benefit, cost, and risk are always linked and if too much expenditure is given to the minimization of risk, unnecessarily, the benefit may disappear. In this context it is important to consider risk aversion.

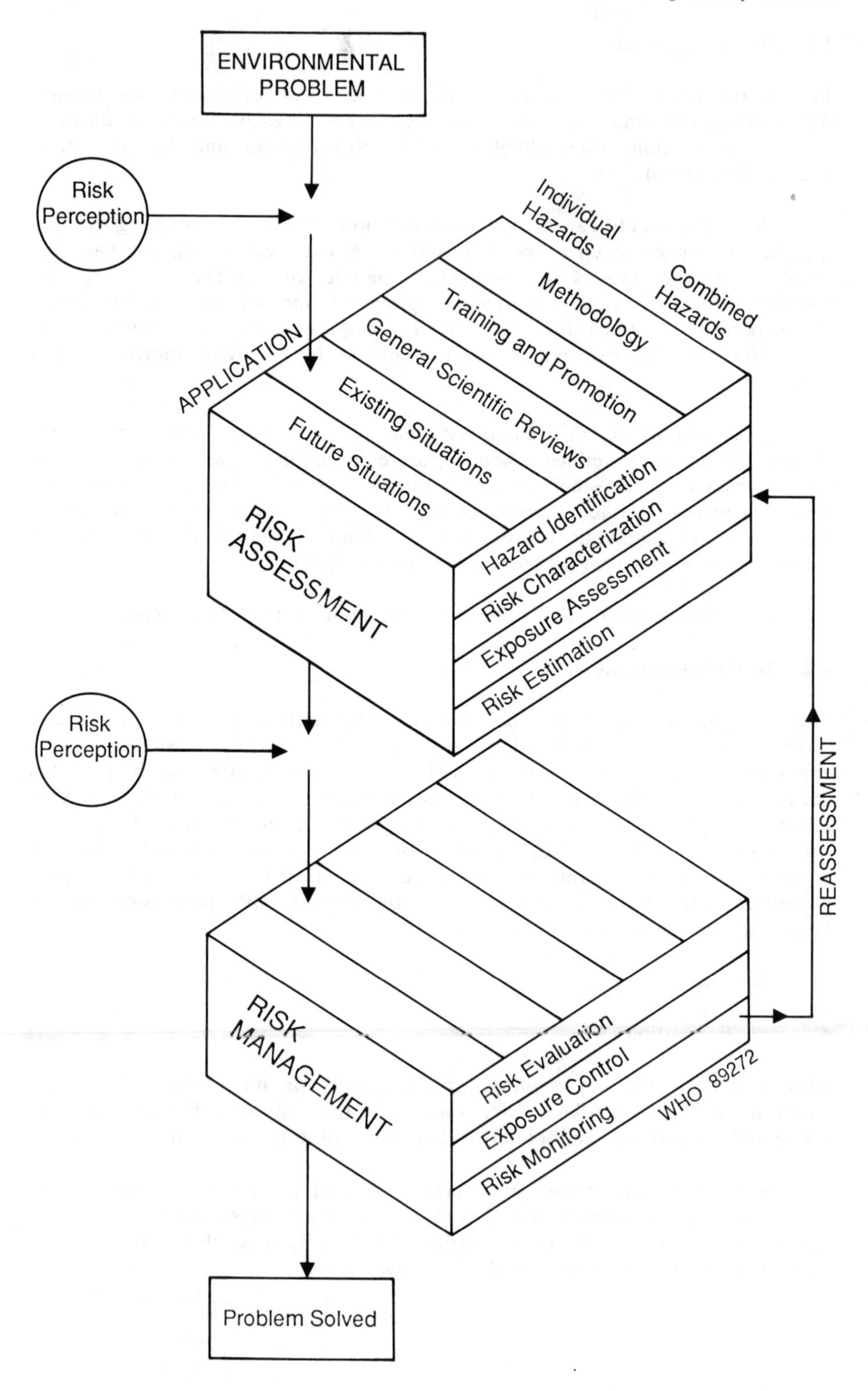
ENVIRONMENTAL PROBLEM
Risk Perception
Individual Hazards
Combined Hazards
Methodology
Training and Promotion
General Scientific Reviews
Existing Situations
Future Situations
APPLICATION
RISK ASSESSMENT
Hazard Identification
Risk Characterization
Exposure Assessment
Risk Estimation
Risk Perception
REASSESSMENT
RISK MANAGEMENT
Risk Evaluation
Exposure Control
Risk Monitoring
WHO 89272
Problem Solved

4.4 Risk Aversion

Risk aversion can be considered to describe the tendency of an individual person to avoid risk. In general terms, it relates to one's acceptance of paying a premium above the expected value to avoid a gamble, *e.g.* stocks and shares offer a higher return than a conventional bank account.[7] However, even rational persons may make decisions or attempt to pressurize governments into making choices based on preferences by exaggerating the risks. This can result in action being taken, which results in increased hazard, for example because a more hazardous substance or procedure is substituted for a safer one as a result of an incorrectly assigned or assessed risk. Such aversions must hence be considered within the true knowledge base of the situation.

In the case of large incidents societal risk also needs to be considered.

4.5 Societal Risk

Societal risk is the total probability of harm to a human population and effect on the future of whole communities. In contrast to individual risk, it takes into account the likely size of the population that will be affected. For example, the individual risk of being affected 1 km from a major hazard, should an accident occur, is the same whether the site is in the middle of a city or in remote countryside, but the societal risk is considerably greater in the city because of the numbers potentially affected. Societal risk includes the probability of adverse health effects to descendants and the probability of disruption resulting from loss of material goods, electricity, water supplies, and sewage disposal, *etc.* With the now rapid transmission of news, fear can be generated locally, nationally, and internationally. The scientific community and the professional societies have a great role to play in coping with this but they must be in a position to react both rapidly and objectively. This leads to safety.

5 SAFETY

Safety can be defined as the practical certainty that injury will not result from a hazard under defined conditions, or the high probability that injury will not result from the use of a substance under defined conditions of quantity and manner of use. It is also proportional to the reciprocal of risk. The safety of a chemical in the context of human health is the extent to which a chemical may be used in the amount necessary for the intended purpose with a minimum risk of adverse health effects.

The techniques described above and in the chapters in this book should ensure that chemical substances can be handled safely.

It is of paramount importance that any conclusions relating to risk and safety are reached by the use of sound data and good science and without confrontation.

It has to be stressed that risks can rarely be eliminated but only minimized, controlled, and managed within defined and acceptable limits. There must be a beneficial outcome from any corrective actions. Chemical substances are vital to our well-being, our lives, our wealth, and not least to our health, and those of other organisms on which mankind depends.

As stated in the International Register of Potentially Toxic Chemicals (IRPTC) Bulletin in 1985[8] 'Chemical safety is a serious matter because as long as risk management operations are inadequate, chemicals in the environment will continue to cause nightmares to many people, and even more than that, they will continue to cause damage, fatalities, and even major disasters'. It is the Editor's sincere wish that the contents of this book will go a long way to redress this position. Also it is hoped that the reader will be more informed and thoughtful of all of the environmental problems that beset mankind.

6 OTHER DEFINITIONS

The problem of defining all the terms involved in risk management is considerable.

Seven further widely used terms are defined below, but it should be remembered that few of such terms, including those indicated, and as exemplified by Koundajkian and Illing and also by Broecker, can be defined in an absolute sense.

The reader will also find alternative definitions in many of the works given in the Reading List.

A number of the terms quoted are currently under review by The International Union of Pure and Applied Chemistry (IUPAC).

i) Acceptable risk is the probability of suffering disease or injury that an individual, group, or society agrees to permit.

ii) Dose is the amount of a substance administered to, or absorbed by an organism. 'Uptake' is preferred to the usual term 'dose' because the precise dose administered is very difficult to measure and because uptake indicates effective exposure.

iii) Dose-effect relationship is the association between dose and the magnitude of a continuously graded effect, either in an individual or a population.

iv) Dose–response relationship is the association between dose and the incidence of a defined biological effect in an exposed population.

v) Individual risk is the probability that an individual person will experience an adverse effect.

vi) Tolerable risk is the probability of suffering disease or injury which can be endured by an individual, group, or society, but which may not be entirely acceptable in ideal circumstances.

vii) Voluntary and involuntary risk can be illustrated by an example. An individual may willingly (voluntary risk) take part in a relatively dangerous activity such as driving a car or motor cycle racing, but the same individual may be unwilling to accept much lower risks from hazards that they are unable to avoid (involuntary risk), such as exposure to pesticides in food.

It should be stressed that there is a difference between *acceptable* and *accepted* risk, *e.g.* smoking can be widely *accepted* but regarded increasingly by the majority to be an *unacceptable* risk.

A person may accept a level of risk based on his past experience and knowledge. Such acceptance is in many cases associated with an inadequate knowledge (*e.g.* ignorance) or even more, a total inability to wish to believe that any significant risk exists.

In contrast, in a scientific or rational manner, the establishment of the acceptance of a risk will depend on the best available estimate of the probability (likelihood) of an incident or activity, and comparing this to the best of one's judgement against its acceptability using suitable data or criteria.

7 REFERENCES

1 D.M. Sanderson, 'Methods of data retrieval – manual', in 'Toxic Hazard Assessment of Chemicals', ed. M.L. Richardson, The Royal Society of Chemistry, London, 1986, pp. 24–28.

2 D.M. Sanderson, 'Methods of data retrieval – computer', in 'Toxic Hazard Assessment of Chemicals', ed. M.L. Richardson, The Royal Society of Chemistry, London, 1986, pp. 15–22.

3 H. Kidd, 'Information sources for chemistry, agriculture and the environment', in 'Chemistry, Agriculture and the Environment', ed. M.L. Richardson, The Royal Society of Chemistry, London, 1991, pp. 81–90.

4 C.R. Krishna Murti (deceased) and D. Nag, 'Human health impact of pesticides in the environment', in 'Chemistry, Agriculture and the Environment', ed. M.L. Richardson, The Royal Society of Chemistry, Cambridge, 1991, pp. 491-510.

5 D.T. Taylor, 'Separating the wheat from the chaff - the selection of appropriate toxicological data from the world literature', in 'Toxic Hazard Assessment of Chemicals', ed. M.L. Richardson, The Royal Society of Chemistry, London, 1986, pp. 51-63.

6 Control of Environmental Hazards - Assessment and management of Environmental Health Hazards, WHO/PEP/89.6, World Health Organization, Geneva.

7 A.L. Nichols and R.J. Zeckhauser, *Reg. Tox. Pharmacol.*, 1988, **8**, 61.

8 Editorial International Register of Potentially Toxic Chemicals Bulletin, 1985, 7 (2), 1.

Editorial Board

The Universe requires an eternity ...
Thus they say that the conservation of this world is a perpetual creation and that the verbs, 'conserve' and 'create', so much at odds here, are synonymous in heaven'.

Historia de la Eternidad
Jorge Luis Borges

1991

Acknowledgements

The Editor, on behalf of The Royal Society of Chemistry, wishes to thank n.v. Procter and Gamble, European Technical Center SA, Strombeek-Bever, Belgium for a donation, so making copies of this work available to the less-developed nations at the United Nations Conference held in Rio de Janeiro, Brazil, in June 1992.

Setting the Scene

1
Introduction

P. P. Koundakjian and H. P. A. Illing

HEALTH AND SAFETY EXECUTIVE, MAGDALEN HOUSE, STANLEY PRECINCT, BOOTLE, MERSEYSIDE L20 3QZ, UK

1 BACKGROUND

Chemicals are an integral part of modern life, and many possess a potential for harm. A harmful effect may result from a direct physical property of the chemical (explosivity, corrosivity) or a toxic property, a deleterious consequence of the interaction of the chemical and a biological system. Toxic effects may be mediated at various levels; they can be on a single organism, a population, an ecosystem, or the biosphere. They may be effects on man, such as death, ill-health, or loss of amenity. They can be on other species or cause detrimental changes to the fauna and flora of a given habitat. They can also be mediated indirectly, *i.e.* via consequences of chemically induced changes to the atmosphere or to weather patterns. For these effects to take place, exposure to the substance is needed.

The risks associated with a potential for harm due to exposure to chemicals need to be identified, assessed, and managed appropriately. This book is intended to address the whole process of risk management associated with the production, use, and disposal of chemicals – a 'cradle to grave' examination of this complex subject.

2 DEFINING THE PROCESS OF RISK MANAGEMENT

Risk management was defined by a Royal Society Study Group[1] as 'the making of decisions concerning risk and their subsequent implementation'. The WHO[2] called risk management 'the managerial, decision making and active hazard control process to deal with those environmental agents for which the risk evaluation has indicated that the risk is too high!' Before attempting to introduce or change risk management procedures it is first necessary to evaluate the risks, based on the available evidence. This process of risk assessment should precede decisions implementing activities aimed at managing risks, and should be repeated iteratively in order to optimize the management process. The next section examines the concept of risk assessment process.

3 RISK ASSESSMENT

Risk assessment is concerned with identifying and characterizing the risks arising from the use of a chemical. IUPAC[3] defined risk assessment as a decision making process that entails consideration of political, social, economic, and engineering information with risk-related information so as to develop, analyse, and compare regulatory options and to select the appropriate regulatory response to a potential health hazard. The Royal Society Study Group Report[1] divides this process into 'risk estimation' and 'risk evaluation'. The Report was concerned with engineering risk, the potential for plant failures, and the consequences which arise therefrom, as well as chemical risk; thus its definitions are very wide ranging. It identified 'risk estimation' as:

i) The identification of the outcomes;

ii) The estimation of the magnitude of the associated consequences of these outcomes;

iii) The estimation of the probabilities of these outcomes.

In terms of the biological effects of chemicals, stage (i) involves a knowledge of the dose-effect or dose-response relationships, and stage (ii) likely exposure. The different definitions evolved by other international bodies for the various stages of biological risk estimation are given in Table 1. There is, as yet, no consistency in the names used for the different stages. In the EC the term risk assessment is used only in circumstances where a definite probability is obtained;[4] others include semi-quantitative or qualitative assessments. The latter are, in EC terms, 'hazard assessments'.

According to the Royal Society Study Groups, risk evaluation is 'the complex process of **determining the significance or value** of the identified hazards and estimated risks to those concerned with or affected by the decision.[1] Sociological and political considerations are involved in this stage, as well as scientific principles. Other bodies omit overt mention of this risk evaluation stage or (as with WHO[2]) consider it the first phase of risk management.

Here, we will divide the process of risk estimation into hazard identification and assessment (including dose-response determination) and exposure assessment and risk characterization. We will include as a distinct entity a discussion on the risk evaluation stage. Two previous books in this series have covered hazard assessment and risk assessment in detail;[6,7] thus only a general overview is given here.

Table 1 *Stages in risk estimation*

EC[3]	*National Research Council (USA)*[5]	*World Health Organization*[2]
Hazard Identification The process of associating a hazard with a particular chemical	**Hazard Identification** Determining whether an agent can cause an increase in the incidence of an effect, *e.g.* a health condition	**Hazard Identification** Identifying the environmental agent of concern, its adverse effects, target populations, and conditions of exposure
	Dose–response Assessment Characterizing the relationship between dose of an agent administered or received and the incidence of an adverse effect	**Risk Characterization** Describing the different potential health effects of the hazard and quantify-ing the dose–effect and dose–response relationships in a general scientific sense
Hazard Assessment* The process whereby the potential for a chemical to cause adverse effects in target species or systems is assessed. The effects of the substances are related to an assessment of (possible) exposure	**Exposure Assessment** Measuring or esti-mating the intensity, frequency, and duration of exposure to an agent	**Exposure Assessment** Quantifying exposure (dose) in a specific population based on measuring emissions, environmental levels, biological monitoring, *etc*.
	Risk Characterization Estimating the incidence of an effect under the conditions of exposure described in the exposure assessment	**Risk Estimation** The process of combining the risk characterization, dose–response relationships, and exposure estimates to quantify the risks in a specific population

*The term Risk Assessment is only employed by the EC when the assessment is carried out in quantitative terms – *i.e.* when the probability that a substance causes adverse effects as a result of its presence in the environment (at a given concentration) is expressed numerically.

3.1 Hazard Identification

Hazard identification is the process of associating a hazard with a particular chemical. This may involve any or all of experimental work, information retrieval, and deductive work based on analogy, and is followed by interpretation of the data. It is the first essential in identifying the chemicals of concern and, usually, the type of concern. The relationship between the dose or concentration of the chemical and the expression of toxicological and/or ecotoxicological effects in the species of concern is normally considered to be important at this stage. Details of principles and methods involved in assessing toxic effects of chemicals are given elsewhere.[6–17]

A dose–response curve is used in cases of stochastic effect, *i.e.* one for which the probability of occurrence, rather than the severity, depends on the absorbed dose.[10] A dose–effect curve is relevant to a non-stochastic process, *i.e.* a process where the severity varies with the exposure level.[10] Genotoxic carcinogenesis is an example of a stochastic effect; most toxic effects are considered non–stochastic. If a threshold exists, as is likely with a dose–effect relationship and may occur with a dose–response relationship, then a 'no–adverse effect concentration' can, in theory, be determined.

3.2 Exposure Assessment and Risk Characterization

The process described in this section requires that real or estimated exposure concentrations in the environmental compartments of concern are compared with the relevant 'no adverse effect concentrations' and, where possible, the dose–response curve (or dose–effect curve).

Exposure information is necessary as well as hazard information for this stage of the risk estimation. Different kinds of exposure can be categorized and exemplified in several ways. One possible scheme is shown in Figure 1. Exposure may be measured, or predicted, possibly using model systems. Sometimes, exposure levels can be estimated by examining use patterns (*e.g.* industrial use categories, food basket information, and/or disposal methods for the material). When examining accidental exposures, the size and type of potential exposure may be estimated using failure rates for various parts of the system, coupled with predicted release sizes and dispersal patterns.

4 RISK EVALUATION: JUDGING THE ACCEPTABILITY OF THE RISK

The combination of the potential for harm (the hazard) and the likelihood (probability) of it happening (the risk) constitutes the information required of the risk estimation. Because of the nature of information on the biological effects of chemicals, this likelihood is often only known in

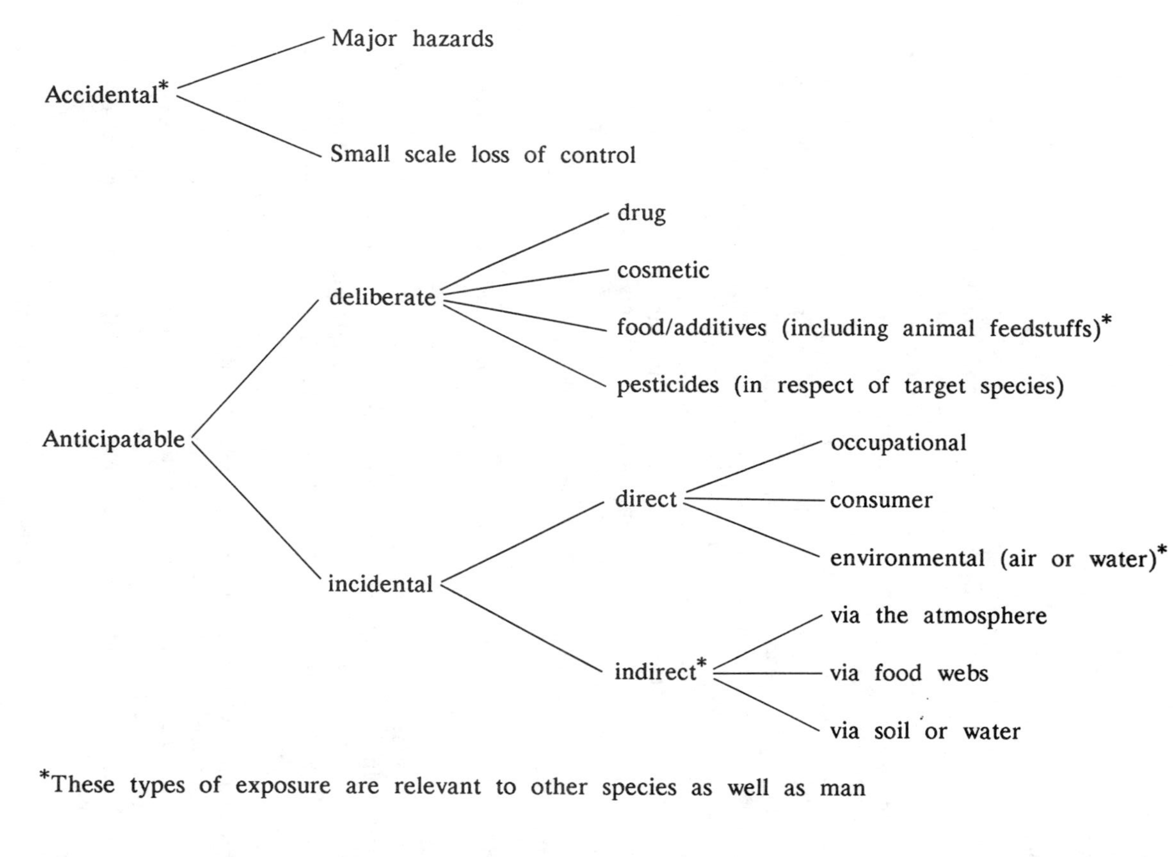

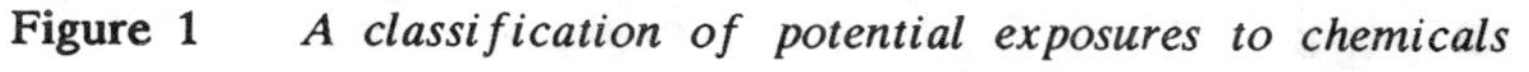

Figure 1 *A classification of potential exposures to chemicals*

qualitative or semi-quantitative terms. Once the levels of risk associated with the uses of the chemical have been determined or estimated, those levels have to be judged against criteria to determine whether further management activity will be needed. At this point one is introducing societal as well as scientific judgements into the risk management processes, and judgements on risk and benefit. The Royal Society Study Group[1] considered this risk evaluation as the last stage of the process of risk assessment; the WHO[2] considered it as the first stage of the risk management process.

4.1 Criteria for Risk Evaluation

Concepts of 'acceptable' (*i.e.* 'tolerable' or 'negligible') and 'unacceptable' risk become important here. The concept of 'acceptable' risk was present in the Royal Society Study Group Report,[1] and the ideas of 'tolerability' of risk are presented, at least in the context of nuclear installations, in a paper published by the Health and Safety Executive.[18] Similar ideas are used in the field of major hazards assessment generally.[19] The meanings of these statements are given in the Preview. The concepts of negligible, tolerable, and unacceptable risk to risk assessment in general have been utilized in legislative approaches to limit setting for far longer.[1,20] These concepts can be summarized in Figure 2. Tolerability of a risk is always associated with a risk-benefit analysis. The criteria against which to judge a risk can sometimes be set out in numerical form and risks compared numerically (quantitative risk assessment; see ref. 19). A judgement is then employed to determine 'negligible' or 'tolerable' risk. For many toxic risks the data are not amenable to closely defined numerical approaches. In these circumstances, qualitative or semi-quantitative approaches, based on professional judgement and 'safety/uncertainty factors' are employed.[15,16]

Three further complications affect the judgement of risk for effects on man, and hence the size of safety factors. The first is that of 'individual' and 'societal' risk (see Preview for definitions). Societal risk becomes important when the potential for major accidents is being considered. In most other circumstances the primary consideration is that of individual risk. As risks to non-human species are usually judged on the basis of populations, judgements concerning them are essentially based on societal risk.

The second is that of 'voluntary' and 'involuntary' risk. These terms relate to human health and are also explained in the Preview. Acceptance of the risk may vary with the 'voluntariness' of the exposure. For example, relatively high risks can be considered tolerable for recreational activities, but would be unacceptable if they were associated with drinking water.

The third complication is caused by the need for a societal judgement and is the difference between an 'acceptable' risk and an 'accepted' risk. Again, the terms are defined in the Preview. Essentially they revolve around public perception of risk and public confidence in scientific judgements and professional advice.

INTOLERABLE/UNACCEPTABLE RISKS

UPPER LIMIT OF RISK ————————————————

TOLERABLE RISK
(Only if risk reduction is impractical or cost grossly disproportionate)

MAY BE JUSTIFIABLE
(Further control required to reduce risk 'so far as is reasonably practicable')

TOLERABLE RISK
(If cost of risk reduction would exceed the improvement gained)

CUT-OFF POINT ————————————————
(NON-ACTION LEVEL)

BROADLY ACCEPTABLE RISK
(Detriment judged to be trivial/negligible)

Figure 2 *Risk and its acceptability (Based on references 1 and 18)*

4.2 Who Judges?

A societal judgement may be taken by Government after specific enquiry (as with, for example, public enquiries prior to authorization of construction or extension of major nuclear or chemical plants or of developments around them). Alternatively, the public, through Government and the legislative process, may delegate the societal judgements via ministers to specially authorized bodies which combine representation from the interested parties and experts. These bodies may, in turn, obtain professional advice from specific expert committees. One process fitting this model is that for setting occupational exposure limits in the UK (Table 2). The scientific part of the judgements is becoming more international, with bodies such as IPCS publishing scientific evaluations of the hazards, and occasionally, as with JECFA, judging 'acceptable' risk. Nevertheless, variations in exposure conditions and differences in the background risk (and consequent risk-benefit analysis), and the need for societal input into decision taking on the acceptance of risk mean that final decisions in some areas (*e.g.* pesticide usage) cannot be completely harmonized.

Table 2 *Summary of the process for setting occupational exposure limits in the UK*

Stage	*Public body*	*Process undertaken*
Societal judgement, taken after public consultation to examine acceptability (tolerability) of proposals	Health and Safety Commission (HSC) and and Ministers	HSC issues consultative document and considers public response before promulgating OES* or recommending MEL to Ministers
Examination of practicability – risk/benefit analysis for 'tolerable'	Advisory Committee on Toxic Substances (ACTS)	Performs risk/benefit analysis to determine 'tolerability' of risk and recommends a MEL and concomitant control requirements
Scientific evaluation of hazard (assumes no difficulty over practicability)	Working Advisory Group on toxic chemical hazards (WATCH)	Recommends 'acceptable' level of exposure for 'negligible' risk as an OES

*OES = Occupational exposure standard; MEL = Maximum exposure limit.

Society, through Government, may also adopt a separate approach, in leaving risk assessment to the individual or company. This is more likely where public concern is most easily reflected back via the market; sales rapidly become inadequate to support a product if it is considered to cause minor ill-health. *'Caveat emptor'* may be true here, or there may be legislation making the seller or provider responsible for the judgements on safety (the risk assessments). These are usually cases where 'acceptable' and 'accepted' risk are reasonably easily linked.

Judgements about the magnitude of risk tend to be scientific and professional; those concerned with acceptance tend to be Governmental, corporate, or individual decisions, and taken cognizance of public opinion.

Associated with these judgements are consideration of the management procedures necessary to handle any remaining risk.

5 RISK MANAGEMENT

Risk management is an active process. The risk recognized as being associated with a chemical is set against defined criteria and rendered acceptable or, at least, tolerable in relation to the benefits associated with that use of the chemical. Essentially one can conceive two sets of circumstances which require management, handling the consequences of loss of control (as with accidents), and controlling anticipatable exposure adequately.

Details of how to handle the former type of management, including how to handle major hazards, can be found elsewhere.[14,21,22] This book is largely concerned with the latter type of management.

Management of the risks associated with the use and disposal of chemicals flows from the risk evaluation. Management involves controlling exposure to appropriate levels and monitoring those levels to ensure that there is a satisfactory reduction in risk.[1,2] Procedures may be set up by Governments, international bodies, companies (including multinational companies), or individuals.

At the legislative level, exposure controls can range from a total ban on manufacture and use, through various intermediate options, such as restricting outlets or uses, setting exposure limits, or process controls, and managing waste disposal, to acceptance of unrestricted manufacture, use, and disposal. Governmental action may also be in the form of legislation to ensure that individuals can take informed decisions such as insisting on appropriate labelling of the chemical or product and ensuring that users understand the implications of that labelling. This can, in turn, lead users to devising acceptable risk management procedures. Such ideas as substitution of one process or chemical for another, use of appropriately designed equipment and ventilation, use of suitable protective equipment, and use of the most appropriate disposal systems come into the decisions.

In order to confirm that exposure has been kept within set levels, some form of monitoring system is required. This may be direct measurements of levels of chemical in the relevant matrix (water, air, food, *etc.*) or a measure based on ascribing a biological end point to a chemical (*e.g.* drug 'side-effect'-monitoring or monitoring of the effects of chemicals on plant/animal life in soil or water). Monitoring will reveal whether control measures are adequate or whether further management activity, leading to lower exposure and reduced risk, is needed.

In risk management, reaction to new information and changed opinion is essential. Thus there is a cycle of assessment, management, and reassessment. It should be obvious that risk management is an extremely complex activity. Potential risks which need management can occur at all stages in the life and use of a chemical (*e.g.* manufacture, transport, use, disposal). The impact of different national and international legislation affects these decisions. Political and societal pressures can also influence the decisions taken. The authors of the various chapters in this book will examine various facets of risk management of chemicals so that the question 'can chemicals be used safely?' may be answered.

6 CONCLUSIONS

In this chapter we have tried to outline general approaches to risk assessment as a preliminary to risk management. Much more detailed discussion of risk management and exemplication of the principles outlined here is contained in the following chapters.

7 DISCLAIMER

Opinions expressed are those of the authors, and should not be considered as statements of HSE policy.

8 REFERENCES

1 Royal Society Study Group, 'Risk Assessment', The Royal Society, London, 1983.

2 WHO, 'Control of Environmental Hazards', Duplicated document, World Health Organization, Geneva, 1989.

3 IUPAC 'Glossary of Terms Used in Toxicology', Prep. J.H. Duffus, Draft 4, 1991.

4 Commission of the European Community, Workshop on Environmental Hazard and Risk Assessment in the Context of Directive 79/831/EEC, Summary Document XI/730/89 rev.3, Commission of the European Community, Brussels, 1990.

5 National Research Council, 'Risk Assessment in the Federal Government: Managing the Process', National Academy Press, Washington, 1983.

6 'Risk Assessment of Chemicals in the Environment', ed. M.L. Richardson, The Royal Society of Chemistry, London, 1988.

7 'Toxic Hazard Assessment of Chemicals', ed. M.L. Richardson, The Royal Society of Chemistry, London, 1986.

8 WHO, 'Environmental Health Criteria 6. Principles and Methods for Evaluating the Toxicity of Chemicals. Part 1,' World Health Organization, Geneva, 1978.

9 WHO/IPCS/JECFA, 'Environmental Health Criteria 70. Principles for the Safety Assessment of Food Additives and Contaminants in Food', World Health Organization, Geneva, 1987.

10 WHO/IPCS, 'Environmental Health Criteria 27. Guidelines on Studies in Environmental Epidemiology', World Health Organization, Geneva, 1983.

11 R.M. Turner and S. Fairhurst, HSE Specialist Inspector Report 21, 'Assessment of the Toxicity of Major Hazard Substances', Health and Safety Executive, Bootle, 1989.

12 J.W. Bridges, 'Identification of Toxic Hazard', in 'Major Chemical Disasters – Medical Aspects of Management', ed. V. Murray, Royal Society of Medicine Services Ltd., London, 1990, pp. 131–139.

13 ECETOC, 'Emergency Exposure Indices for Industrial Chemicals', Technical Report No. 43, ECETOC, Brussels, 1991.

14 H.P.A. Illing, 'Toxicology and Disasters', in 'A Textbook of Basic and Applied Toxicology', ed. B. Ballantyne, T.C. Marrs, P.M. Turner, MacMillan, Basingstoke, 1992.

15 H.P.A. Illing, *Ann. Occup. Hyg.*, 1991, (in press).

16 V.J. Feron, P.J. van Bladeren, and R.J.J. Hermus, *Food Chem. Toxicol.*, 1990, **11**, 783.

17 'Toxic Substances and Human Risk; Principles of Data Interpretation', ed. R.G. Tardiff and J.V. Rodericks, Life Science Monographs, Plenum Press, New York, 1987.

18 HSE, 'The Tolerability of Risk from Nuclear Power Stations', HMSO, London, 1988.

19 HSE, 'Quantified Risk Assessment: its Input to Decision Making', HMSO, London, 1989.

20 H.P.A. Illing, *Human Exp. Toxicol.*, 1991, **10**, 215.

21 V.C. Marshall, 'Major Chemical Hazards', Ellis Horwood, Chichester, 1987.

22 'Major Chemical Disasters – Medical Aspects of Management', ed. V. Murray, Royal Society of Medicine Services Ltd., London, 1990,

2

The Chemical Safety Programme of the ILO

K. Kogi and S. Machida

OCCUPATIONAL SAFETY AND HEALTH BRANCH, WORKING CONDITIONS AND ENVIRONMENT DEPARTMENT, INTERNATIONAL LABOUR OFFICE, 1211 GENEVA 22, SWITZERLAND

1 INTRODUCTION

Internationally, control of chemical risks has become a matter of increasing concern in safety and health programmes for both general and working populations. Awareness is growing that a series of comprehensive measures are necessary for effective control of chemical risks. Increasing importance is placed on the supply and use of hazard information, the routine control of exposure, and the application of practical improvement measures through improved risk management and training. There is also a need for direct involvement of employers and workers in the assessment of risks and planning and implementation of improvements.

These trends are clearly reflected in the International Labour Office's (ILO) chemical safety programmes. As a part of its activities concerned with occupational safety and health and the improvement of working conditions, the ILO programme in chemical safety emphasizes:

i) Standards setting for basic principles in chemical risk assessment and control;

ii) Promotion of coherent national policies and programmes on chemical safety and major hazard control (prevention of industrial disasters);

iii) Development of guides and training programmes;

iv) Collection and dissemination of information; and,

v) Technical co-operation.

2 STANDARDS SETTING

2.1 Conventions and Recommendations

Since the inception of the ILO in 1919, the International Labour Conference, under its tripartite structure involving the delegates from governments and employers' and workers' organizations, has adopted numerous international labour standards, namely Conventions and Recommendations concerning working conditions and the working environment.[1-11] At its Session in November 1990, the Governing Body of the ILO decided to place an item on the prevention of industrial disasters on the agenda for the 1992 International Labour Conference with a view to adopting new international standards. These new standards on the prevention of industrial disasters are envisaged for adoption in 1993 after a two year period for consultation after the Conference.

The Occupational Safety and Health Convention[1] and Recommendation[2] laid down the foundation for a policy extending to an undertaking to introduce a comprehensive and coherent system for the prevention of occupational accidents and diseases. The Occupational Health Services Convention[3] and Recommendation[4] sought the establishment of occupational health services which embody the assessment for the health of workers and monitoring of the working environment. These have an advisory function for implementation of the necessary improvements. These four instruments[1-4] provide an important basis for both national and enterprise level action in chemical safety.

2.2 Principles for the Control of Chemical Risks

The Chemicals Convention[7] and Recommendation[8] paved a new path for control in the use of chemicals, linked with environmental considerations. The focus for the controlling principles is the classification and labelling of chemicals, provisions of chemical safety data sheets (CSDS), minimizing the exposure of workers, the provision of information and training to workers, rights of workers, and ensuring co-operation of all parties.

2.2.1 Classification Systems. The systems and specific criteria appropriate for the classification of all chemicals, according to the type and degree of their intrinsic health and physical hazards and for assessing the relevance of the information required to determine whether a chemical is hazardous, should be established by the competent authority or by an approved body. The hazardous properties of mixtures can be determined by calculation. In the case of transport, systems and criteria shall take account of the UN Recommendations on Transport of Dangerous Goods. The classification system and its application shall be progressively extended.

2.2.2 Labelling and Marking. All chemicals should be marked to indicate their identity, particularly noting local language requirements. Hazardous chemicals shall, in addition, be labelled, in a manner which is easily understandable to workers, so as to provide essential information regarding their classification, the hazards they present, and the safety precautions to be observed. In the case of transport, due account should be taken of the UN Recommendation.

2.2.3 Chemical Safety Data Sheets (CSDS). For hazardous chemicals, chemical safety data sheets (CSDS) should be provided to employers, and contain detailed essential information regarding their identity, supplier, classification, hazards, safety precautions, and emergency procedures. Criteria for the preparation of CSDS are to be established by the competent authority or by an approved body. The chemical or common name used to identify the chemical on the CSDS shall be the same as that on the label.

The criteria for the preparation of CSDS for hazardous chemicals should ensure that they contain essential information including:

Chemical product and company identification (including trade or common name of the chemical and details of the supplier or manufacturer);

Composition/information on ingredients (in a manner that clearly identifies them for the purpose of conducting a hazard evaluation);

Hazards identification;

First–aid measures;

Fire–fighting measures;

Accidental release measures;

Handling and storage;

Exposure controls/personal protection (including recommended methods of monitoring workplace exposure);

Physical and chemical properties;

Stability and reactivity;

Toxicological information (including the potential routes of entry into the body and the possibility of synergism with other chemicals or hazards encountered at work);

Ecological information;

Disposal considerations;

Transport information;

Regulatory information; and

Other information (including date for the preparation of the chemical safety data sheet).

<u>2.2.4 Responsibilities of Suppliers</u>. Suppliers of chemicals, whether manufacturers, importers or distributors, are to ensure that:

i) Chemicals have been classified on the basis of knowledge of their properties and a search made for available information, or be assessed;

ii) Such chemicals are marked so as to indicate their identity;

iii) Hazardous chemicals which they supply are then labelled and CSDS are prepared for such hazardous chemicals and a copy provided to their customers. Furthermore, suppliers of chemicals which have not been classified are to identify these chemicals, and assess the properties on the basis of a search of available information and then determine whether they are hazardous.

<u>2.2.5 Action at Workplace</u>. Employers should ensure that all chemicals have been labelled or marked and that the CSDS have been compiled for hazardous chemicals; the CSDS are also to be made available to the workers concerned. If chemicals have not been labelled or marked, or CSDS have not been provided, employers should obtain the relevant information from the supplier or from other sources, and should not use the chemicals until such information has been obtained.

Employers should ensure that when chemicals are transferred to other containers, the contents are shown so as to make their identity known to the workers; any hazards associated with their use and any safety precautions to be observed should also be clearly shown.

Employers should:

i) Ensure that workers are not exposed to chemicals to an extent which exceeds exposure limits established by the competent authority or by an approved body;

ii) Assess the exposure of workers to hazardous chemicals;

iii) Monitor and record the exposure of workers to hazardous chemicals when this is necessary or as may be prescribed by the competent authority; and

iv) Ensure that the records are kept for a prescribed period of time and that these are accessible to the workers and their representatives.

Employers should assess the risks from the use of chemicals at work, and protect workers against such risks by appropriate means such as:

i) The choice of chemicals that eliminate or minimize the risk;

ii) The choice of technology that eliminates or minimizes the risk;

iii) The use of adequate engineering control measures;

iv) The adoption of working systems and practices that eliminate or minimize the risk;

v) The adoption of adequate occupational hygiene measures; and

vi) Where recourse to the above measures does not suffice, the provision and proper maintenance of personal protective equipment and clothing. These should be supplemented by precautions to limit exposure of workers to hazardous chemicals, to provide first-aid, and to make arrangements to deal with emergencies.

Further, employers should:

i) Inform the workers of the hazards associated with exposure to chemicals used;

ii) Instruct the workers how to obtain and use the information provided on labels and CSDS;

iii) Use the CSDS along with information specific to the workplace, as a basis for the preparation of instructions to workers; and

iv) Train the workers on a continuing basis.

2.2.6 Rights of Workers. Workers have the right to remove themselves from danger resulting from the use of chemicals when they have reasonable justification to believe that there is an imminent and serious risk to their safety or health, and should inform their supervisor immediately. Workers who remove themselves from danger, or who exercise any other rights under this Convention shall be protected against undue consequences. (See also chapters by P. Jacques and A. Rice.)

Workers and their representatives shall have the right to:

i) Information on the identity of chemicals, precautionary measures, education, and training;

ii) Information contained in labels and markings;

iii) CSDS; and

iv) Any other relevant information.

When disclosure of the specific identify of an ingredient in a mixture to a competitor would be liable to cause confidentiality problems for the employer's business, the employer may, in providing the above necessary information, protect that identity in a manner approved by the competent authority.

3 PROMOTION OF COHERENT NATIONAL POLICIES AND PROGRAMMES ON CHEMICAL SAFETY AND MAJOR HAZARD CONTROL

3.1 ILO Activities

ILO activities are designed to stimulate and to reinforce national action in occupational safety and health. They are now promoted through the 'International Programme for the Improvement of Working Conditions and Environment' (PIACT) launched in 1976.[9] Within the framework of PIACT, a series of activities should be undertaken to promote national and workplace level actions for preventing accidents and diseases eminating from the use of chemicals and good management of chemicals. The support is provided usually to member States by means of Conventions and Recommendations, codes of practice, guides and manuals, technical advice and information, and technical co-operation projects. Particular attention should be paid to the following three aspects.

<u>3.1.1 The Implementation of Principles Set Out in the Chemicals Convention and Recommendations</u>. The Chemicals Convention[7] and Recommendation[8] established basic principles for national policies for the promotion of chemical safety at workplaces. The Convention[7] will provide a binding obligation to the member States which ratify it. The provision of the Convention and Recommendation[7,8] gives guidance to all member States irrespective of whether they ratify the Convention or not. The ILO activities for chemical safety carried out within the PIACT[9] programme, including preparation of codes, guides and manuals, provision of advice and information, organization of seminars and workshops, and technical co-operation projects, aim at promoting the implementation of principles set out in the Chemicals Convention and Recommendation.[7,8]

<u>3.1.2 Harmonization of Systems of Classification</u>. The ILO is to play a leading role in promoting the harmonization of systems of classification in a manner that contributes to safety in the use of chemicals while promoting international trade. The 1989 International Labour Conference adopted a resolution concerning the harmonization of systems of classification and labelling for the use of hazardous chemicals at work.

This is because the Chemicals Convention and Recommendation[7,8] require national criteria and systems to be established for the classification of chemicals according to their intrinsic hazards. The ILO's work in this field will help to eliminate contradictions found between different systems, avoid duplication of effort, and reduce the unnecessary use of animals for laboratory testing. A review of existing national, regional, and international classification systems has been commenced. Criteria for an international system of classification will be prepared within the framework of the UNEP/WHO/ILO 'International Programme on Chemical Safety' (IPCS). Proposals will then be made for the harmonization of the classification used to rank the hazards of chemicals. This work will be undertaken in co-operation with the Commission of the European Communities, the OECD, the United Nations Committee of Experts on the Transport of Dangerous Goods, and various national and international institutions.

<u>3.1.3 Prevention of Industrial Disasters (Major Hazard Control)</u>. Based on the discussions on the issues raised by the disaster at Bhopal,[10,11] the International Labour Conference adopted, in June 1985, a resolution concerning the promotion of measures against risks and accidents arising out of the use of dangerous substances and processes in industry. In response to the resolution, the ILO convened a Tripartite *Ad Hoc* Meeting of Consultants on Methods of Prevention of Major Hazards in Industry, in October 1985, in order to obtain guidance on future action in the field of major hazard control. To give effect to the recommendations of this meeting, the ILO has taken a series of actions.

The ILO has assisted member States in providing advisory services for establishing a national system for preventing industrial disasters. ILO action relating to major hazard control has placed particular emphasis on technical co-operation activities. The aim of these activities is to install a national system for identification of potentially hazardous industries, for analysis and control of these industrial activities, and for emergency operations in case of major accidents. Specifically, the ILO has been executing technical co-operation projects in India, Indonesia, and Thailand aiming at establishing a central national unit for major hazard control advisory services to the industries and government inspectors, strengthening the chemical inspection capabilities of inspectorates, and training of management, supervisors, and workers concerned.

A manual on Major Hazard Control was published in 1988 by the ILO for general use in all countries which have industrial activities with major accident potential. This Manual reflects the experience of several Western European countries which practice major hazard control systems.

A Code of Practice on the Prevention of Major Industrial Accidents was adopted by the Meeting of Experts held in October 1990 and was published in May 1991. The main components of a major hazard control system described in the Code are identification of major hazard installations, assessment of major hazards and their control by the management, emergency planning, safety report, and siting and land-use planning. Nuclear hazards and those of military nature as well as transportation of hazardous chemicals are excluded from the scope of the Code.

It should be emphasized that a major hazard control system can function effectively only in parallel with the smooth functioning of an overall programme for chemical safety. The principles set out in the new international standards on Safety in the Use of Chemicals at Work,[7,8] are an important basis for national and workplace-level action in this direction. There is a need for the development of a system for prevention of industrial disasters in addition to the system for safety in the use of chemicals in all kinds of activities.

4 CODES OF PRACTICE, GUIDES, AND MANUALS

The ILO has been supplementing Conventions and Recommendations on occupational safety and health by issuing codes of practice, manuals, and technical guides. These are intended to stimulate action at both national and operational levels in a given area. Examples of publications related to chemical safety include the 'Code of Practice on Safety and Health in Agricultural Work', the 'Occupational Exposure to Airborne Substances Harmful to Health', the 'Code of Practice on Occupational Safety and Health in the Iron and Steel Industry', and the 'Code of Practice on Safety, Health and Working Conditions in the Transfer of Technology to Developing Countries'. The ILO 'Encyclopaedia of Occupational Health and Safety' is also an important reference book which contains a great deal of information of chemical safety. The preparation of a new edition of the Encyclopaedia is now under way.

Other recent examples are 'Major Hazard Control – A Practical Manual' (1988), 'Safety in the Use of Mineral and Synthetic Fibres' (1990), 'Code of Practice on the Prevention of Industrial Disasters' (1991), 'Safety and Health in the Use of Agrochemicals – A Guide' (1991). A training manual on 'Safety in the Use of Chemicals at the Workplace' is in preparation.

To promote the application of standards by member States, a draft Code of Practice on 'Safety in the Use of Chemicals at Work' was prepared during 1990–91 and will be submitted to a Meeting of Experts in 1992. This Code of Practice will provide guidance on the safe use of chemicals with specific reference to their classification, labelling, transport, storage, handling and use, and waste disposal. It will assist member States in the formulation of appropriate national legislation and the establishment of a national infrastructure to deal with chemical safety; it may also be used as a training manual.

5 COLLECTION AND DISSEMINATION OF INFORMATION

5.1 The 'International Occupational Safety and Health Information Centre' (CIS)

CIS makes available to users comprehensive computerized databases in the field of safety and health. Special attention is paid to chemical safety by including information on detailed elements of safe practice obtained from various chemical safety data sheets (100,000 currently available) developed by industry, governments, and safety and health institutions. The Centre operates through national centres throughout the world. A recent and effective means of disseminating the CIS database has been its inclusion on CD-ROM or laser discs capable of storing the whole database together with some other relevant databases. Currently more than one hundred safety and health databases are available from CIS.

5.2 CD-ROMs

The OSH-ROM Produced by SilverPlatter Information Services contains the three major safety and health databases: 'CISILO', 'NIOSHTIC', and 'HSELINE', as well as the database on major hazards 'MHIDAS', thus placing in excess of 250,000 abstracts of the world literature at the disposal of the user. The 'CCINFODISC', produced by the 'CIS national centre in Canada. The Canadian Centre for Occupational Health and Safety (COOSH), contains, in addition to 'CISILO', 60,000 chemical safety information sheets in full text, as well as other databases such as 'NIOSHTIC', 'RTECS' (Registry of Toxic Effects of Chemical Substances), Canadian Standards, *etc.* Annual subscriptions to both disc services include quarterly updates. Reading and retrieving information from these discs requires relatively inexpensive equipment and no special computer training. Basic equipment includes a microcomputer, a 'CD-ROM' reader, and a printer.

5.3 Information Exchange

The need for the exchange of experiences and information to promote practical safety and health measures at the workplace is growing rapidly. Requests concerning work hazards, chemicals, and preventive measures by governments and employers' and workers' organizations are increasing both in number and in complexity. Particularly important are requests for information concerning safety and health legislation, effective safety and health methods, and training materials. Thus the collection and dissemination of priority information concerning safety and health legislation, training courses, guides, and reports will continue to be expanded. The periodical ILO-CIS Bulletin, 'Safety and Health at Work', is distributed worldwide to subscribers. It now contains information on over 20 specific priority subjects, determined annually under the guidance of the national CIS centres. Current priorities include new technologies, biotechnology, carcinogenic and other hazardous substances, agents such as asbestos and man-made mineral fibres, the chemical industry, and agriculture. On the basis of the experience gained through the 'CIS' activities, technical assistance will continue to be provided to developing countries for the establishment and improvement of safety and health information centres.

5.4 The International Occupational Safety and Health Hazard Alert System

This became operational in 1982. It can disseminate rapidly information about specific hazards to associated centres in member States. The use of the System is extended to international exchanges on new issues and information related to new technologies and major hazard installations. Information on major accidents is included. In collaboration with the participating agencies in 100 member States, the System will act as a clearing-house for information on specific hazards or dangerous processes of particular interest to member States and on newly developed improvement measures.

6 TECHNICAL CO-OPERATION

Over 60 countries have received ILO technical assistance on health and safety matters through the ILO's technical co-operation activities. This can take various forms. These activities provide assistance to countries or regions in setting up or strengthening systems of technical inspection services or occupational safety and health centres and upgrading training capabilities of government, employers', and workers' organizations. A number of projects to strengthen the national capability for the improvement of occupational safety and health are operational and chemical safety is one of the important fields included in these projects. In addition, there are currently three projects aiming at establishing a national system of major hazard control. Further similar activities are under consideration.

Two major regional projects were started in 1991 for Africa and for Asia. Both are funded by Finland and aim at strengthening the national information collection and dissemination infrastructure of participating countries. The promotion of chemical safety programmes is an important element of the projects. The projects will improve enforcement practices and advisory services and will set up training systems for safety and health specialists, employers, and workers. The projects will play also a key role in reinforcing the national and regional exchange of practical safety and health information.

7 ACTION PROGRAMMES TO MEET LOCAL NEEDS IN CHEMICAL SAFETY

It is essential to ensure that an action programme for chemical safety meets the local needs of different working populations. Priorities should not be determined by merely applying a model or other criteria established outside the country or region. The identification of hazards and knowledge of standards are important but are often not sufficient for solving local problems. It is important to take measures, on a step-by-step basis, which are practical in the local context. Due account has to be given to the availability of technical expertise, resources, and awareness of safety and health issues. Participatory approach to encourage local people to form work groups to investigate potentially available options for the improvement is recommended as an effective way for finding practical solution.

Technologies transferred to developing countries have brought new hazards, but technical equipment and chemicals are often used without protective measures and without the workers being informed of the risks likely to be encountered. The transfer of machinery, chemicals, and processes should be accompanied by the knowledge detailing the effect which machinery, chemicals, and processes may have upon the safety and health and working life of those who operate or work with them. For the technology to be transferred safely, any appropriate or necessary adaptations should be made to the original technology to ensure that the processes, plants, and equipment take adequate account of the differences between the receiving country and the supplying country. The ILO Code of Practice on Safety, Health, and Working Conditions in the Transfer of Technology to Developing Countries published in 1988 provides a guidance in this regard.

Recent experiences in the ILO activities show that effective services in occupational safety and health have some common features. These features directly relate to chemical safety and major hazard control activities. They include: to be based on local traditions and culture and on local practice; to provide positive guidance and feedback for better safety and management goals; and to facilitate active participation of employers and workers. It is necessary to learn from these effective activities and provide support for organizing workplace action aimed at local solutions, while priorities for action may vary according to the extent of problems and available means of action.

Future action in chemical safety should be given support by the harmonization of existing classification systems, the establishment of a common labelling system, the dissemination of chemical safety data sheets and the development of practical training programmes. The ILO will collaborate with the ILO/UNEP/WHO 'International Programme on Chemical Safety' (IPCS). In particular, technical assistance in strengthening regulatory measures, surveillance programmes, and organizing training activities, based on such international support and tripartism, will be the key for the future action in controlling the use of chemicals in all occupations.

8 REFERENCES

1 The Occupational Safety and Health Convention No. 155, 1981, International Labour Office, Geneva.

2 The Occupational Safety and Health Recommendation No. 164, 1981, International Labour Office, Geneva.

3 The Occupational Health Services Convention No. 161, 1985, International Labour Office, Geneva.

4 The Occupational Health Services Recommendation No. 171, 1985, International Labour Office, Geneva.

5 The Asbestos Convention No. 162, 1986, International Labour Office, Geneva.

6 The Asbestos Recommendation No. 172, 1986, International Labour Office, Geneva.

7 The Chemicals Convention No. 170, 1990, International Labour Office, Geneva.

8 The Chemicals Recommendation No. 177, 1990, International Labour Office, Geneva.

9 'Improving Working Conditions and Environment: An International Programme (PIACT)', 1984, International Labour Office, Geneva.

10 'Major Hazard Control: A Practical Manual', 1988, International Labour Office, Geneva.

11 'Prevention of Major Industrial Accidents', 1991, International Labour Office, Geneva.

3

Risk Management of Chemicals – An Industry Point of View

B. Broecker

HOECHST AKTIENGESELLSHAFT, ABT. UCV – D787, POSTFACH 80 03 20, BRÜNINGSTRASSE 50, 6230 FRANKFURT AM MAIN 80, GERMANY

1 INTRODUCTION

There are a large number of publications concerned with the topic of risk management or, to put it more precisely, the reduction of the risk from chemicals to an acceptable level for humans and the environment.[1-5] The strategies recommended in many of these publications are mostly very similar. Some of these publications, however, have the disadvantage that they are based on purely theoretical scientific principles which cannot or only very rarely can be applied in practice. Their understanding is also often confused by non-consistent definitions of basic terms especially of hazardous properties, of hazard, and of risk.[1-7] Without disputing these various definitions, it should be stressed that in this chapter the following definitions are used which have been adapted from the results of a workshop on environmental hazard and risk assessment in the context of Directive 79/831/EEC held in October 1990 (see Table 1). (see also Preview and chapter by Koundajkian and Illing). This workshop dealt only with environmental hazard but the proposed definitions could be used also with minor modifications for health hazard.

Table 1 *EEC definitions*

Hazard identification:

The hazard identification aims at a first stage to identify a substance as a substance of concern. The hazard identification takes into account the hazardous properties **inherent** in a substance.

Hazard assessment:

A hazard assessment is a means to assess the potential for a substance to cause adverse effects on an environmental species and/or on man. Therefore, the hazard assessment requires information on environmental exposure (the environmental compartments of concern, quantities) and effect data (with reference to the environmental compartment of concern) and is normally expressed by a comparison of the (predicted) environmental concentration with the (predicted) no effect concentration for concerned species or ecosystems.

Risk assessment:

The term risk assessment is often confused with the term hazard assessment; it is frequently used as a comprehensive term to cover all kinds of evaluation of substances. **As a risk usually is expressed as the probability of the occurrence of an adverse effect**, the term 'risk assessment' should not be used if no probabilities are calculated. The risk assessment is a means to estimate the probability that a substance can cause adverse effects as a result of the (at a given concentration) presence of that substance in the environment.

Risk management:

The term risk management means to take measures which are appropriate at least to diminish significantly the presence (and thus the hazard) of a substance in the environmental compartments of concern. The measures which are appropriate for the purpose should be considered case by case; they can include total banning of a substance, restriction of special uses, or control measures.

Consequently, an attempt will be made to describe how, as a rule, the hazard potential of chemical products is assessed currently in the chemical industry and those measures which can be taken to limit this hazard if necessary to an acceptable level.

2 PRINCIPLES

The risk management of chemical products is an iterative process which proceeds through the steps described in Table 2. According to this Scheme one must first determine, by means of suitable experimental tests, whether a chemical product has any dangerous properties. If this should be the case, parallel investigations must determine whether these properties are of any significance whatsoever in practice because either both humans and/or the environment are exposed to this substance.

Next the hazard potential of chemical substances must be assessed by correlation of the maximum tolerable concentrations of the chemical for man or environment and the exposure levels to be anticipated. If the above procedures yield a hazard potential, then as a subsequent step (probably the most difficult of all) an evaluation must be made as to whether or not this hazard potential should be regarded as acceptable to humans and the environment. If not, suitable measures must be taken to reduce the potential of hazard respective to the risk to an acceptable level. Each of the steps in Table 2 has its specific problems in industrial practice and these problems are summarized in Section 3.

Table 2

Hazard identification by testing of biological effects, physico-chemical properties and collection of exposure information

↓

Hazard assessment by correlation of biological effects and exposure

↓

Assessment of acceptability of hazard in respect of the risk

↓

Reduction measures if the hazard associated with the risks is not acceptable

3 IDENTIFICATION OF HAZARDOUS PROPERTIES

Hazardous properties of chemical substances can consist of, on the one hand, their physico-chemical properties such as their inflammability or explosiveness; and, on the other hand their capacity to cause injurious biological effects in humans and/or the ecological system when humans and the environment come into contact with these substances.

Whereas, currently, the determination of physico-chemical properties presents no special experimental difficulties, unfortunately, this is not the case with regard to the biological effects. The first problem is to decide at the outset which biological end points must be experimentally identified.

If one wanted to investigate a chemical substance in detail for all the biological effects known today, it would cost several million dollars and take 3 to 5 years. But even then it would not be possible to exclude the possibility that certain biological effects had been overlooked.

In this respect a compromise must be made between scientific completeness and economic feasibility. Hence, currently, it is general practice to restrict experimental testing to the most important biological end points. OECD, within its Chemicals Programme[8] came to an agreement that for initial hazard assessment data - a screening information data set (SIDS) - should be available as indicated in Table 3.

Table 3 *Screening information data set*

Chemical identity

CAS number
Name of the substance
Structural formula

Physical–chemical data

Melting point
Boiling point
Vapour pressure
Partition coefficient (n–octanol/water)
Water solubility

Sources

Production ranges
Categories and types of use

Environmental fate and pathways

Aerobic biodegradability
Abiotic degradability
Estimates of environmental fate, pathways, and concentrations

Ecotoxicological data

Acute toxicity to fish
Acute and prolonged toxicity to daphnids
Toxicity to algae
Toxicity to terrestrial and avian organisms

Toxicological data

Acute toxicity
Repeated dose toxicity
Genetic toxicity
Reproductive toxicity

Whilst OECD believes that these data should be available as a minimum, it is customary in industry to take a step–by–step approach depending on the exposure potential. That is to say, for substances that are handled only in closed systems, information is restricted essentially to acute effects on humans and the environment. Only those substances to which humans are actually exposed for extended periods or which are released into the environment in appreciable quantities are required to be investigated for chronic effects on humans and the environment.

This shows that the parallel steps of exposure determination and the subsequent hazard assessment cannot proceed separately but are processes that run iteratively and mutually influence one another.

4 EXPOSURE DETERMINATION

The assessment of the exposure potential is also very complex and in practice a difficult process. First, it must be pointed out that it is by no means always an easy matter for a manufacturer of chemical substances to estimate what the exposure profile of his chemicals at his customer's plant is really like, since he does not know all the purposes for which his substances are used. Consequently, he can only reasonably include in his considerations of this problem those use patterns for which he recommends his product. Nevertheless, in order to prevent customers from relying on the evaluation made by the manufacturer, the latter would have to indicate to his customers the conditions under which he arrived at this evaluation more clearly than is now the usual practice.

In fact, however, for many chemical products, exposure of humans and the environment occurs primarily during their use and not during their manufacture. It must be remembered in this connection that in the final analysis all chemical products except intermediates enter the environment finally when they have become wastes. But even when all areas of application are sufficiently well known, in many cases factors such as the type and number of humans who will be exposed to the substances and the extent and frequency of the exposure are so difficult to estimate accurately that a quantitative exposure estimate can be made only very rarely.

5 HAZARD ASSESSMENT

Once the biological activity potential and the exposure potential have been determined, the hazard potential must be assessed by correlating these two factors. This process is also very complex in practice and is rendered more difficult by our limited knowledge of certain biological effects.

The greatest problem is the necessity of extrapolating from the results of animal studies to humans or, in the case of ecological systems, from the results obtained with one species to entire ecological systems. In the area of human toxicology there is now, fortunately, considerable scientific experience available, which does not mean, however, that we will not again experience unpleasant surprises in isolated cases.

Standard formulae for determining the concentrations that are tolerable to humans from the results of animal experiments have only limited scientific applicability.[9] Far more important are the specific properties and exposure profiles of the respective substances. A further problem is the difference in sensitivity of the persons affected, especially the high–risk groups such as the elderly, children, pregnant women, *etc.*

There is a special problem with regard to the effect end points, where in accordance with current scientific knowledge a no–effect level can be determined, especially carcinogenic, mutagenic, and embryotoxic effects. Here assistance can be provided by application of conventions. For example, with regard to carcinogenic substances this involves the establishment of Limit Values that are technically feasible and below which available toxicological and industrial medical knowledge make it likely that no appreciable hazard potential will appear. The matter becomes even more difficult in the area of ecological toxicity. According to present knowledge it cannot be assumed that one can extrapolate from one species, *e.g. Daphnia*, to another species, *e.g.* fish, not to mention to the great number of organisms that occur in different environmental compartments. So a Workshop organized by OECD on Ecological Effects Assessment in 1988[10] came to the conclusion that there is uncertainty at each stage in an ecological risk assessment. The prediction of effects is based on test data which themselves are subject to statistical variability. The extrapolation of these results from acute to chronic effects, from one species to another, and to other taxonomic groups, introduces further levels of uncertainty. There is further uncertainty in extrapolating laboratory test results to effects in the real environment. The estimation of environmental exposure for most types of chemicals is also very imprecise and measured concentrations are subject to sampling errors.

However, due to the fact that for ecological effects, if there are only primarily acute data on a few species available, at least for the time being, there is no alternative but to undertake a rough extrapolation from these data. For chemicals which intentionally or unintentionally are discharged in large quantities into the environment a better database especially for chronic effects is, therefore, urgently needed. Here again, the only practical solution is to set up a convention by defining and examining certain biological effects in certain key organisms and orienting oneself to these data.

6 QUANTITATIVE ASPECTS OF HAZARD AND RISK ASSESSMENT

Whilst, in principle, hazard assessment is quantitative, because quantitative figures for no–effect concentrations and exposure levels need to be compared it is often (incorrectly) described as qualitative. Due to uncertainty involved in both figures the results, however, have only very qualitative character. Therefore, in most cases attempts to transfer hazard assessment into risk assessment by including the probability of the manifestation of injuries to health or environment have no real basis.

It is true that there have been many attempts to derive a quantified risk potential from the results of animal studies, for example, to quantify carcinogenic activity. But the results of such estimates are extremely dependent on the model on which they are based, and different models can yield results differing by several orders of magnitude (see Table 4). This uncertainty is multiplied by the use of uncertainty or safety factors to be applied to the correlating effects and exposure. For these reasons a quantitative risk estimation can be made only in rare cases where sufficient data on relevant effects and for exposure are available. In the large majority of cases, the results of hazard assessment will have to be described as follows:

i) Hazard potential low due to large safety margin between exposure and maximum tolerable effects concentrations.

ii) Hazard potential may exist due to low safety margin between exposure and maximum tolerable effects concentrations.

iii) Hazard potential likely because maximum tolerable effects concentrations exceed exposure levels.

Table 4 *Virtually safe doses of formaldehyde in rats for selected excess risks of squamous cell carcinoma of the nasal cavity predicted from adjusted proportions using the probit, gamma multihit, logit, Weibull, multistage, and linear extrapolation models*

	Formaldehyde virtually safe dose (p.p.m.)			
Model	1 in 10^5	1 in 10^6	1 in 10^7	1 in 10^8
Probit	2.60	2.14	1.79	1.52
Multihit	2.05	1.52	1.14	0.86
Logit	1.41	0.89	0.56	0.35
Weibull	1.26	0.76	0.46	0.28
Multistage	1.26	0.65	0.37	0.21
Linear	0.0059	0.00059	0.000059	0.0000059

7 ASSESSMENT OF THE ACCEPTABILITY OF THE HAZARD POTENTIAL

When it can be concluded from the above considerations that the possibility of a hazard potential to humans and to the environment from certain chemical substances cannot be completely excluded, the next step is probably the most difficult task, namely estimating whether this hazard potential can be accepted or not. In this connection various publications have proposed first to quantify the hazard potential in the form of the risk arising from the substances and then to compare it with other risks that are generally considered acceptable. As outlined above, in most cases this is pure theory.

Definitions of acceptability of a specific hazard potential by society are additionally complicated because different sectors of society have different ideas regarding the benefits of chemical products. A cigarette smoker will judge the risk to his fellow men who are exposed to passive inhalation of tobacco smoke very differently from a fervid non–smoker. Similarly, a driver who has to drive through streets that are slippery with snow and ice will assess the risk to the environment caused by road de–icing salt quite differently from someone who uses public transport only.

The above illustrates that we are concerned not with scientific but with political evaluation processes, with which industry of necessity has significant difficulties. The situation is made even more complicated by the introduction of the so–called minimization principle which means the requirement to minimize emissions of chemical substances as far as possible even if they are not hazardous. (See also chapters by B. Samimi–Eidenbenz and G. Vonkeman.)

8 RISK MANAGEMENT BY RISK REDUCTION

When the decision has been reached, for whatever reason, that the risk caused by certain chemical products must be considered unacceptable for humans and the environment, measures must be taken to reduce this risk accordingly. A wide spectrum of possible strategies is available for this purpose.

When the hazard potential at the workplace is considered to be too high, the exposure must be reduced primarily by technical protective measures or, if this is not possible to a sufficient degree, by use of personal protective measures where possible.

It is inconceivable that such actions could be instituted for the end user. Here the risk of exposure must be entirely eliminated by discontinuing the use of these substances in the consumer sector, or the proportion of the substances in consumer products must be diminished to such an extent that an increased risk of exposure does not arise from normal use.

With regard to substances that enter the environment, it must be borne in mind that once these substances are released into the environment as a rule they are not recoverable. And only certain product groups, such as plastics and the like, can be recovered and recycled. In all other instances either the amount of the substance that enters the environment must be so greatly reduced by suitable precleaning methods that the environment is not endangered or, if this cannot be ensured, the use of the substance must be discontinued.

9 CURRENT PRACTICE OF RISK MANAGEMENT BY INDUSTRY

Identification of Hazardous Properties

As described above, the first step to adequate risk management is hazard identification. The data available on hazardous properties are very often incomplete especially so far as chronic effects are concerned. Industry, partially due to legislative pressure and partially on the basis of voluntary programmes, *e.g.* the OECD Existing Chemicals Programme, is currently closing those gaps by testing. For chemicals which have a low use pattern the hazard assessment and evaluation for chronic effects can be given a lower priority.

Due to financial implications for R and D substances, the identification of hazardous properties will be restricted primarily to acute effects. In those cases, however, customers will be recommended to handle these chemicals with all necessary caution.

10 ASSESSMENT OF ACCEPTABILITY OF HAZARD

10.1 Application of Thresholds

The most efficient way to assess the acceptability of a specific hazard is the acceptance of threshold values below which a hazard for man and environment is unlikely to occur. Even if available data on effects allow the deduction of such values, this approach is criticized increasingly on the grounds of the uncertainty involved in setting such standards and of the lack of incentive to industry to minimize emissions as far as possible even if it is not necessary to keep to these levels.

However, realizing that industry must be interested to use its resources in the most efficient way, threshold values as guides for the required emission reduction remain crucial.

Due to this situation industry attempts increasingly to set its own health standards for application (*e.g.* see Reference 11), at least as far as the workplace is concerned. This also applies to health effects for which the present state of science does not permit definition of no-effect-concentration standards on the basis of pragmatic conventions.[12] To set safe levels for the environment is much more difficult because in most cases adequate data on chronic ecotoxicological effects are not available. There is also no general agreement on trophic levels and the number of species within respective levels for which data should be available, not to mention safety factors necessary between effects data and exposure levels. An approach very often used is to apply a safety factor of 1000 between acute LC_{50} for fish or *Daphnia* and concentration in the aqueous environment.

This approach, however, is feasible only for permanent discharges which might lead to chronic effects and not for accidental spillages or intermittent discharges for which acute effects are of prime relevance. These considerations can be used as a rough guide for acceptability of discharges, but only for point sources for which the environmental concentration can be predicted with some accuracy. For diffuse discharges the need for reduction measures has to be decided primarily on the basis of their inherent properties.

10.2 Termination of Production for Chemicals with Irreversible Effects and Uncontrolled Emissions

The use of chemicals with irreversible effects in areas where exposure cannot be controlled, *e.g.* in the consumer–sector or in applications with significant discharge into the environment, has to be ceased as far as possible. The same must be true for chemicals discharged in significant amounts into the environment which may cause chronic ecotoxic effects due to their persistence, bioaccumulation, and ecotoxicity. In both cases continuation of use in those areas can hardly be justified even if benefits to society are considered to be significant.

A large grey area of chemicals remains, for which, due to lack of precise data on exposure and effects, it cannot be completely excluded that under accidental or temporary situations manifestation of hazardous effects on man or environment will take place for a limited time period. Another grey area concerns chemicals for which data on hazardous effects are dubious or equivocal. In these cases it depends upon the severity and reversibility of hazardous effects in relation to the benefits of the substance whether the hazard is acceptable or not.

In general the philosophy used by industry can be summarized in a simplified form as shown in Table 5. If substitutes with lower hazard potential are available they should be used. In practice this often creates problems because substitutes are often less technically efficient or investigated less adequately with their hazardous properties.

11 CONCLUSIONS

It is evident from the above discussion that risk management represents a difficult problem for industry. On the one hand it is influenced by the limitations of the state of the art of the science and on the other by political evaluative processes. Hence, it is inevitable that industry makes different decisions from those considered essential by certain population groups. It is foolish and unrealistic to place the blame for this dilemma at industry's door.

Table 5 *Summary of recommendations*

	Closed systems	*Controlled industrial use with limited exposure*	*Uncontrolled industrial use or consumer use*
Substances with irreversible health effects for which no-effect-levels cannot be defined	Yes	No (unintended impurities may be permissible)	No (unintended impurities may be permissible)
Substances with irreversible health effects with no-effect-levels	Yes	Yes (if exposure below NOEL)	No (if exposure certainly below NOEL use may be possible)
Substances with reversible health effects	Yes	Yes (if exposure normally below NOEL)	Yes (if exposure normally below NOEL)
Substances causing chronic ecotoxic effects due to persistency, accumulation and ecotoxicity	Yes	Yes (if emissions are controlled)	No

It is much more important to make risk evaluations more objective through the creation of a suitable political framework and, especially, conventions so that industry can orientate itself according to these rules. Whilst this is not undertaken, the conflicts over the threat posed by chemical products to man and the environment will continue.

12 REFERENCES

1. D.P. Lovell, 'Risk Assessment – General Principles', in 'Toxic Hazard Assessment of Chemicals', ed. M.L. Richardson, The Royal Society of Chemistry, London, pp. 207–209.

2. J.J. Cohrssen and V.T. Covello, 'Risk Analysis: A Guide to Principles and Methods for Analyzing Health and Environmental Risks', The National Technical Information Service, US Department of Commerce, 5285 Port Royal Road, Springfield, VA 22161, 1989.

3. P.F. Ricci and M.C. Cirillo, *J. Hazardous Mater.*, 1985, **10**, 433–447.

4. World Health Organization, 'Environmental Toxicology and Ecotoxicology', Proceedings of the Third International Course, Edinburgh, UK, 6–13 September 1985, WHO Regional Office for Europe, Copenhagen, 1986, 342.

5. 'Risk Assessment of Chemicals in the Environment', ed. M.L. Richardson, The Royal Society of Chemistry, London, 1988, pp. 543–558.

6. 'Nomenclature for Hazard and Risk Assessment in the Process Industries', The Institution of Chemical Engineers, Rugby, UK, pp. 10–13.

7. The Royal Society, 'Risk Assessment', Report of a Royal Society Study Group, The Royal Society, London, 1983, pp. 22–23.

8. OECD Decision–Recommendation of the Council on the Co–operative Investigation and Risk Reduction of Existing Chemicals (adopted by the Council at its 750th Session on 31 January 1991) 32132.

9. ECETOC Technical Report No. 10. 'Considerations Regarding the Extrapolation of Biological Data Deriving Occupational Exposure Limits', ed. L. Turner, ECETOC, Brussels, 1984.

10. OECD Environmental Monographs, No. 26, 'Report of the OECD Workshop on Ecological Effects Assessment', May 1989.

11. Bundesministerium für Arbeit, Gefahrstoffe. Vorläufige Arbeits–platzrichtwerte. Bek. des BMA vom 23 Januar 1991–III b 4–35125–5, Bundesarbeitsblatt, 1991, **3**, 69–70.

12. Deutsche Forschungsgemeinschaft, 'Maximale Arbeitsplatzkonzen–trationen und Biologische Arbeitsstofftoleranzwerte 1990', Deutsche Forschungsgemeinschaft, VCH Verlagsgesellschaft mbH, 1990, p. 83.

4

Some Current Issues: The Trade Union View

Peter Jacques

TRADES UNION CONGRESS, CONGRESS HOUSE, GREAT RUSSELL STREET, LONDON WC1B 3LS, UK

1 BACKGROUND

Recent estimates indicate that between 15,000 and 30,000 chemicals are currently in major use everyday.[1] Many new chemicals are introduced each year. Without adequate risk evaluation and control many could be harmful or dangerous. Trade unions have long been aware of the need to control damage to health arising from use of such substances. Unlike accident hazards, where the hazard is often visible and the effects are generally immediate, the harmful effects of toxic substances are not always so obvious. For thousands of workers, the 'hidden hazards' of chemicals and other substances hazardous to health present a real threat every day of their working lives. Current estimates of work related ill health vary considerably. Early death from past exposure to hazardous agents, including toxic substances, is estimated conservatively to be at least an order of magnitude greater than death due to accidents.[2]

This chapter is intended to help readers understand aspects of the approach of the Trades Union Congress (TUC) and unions to the dangers of toxic substances used at work. It also describes how unions seek to secure improvements in the way substances are used and their role in checking that control measures are being properly applied at the workplace.

2 TRADE UNION INVOLVEMENT

The trades union movement has a central role to play in securing the protection of people from the hazards of their work. Since the introduction of the Health and Safety at Work Act in 1974, the TUC has continued to pursue and develop its general strategy for the improvement of occupational safety and health. Four key elements in this strategy are:

i) Securing the effective involvement of unions in health and safety issues – at national, industry, and workplace levels; for example through well understood and effective procedures for consultation;

ii) Defining problems and priorities – for example by the promotion of research;

iii) Securing improved health and safety standards in legislation; codes and guidance; and

iv) Ensuring compliance by employers with such standards and the adoption of safe systems of work – through the development of health and safety policies and services at the workplace and effective enforcement of standards by an adequately staffed inspectorate.

At national level the TUC is involved as a key participant in the Health and Safety Commission (HSC) and the Commission's Advisory Committee on Toxic Substances (ACTS). The TUC also participates in the ACTS WATCH (Working Group on the Assessment of Toxic Chemicals) panel which undertakes scientific evaluation of the toxicity of substances hazardous to health and advises ACTS about possible control measures. Unions are also involved in dealing with toxic substances issues via HSC Industry Advisory Committees (IACs) many of which are confronted with specific substances related occupational health problems.

In addition to their role in the policy processes, unions are also guaranteed a key role in securing compliance with health and safety standards at workplace level through the safety representatives and safety committees system.

Under the Safety Representatives and Safety Committees (SRSC) Regulations[3] of 1977, safety representatives appointed by recognized trades unions have legal rights to:

Represent their members;

Investigate potential hazards, complaints by members, dangerous occurrences, cases of industrial disease, or accidents at work;

Make formal workplace inspections at least once every three months;

Liaise with Factory Inspectors or other enforcement officers;

Require the formation of a joint union/management health and safety committee.

Employers must let safety representatives have:

Time off with pay to carry out their job;

Time off with pay for trade union approved training;

Facilities and assistance (*e.g.* access to a telephone, photocopying, filing facilities, *etc.*); and

Health and safety information.

The TUC and unions seek to use opportunities open to the at both national and local level to tackle occupational risks – including risks arising from exposure to chemicals. At present there are some 90,000 safety representatives in the UK. The TUC has a major safety representatives training programme and produces extensive advisory literature – including, for example, on COSHH.[4]

2.1 Trade Union Approach to Occupational Risk

As with occupational exposure to all harmful factors which are capable of causing potentially disastrous health effects, the TUC favours the adoption of a very cautious approach to control of exposure to toxic substances – particularly given widespread uncertainties about dose/effect relationships.

The aim of the TUC is to ensure that, at any particular time, risks to workers' health are reduced to the lowest level possible. This does not mean, however, that the TUC accepts that all efforts for risk reduction should cease when an 'acceptable' level of risk has been reached. No level of risk to health can be considered as entirely insignificant. The fact that workers continue to work in the face of a perceived level of risk should not be taken to imply that they accept that level of risk as an inevitable penalty to be paid for the advantages of employment and income which work brings. Workers who are expected to take the risks of working do so, in the main, not because they believe such risks to be unimportant but because, in the short term, they have little choice. The TUC view therefore is that, although work related risks may have to be tolerated, they must continue to be reduced as scientific knowledge, techniques, technology, and aspirations advance.

2.2 Interests and Perceptions

From this point of view, the TUC has argued[5] that risk assessment and control is a dynamic social process involving the interplay of a number of constantly developing factors including evidence of harm, practicability of controls, availability of resources, and expectations of those exposed to risk.

Perceptions of these factors will also vary according to the interests of those involved. For example, in relation to the strength of evidence required to demonstrate the level of a particular risk, those anxious to control the cost of associated preventive action may seek evidence at a level that is 'beyond all reasonable doubt'. On the other hand, those concerned to err on the side of caution by maximizing the impact of preventive action will base their case on evidence which is pitched at the lower 'balance of probabilities' level.

The TUC's view is that the consequences of past errors in risk evaluation (*e.g.* in relation to harmful agents such as asbestos, where tens of thousands of asbestos-related deaths due to past exposure are still awaited) should henceforth dictate a presumption in favour of harm. The opposite approach, namely assuming agents to be harmless until proven otherwise by predominantly human evidence, is indefensible on both moral and political grounds. The TUC's approach to risk evaluation and control therefore demands that more weight should be given to evidence which points in the direction of harm than to negative epidemiology.

Besides recognizing that quantitative evaluation of risk is an imperfect science, it should also be remembered that the character of risks (*e.g.* whether they involve major catastrophe, latency, involuntary risk, *etc.*) influences the way in which risks are perceived and thus the extent to which they will be tolerated at any particular level of probability or consequence. In responding to recent Health and Safety Executive discussion documents on risk[6] the TUC has argued that it is unrealistic to expect the process of risk assessment and 'acceptance' at work to be based on a uniform index of harm encompassing not only risk levels in other occupations but risks encountered in leisure and social activities such as transport or smoking.

3 COST/BENEFIT

The TUC has also argued[7] that it is unrealistic to assume that decision making about levels of risk control can be based on a purely cost/benefit approach. Although costs and benefits are clearly important factors, they are not easy to compare. This is because while costs of control measures are overwhelmingly quantitative and are capable of expression in monetary terms, the benefits (and certainly the most important benefits as far as workers are concerned) are overwhelmingly qualitative and hence are not capable of being expressed in the same way. In the area of risk control therefore, cost/benefit approaches must be seen as an aid to judgement and no more.

The TUC sees no alternative to judgements in the area of occupational risk control being based on an essentially pragmatic approach representing the outcome series of countervailing pressures and considerations. The complexity and interaction of these (unions, employers, national and international regulatory authorities, scientific evidence, *etc.*) defy expression within the relatively crude decision aiding models which have been developed so far in the field of risk assessment and 'acceptance'.

3.1 The WATCH/ACTS Process

These approaches have informed the TUC's general approach to participation in policy development in relation to toxic substances via the HSC's Advisory Committee on Toxic Substances (ACTS). Increasingly ACTS work links with related initiatives of the Commission of the European Communities (CEC) which concern work on toxic substances.

ACTS consists of 16 members, including four TUC and four Confederation of British Industry (CBI) nominees, as well as five independent Health and Safety Executive (HSE) nominees.

ACTS, which was established in 1975, continues to have a major work programme. During the period January 1989 to March 1991 ACTS undertook a range of tasks including: reviewing certain substances to reassess earlier Occupational Exposure Limits (OELs); a further ongoing programme of substances reviews working through the ACTS WATCH Group; and other broad issues associated with development of the Control of Substances Hazardous to Health (COSHH) package.

WATCH is the main standing working group of ACTS and provides for the assessment of the scientific evidence on which OELs are based. It is an expert, tripartite group which reports directly to ACTS and plays a particularly important role in recommending the appropriate values for Occupational Exposure Standards (OESs). However, consideration of socio–economic issues that may be involved in limit setting remains a matter for the main Committee.

3.2 Examples of ACTS Work

Between January 1989 and March 1991 ACTS met seven times. Their main concern was to implement the new procedures agreed following the introduction of the COSHH Regulations and to take forward work on exposure limits and the development of effective control and monitoring strategies. In this period, ACTS completed 31 reviews, making 12 proposals for Maximum Exposure Limits (MELs) and 19 proposals for OESs to the HSC, and also six proposals for changes to entries in the Appendix to the COSHH Carcinogens Approved Code of Practice (ACoP). The proposals for MELs and the changes to the Carcinogens ACoP were part of an ongoing role in this annual procedure for recommending changes to COSHH. (The TUC plays an important consultation role by circulating copies of consultative documents to unions for comment and collating a co–ordinated TUC response.) In addition, ACTS have considered some 20 other issues which have not resulted in proposals for new OELs (the generic term for MELs and OESs), *e.g.* Pharmaceuticals.

In addition to these achievements, in the same period ACTS covered a number of other areas where important progress was made including:

i) A review of Indicative Criteria: the ACTS/WATCH guidance for recommending the adoption of OELs;

ii) Updating Guidance Note EH 40: the official source of OELs;

iii) A review of Criteria Documentation: In response to developments in Europe (discussed later), and also to improve resource usage in WATCH substance reviews, a standard format for substance review criteria documentation was introduced. These documents now form an important source of information on the case for the adoption of particular limit values and are to be open to public scrutiny. The intention is to prepare and publish a short news-sheet for each review summarizing the main points of interest. Arrangements have been made to publish summary sheets for the approximately 60 substances which have been reviewed during this period.

3.3 The Indicative Criteria

Traditionally, given the preference of the trades union movement for a precautionary approach to risk assessment of chemicals and for a presumption in favour of harm, trades unions have always favoured control strategies which seek to reduce workplace exposures to the lowest levels practicable. This has always been particularly so in the case of substances giving rise to serious irreversible health effects such as cancer, reproductive effects, respiratory sensitization, neuropathy, permanent CNS damage, *etc.* In practice unions seek reductions in exposure well below OELs to provide a high level of assurance of health protection. Inevitably, however, it has been impossible for unions to avoid participation in dialogue about approaches to standard setting or indeed in agreeing compliance targets. In recent time this debate has resurfaced within the ACTS framework in relation to ongoing discussion of the 'indicative criteria' used by ACTS/WATCH to decide on which kind of OEL should be set for particular substances - either an OES which can provide a 'safe' level of exposure for 'almost all workers' on a daily basis, or an MEL representing an outer boundary of tolerable exposure.

Following the latest review of these criteria they are now expressed as follows:[9]

3.3.1 Occupational Exposure Limit

i) The available scientific evidence allows for the identification, with reasonable certainty, of a concentration averaged over a reference period, at which there is no indication that the substance is likely to be injurious to employees if they are exposed by inhalation day after day to that concentration; and

ii) Exposures to concentrations higher than that derived under criterion 1, and which could reasonably occur in practice, are unlikely to produce serious short or long-term effects on health over the period of time it might reasonably be expected to take to identify and remedy the cause of excessive exposure; and

iii) The available evidence indicates that compliance with the OES, as derived under criterion (i), is reasonably practicable.

3.3.2 Maximum Exposure Limit

i) The available evidence on the substance does not satisfy criteria i) and/or ii) for an OES and exposure to the substance has or is liable to have serious implications for the health of workers; or

ii) Socio-economic factors indicate that although the substance meets criteria i) and ii) for an OES, a numerically higher value is necessary if the controls associated with certain uses are to be regarded as reasonably practicable.

Debate both within ACTS and the wider scientific community about the application of their criteria continues to be intense – particularly given the legal and other implications of substances being given MEL status. The TUC has continued to argue that the criteria are aids to judgement and no more. While the criteria help to structure the judgement process they do not obviate inevitable and necessary debate about key issues such as 'What constitutes a serious and inevitable health effect?' 'What constitutes adequate certainty?', or 'How is it to be decided if control to an OES is not reasonably practicable?' These and other, general issues in the assessment of toxic substances remain before ACTS and illustrate why it is vitally important that trades unions must be involved in contributing the views of their constituents to the policy process.

3.4 'Reasonable Practicability'

Because of the way in which COSHH duties are constructed, the same risk/cost dilemmas which are present in limit setting at the policy level also appear sharply at industry and workplace levels in relation to compliance performance. Again the TUC's view is that unions have a key role to play in ensuring that high-quality judgements are reached as part of the COSHH assessment process.

As with the general duties imposed on various parties by the Health and Safety at Work Act the majority of duties in subsidiary legislation such as COSHH are subject to the qualification of 'reasonable practicability'.

The meaning of what is reasonably practicable to do to comply with health and safety legislation is defined with reference to a given body of decided case law. Essentially the effect of this qualification is to limit an employer's duty to control exposure of employees to sources of harm by not requiring further risk control measures beyond a point of 'gross disproportion' between the level of residual risk and the amount of effort and resources necessary to achieve further risk reduction. Determination of such a point of gross disproportion is necessary in each set of circumstances in which such qualified duties apply. In practice this requires the exercise of structured and informed judgement. The validity of such judgement may ultimately be tested in a court of law. It should be noted, however, that the judgement of what is 'reasonably practicable to do' does not relate to the financial circumstances of the party on whom the duty falls but is a social judgement.

At present, the amount of official guidance to employers and others on the information necessary for and the procedures to be followed in making judgements about reasonable practicability is limited. For example, guidance[10] to the COSHH Regulations in relation to reducing exposure below Maximum Exposure Limits (MELs) under Regulation 7 (the duty to control exposure) states:

> 'The extent to which it is required to reduce exposure further below the MEL will depend on the nature of the risk presented by the substance in question, weighed against the cost and the effort involved in taking measures to reduce the risk'.

The same advice is repeated in paragraph 16 of HSE Guidance Note EH40 Occupational Exposure Limits.

HSE guidance on COSHH Assessments provides useful guidance on elements to be considered in complying with the assessment duty under Regulation 6 but little guidance is given on how to determine the 'point of gross disproportion'. Similarly, COSHH guidance, including the general Approved Code of Practice (ACoP), provides little specific guidance on assessing reasonable practicability in relation to preferred options to achieve adequate control (Regulation 7(2)) or time allowable for compliance with Occupational Exposure Standards (OESs) (Regulation 7(5)(b)). This problem, however, is not unique to COSHH. There is a similar lack of guidance in relation to duty to control exposure to ionizing radiations under the Ionizing Radiations[11] (IRRs) (Regulation 6). At the suggestion of the Health and Safety Commission's Working Group on Ionizing Radiations (on which the TUC is represented), further guidance is being prepared on this issue to accompany the new Part 4 of the ACoP to the IRRs.

The application of reasonably practicable controls can and does lead in practice to different levels of legally permissible risk exposure for workers confronted by the same harmful agent. In practice it is often hard, or indeed impossible, for workers and/or their trade union representatives to enter into dialogue with employers about the judgements involved. Indeed, even where a written assessment has been made - for example to comply with Regulation 6 of COSHH - employers themselves may not have thought carefully about the information and procedures necessary to arrive at satisfactory judgements. This inevitably contributes to a lack of transparency in relation to judgements of what is reasonably practicable. This in turn can undermine confidence not only in specific risk control judgements but in the control regime generally. For these reasons high-quality professional support should be available to safety representatives to help them reach balanced judgements about the adequacy or otherwise of assessments.

In contrast to a minority of agents where a reasonably well quantified dose/response relationship has been established (*e.g.* radiation or noise), for the majority of sources of harm to workers' health there is insufficient information to enable the application of precise cost/benefit of optimization techniques. Nevertheless some general and qualitative judgements about risk levels and associated resource allocation are still possible and in practice these can be useful in helping to evaluate the quality of intuitive judgements about what is reasonably practicable to undertake in specific circumstances. In this respect, it may be worth noting the more qualitative approach to risk assessment and control which is being considered in the area of machinery safety by the European Standards Body (CEN).

On the question of assessment of costs, including relevant quantities to be identified and how cost comparisons should be made, little authoritative advice is currently available. At present, there is no official suggestion that information on costs should be available in support of written COSHH assessments or provided in a form which is open, where appropriate, to 'third party' scrutiny. Such information is clearly required, however, not only by employers but by enforcing authorities as well as unions, to help assess the adequacy of compliance with reasonably practicable duties. The overriding concern must be to help employers, enforcing authorities, and unions, to arrive at and have confidence in judgements about the meaning of the law in particular circumstances. This is likely to assume greater importance in the light of new general legislation on risk assessment to be introduced in the UK to implement relevant sections of the EC Framework Directive for health and safety.

In view of the above arguments, the TUC has suggested that further guidance is needed on how to make judgements about reasonable practicability more transparent within the context of written assessment procedures such as those required under COSHH and similar legislation. In short, where appropriate, assessments should attempt to describe dose/effect and control/cost relationships and in a way which is open to critical evaluation by all parties to the health and safety process – particularly safety representatives.

4 OCCUPATIONAL CANCER: A CASE STUDY

The true extent of cancer by exposure to harmful agents at the workplace remains unclear. Nevertheless, evidence available from a number of sources suggests that Britain, together with other industrialized countries, faces a major health tragedy involving the needless loss of many thousands of lives annually. The TUC has sought to raise awareness amongst unions of cancer risks at the workplace – focusing particularly on the identification of carcinogens and the application of key duties in the COSHH Regulations. Despite advances in cancer treatment and screening, the majority of malignant cancers still result in early death and the course of disease is often extremely distressing not only for sufferers but for friends and relatives and health carers as well.

Estimates of the number of cases of cancer which are caused by exposure to harmful agents at work vary enormously.[12] Epidemiologists such as Sir Richard Doll and Dr. Julian Peto have suggested that between 2% and 8% of cancer deaths each year could be prevented if all workplace hazards were removed. Other commentators in the USA have suggested that up to 30% of cancer deaths may be occupationally related.

Whichever estimate is the more accurate it is clear that literally thousands of workers die every year from cancers that are work related – as opposed to only some 400–500 in industrial accidents. On the other hand, whereas the effects of accidents are often immediate and clearly visible, for most cancers the effects tend to appear only years after exposure to the original hazard has ceased. In addition, some types of cancer, such as lung cancer, are very common throughout the population as a whole and this can mask an excess of such cancers among particular groups which have an occupational cause. The fact that factors such as smoking, diet, or life style generally can contribute to cancer formation also complicates the picture. Often cancer is not properly recorded on death certificates as the true cause of death and seldom are records available of substances to which cancer sufferers have been exposed in the past while at work.

In recent years more information has become available about carcinogens either from tests which have been developed to assess the carcinogenic potential of substances or by epidemiological research – looking for excesses of cancer amongst workers who have been exposed to particular substances in the workplace. Despite the fact that there are major difficulties in screening substances – including the sheer number of substances in daily use – and the fact that resources for new epidemiology are limited, there is now a growing number of substances to which varying degrees of suspicion are attached. Whilst this does not mean that all other substances can be regarded as 'safe', it does offer the opportunity to target preventive measures specifically at particular substances and processes where there is good reason to believe that a cancer risk may be involved.

Judging whether or not a substance may cause or promote the development of cancer is beset with a wide range of technical, scientific, and indeed political difficulties. A considerable number of bodies currently carry out assessment of the strength of evidence as to whether or not specific carcinogenic substances are carcinogenic in humans.

Essentially three sources of evidence are used:

i) The potential for substances to cause changes in genetic material in cell cultures – *in vitro* testing;

ii) Animal studies in which cancer incidence in laboratory animals exposed to a substance is compared with the incidence of cancers in an identical group of animals which are not exposed – the 'control group'; and

iii) Epidemiological studies – that is comparing cancer incidence in workers who have been exposed to a suspect carcinogen to a similar group of workers who have not.

There are enormous difficulties in predicting the cancer-inducing potential of substances by any of the above methods. Where significant excesses of cancer can be shown from human epidemiological studies, evidence is usually taken to be fairly conclusive and substances thought to be responsible are usually allocated the status of 'known human carcinogen'. Nevertheless, it needs to be remembered that in such cases the price of such certainty has been high and has been paid by workers who have developed a fatal disease. Clearly therefore every effort has to be made to assess evidence of carcinogenicity by other means and before workers are exposed to such substances.

As stressed earlier, because levels of certainty about carcinogenicity can have major economic consequences those interests anxious to limit the cost of associated preventive measures will usually seek a high level of proof. On the other hand, those anxious to pursue a precautionary approach – particularly unions who represent workers liable to be exposed – will seek action at a lower level. Bearing in mind the wide range of variables involved in the assessment of carcinogens, the TUC view is that the control of exposure to all carcinogens – including suspect carcinogens – should be equally rigorous. In other words workers liable to be exposed to a particular substance should be given the benefit of any scientific doubt about its potential to cause cancer.

Once carcinogenicity has been determined, the policy focus shifts inevitably to substitution or control of exposure under COSHH. The COSHH Regulations offer a comprehensive legal mechanism for controlling cancer risks at work. These Regulations are shortly to be backed by a new EC Directive on cancer which will require employers to take further steps to control carcinogenic risks.

The COSHH ACoP on carcinogens gives additional advice to employers on how to meet their duties under the Regulations when requiring workers to work with cancer-causing agents. The ACoP applies to any case where workers are exposed, or are liable to be exposed, to substances which have been assigned the 'risk phrase' R45 'May Cause Cancer' under the Classification, Packaging and Labelling (CPL) of Dangerous Substances Regulations and any substances and/or processes listed in Appendix 1 to the ACoP. Some of the more significant advice it contains can be summarized as follows:

Substances which were banned in previous health and safety legislation remain banned (2-naphthylamine, benzidine, 4-aminodiphenyl in 4-nitrodiphenyl, salts of these substances, or any substance containing more than 0.1% of these substances);

Minimum information has to be included in all COSHH assessments which should be written documents;

Employers have to consider substitution by less hazardous substances wherever possible (The Code says that carcinogenic substances or processes should not be used or carried on where there is an equivalent or less hazardous substitute);

If carcinogenic substances do have to be used they should be used in closed systems or with use of high-efficiency local exhaust ventilation.

The key to tackling workplace cancer risks lies in ensuring full compliance of employers with the COSHH assessment duty. The TUC has published special guidance on this new legislation in its booklet 'Clean up with COSHH!' which also includes supplementary notes on occupational cancer.

In order to allow assessments for carcinogens to be discussed meaningfully - for instance, between management and safety representatives via joint safety committees - the TUC has advised that unions should seek the following undertakings from employers:[13]

i) The primary COSHH inventory of substances and potential exposures at the workplace should identify all COSHH carcinogens and suspect carcinogens. A special carcinogens sub-inventory should be prepared.

ii) All COSHH carcinogens and suspect carcinogens should be replaced by less hazardous alternatives wherever practicable. Where not immediately possible, target dates for eventual substitution or phasing out should be agreed. Full data on all possible alternatives should be made available to safety representatives.

iii) Where carcinogens have to be used, control should be by total enclosure or other equally effective engineered means. Assessments should explain standards of 'reasonably practicable' control of exposure including information on all cost/risk judgements.

iv) Environmental monitoring arrangements should be discussed with workers and their trade union representatives.

v) Whenever control measures fail, there should be comprehensive and mandatory contingency plans including evacuation procedures and other essential measures to limit exposure.

vi) Appropriate health surveillance procedures should be agreed - with full access of safety representatives to general information arising from such programmes; where necessary facilities for surveillance to be provided for workers after exposure has ceased or they have left employment.

vii) Special information and training programmes should be provided for all those working with carcinogens or suspect carcinogens – with full opportunities for trade union participation.

5 ISSUES FOR THE FUTURE

While the TUC will continue to create maximum pressure for improving arrangements for the evaluation of toxic risks of substances and the setting of appropriate OELs, it is also highly conscious of the danger of a growing gap emerging between legal standards on the one hand and industrial practice on the other. In particular, the continuing growth in the number of small firms[14] has meant that more and more workplaces are ill equipped to deal with occupational hygiene issues. In the coming period the TUC intends to focus on major toxic risk issues – for example as part of its contribution to the 1992 European Community 'Year of Safety Hygiene and Health Protection at Work' which is scheduled to run from mid-1992 to mid-1993. A major theme in the year will be Clean Air at Work. Issues on which the TUC will be focusing particularly include: occupational cancer; reproductive hazards; sensitization; irritancy; solvent reduction; and dust control.

6 REFERENCES

1 'Hazards at Work': TUC Guide to Health and Safety, TUC Publications, London, 1988.

2 Health and Safety Commission Annual Report (1989/90), HMSO, London.

3 Safety Representatives and Safety Committees Regulations (1977), HMSO, London.

4 'Clean Up with COSHH!', TUC Publications, London, 1989.

5 Report of TUC Nuclear Energy Review Body, TUC Publications, London, 1988.

6 'Quantified Risk Assessment: its Input to Decision Making', Health and Safety Executive, HMSO, London, 1989.

7 'Workers Safety and Health', Bulletin No. 110, National Radiological Protection Board, Chilton, March 1990.

8 HSC Plan of Work 1991/92 and beyond, HMSO, London.

9 HSC Toxic Substances Bulletin, Health and Safety Executive, London.

10 COSHH General Approved Code of Practice, HMSO, London, 1990.

11 Working Group on Ionizing Radiations: Report 1987–88, Health and Safety Commission, HMSO, London, 1989.

12 'COSHH Carcinogens ACoP Appendix: Background – Occupational Cancer', HMSO, London, 1989.

13 TUC Health and Safety Bulletin No. 10, TUC Publications, London, 1990.

14 'Health and Safety in Small Firms: the Case Against Deregulation and the Need for a New Approach', TUC Publications London, 1987.

5
Education and the Employee

Annie Rice

INTERNATIONAL FEDERATION OF CHEMICAL, ENERGY AND GENERAL WORKERS' UNIONS (ICEF), PO BOX 472, 1211 GENEVA 19, SWITZERLAND

1 INTRODUCTION

Chemical hazards and understanding them are among the biggest problems facing workers today. New chemicals are being introduced into workplaces at an ever-increasing pace, whilst many chemicals already in use may pose significant long-term threats. Although the more blatantly unsafe practices have generally ceased, the death and disease is a continuing testament that working conditions still fall short of acceptable standards in many workplaces.

Dr. Jorma Rantanen, director of the Institute of Occupational Health in Finland, estimates that there are about 3.8 million new cases of occupational disease due to exposure to chemical substances in the world each year.[1] Acute pesticide poisoning accounts for about 3 million cases, of which 220,000 are fatal (including suicides).[2,3] Ninety percent of the poisonings and 99% of the deaths occur in developing countries. Acute pesticide poisoning is a work-related risk which frequently threatens not only the worker but also family members.

The risks arising from chemical hazards are not solely related to acute effects. Long-term exposure to low levels of certain chemicals which may give rise to cancer is another important area of concern. The **reported** incidence of cancer in the world is 5.9 million $year^{-1}$. This means that there are about 26 million cancer patients at any one time. Using Doll and Peto's estimate that 4% of cancer is occupationally induced, then the global number of cases of occupationally induced cancers could exceed one million. (See also chapter by P. Jacques.)

The past ten years have also seen a disturbing increase in both the number and size of large industrial accidents worldwide. According to Smets,[4,5] during the period 1920–1978 there was approximately one accident resulting in more than 50 deaths every five years (on average), and he puts the current average at two per annum, *i.e.* a tenfold increase. This increase has been corroborated by other risk research. J. Theys,[6] for example, estimates that both the rate and the severity of industrial, mainly chemical, accidents increased between 1940 and 1985 at the rate of 3–4 serious accidents every five years up to 1970, then 15 for the following 5 years, and 30 every five years since then.

Are the statistics – the deaths, diseases, and disasters – the inevitable consequences of modern industrial chemical use, the price that must be paid for technological development, or can they be reduced to a more acceptable level?

2 MANAGING RISK: THE SAFE WORKER APPROACH OR REDUCTION OF HAZARDS?

It is the employers' basic duty to provide a safe and healthy working environment for workers. Thus it is preferable to eliminate or minimize the risks in using chemicals at source. The alternative is to train the worker to be careful, and provide personal protective equipment. This latter is called the safe worker approach. While the latter may often be simpler, quicker, and, in the short run, cheaper, it is also less effective in reducing workplace accidents and diseases.

In this way, it may be argued that the implication in the title of this chapter, Education and the Employee, as a means of reducing risk is false. The very fact of intervention by the law to set a minimum standard of safety is a recognition that safety lies, in part, in the system of work. However, it also depends on the behaviour of all working with particular chemicals. Accidents and diseases do not necessarily, or even usually, occur because of apathy, and rarely because of carelessness, stupidity, or laziness on the part of workers, but rather because of unsafe and unhealthy methods, systems, and processes. Education is a part of the process of inducing a safety minded culture in which to work.

Putting the blame for occupationally induced illnesses on the victims themselves is rarely stated explicitly, but may be present in many different guises. It may lie behind an approach to disease prevention based on worker awareness – using posters to remind workers to wear their protective equipment. It may lie behind the collection of statistics where the emphasis is on identifying workers or groups of workers who are 'more susceptible' to adverse outcomes of chemical exposure. It could also lie behind safety incentive schemes that are becoming more and more common whereby workers are awarded a bonus if they can work a specified number of hours with no time off.[7] They can all be taken to transfer the onus onto workers to handle chemicals safely. Process design, dangerous operating procedures, poor choice of chemicals used, inadequate work organization, supervision, *etc.*, may all play a (larger) role in chemical exposures. The World Bank supports these views in a study of industrial accidents that showed that 80% resulted from shortcomings in management attention to safety. The other 20% could be related to equipment failure or worker error.[8]

Therefore, an effective preventive strategy aimed at reducing chemical risks needs to focus on underlying work systems and organizational behaviour. Solely relying on making users aware through training and education is clearly inadequate.

3 A WORKPLACE STRATEGY TO CONTROL RISKS

The strategic aim then is to induce an organizational culture which regards safety highly and applies a high priority to safe design, safe management and safe working when manufacturing or using chemicals. One requirement for the operation of this strategy is risk management, which requires identification of the hazards, analysis of the nature of the risks, *etc.* The second part of the strategy is to back up procedures with education, training and technical support in order to induce a culture in which safety is a feature of the organization.

The aim in all education and training then, is to have enough knowledge to understand risks in order to improve existing conditions. Active participation of both management and workers is essential when finding practical solutions for risk management. This requires, above all, co-operation between the two parties. While the primary responsibility to provide a safe and healthy working environment rests with the employer, a joint management/worker committee helps everyone work together for improvement. Attitudes to safety will grow if people participate. The aim, then, is for fully informed, mutually-determined controls at the workplace - or safety through co-operation rather than confrontation.

This approach to safety in the use of chemicals represents a sharp break from former attitudes that occupational health was to be seen as an exclusively management prerogative or the province of technical experts. It also raises the question of education as opposed to training. Management often sees worker safety as a function of **training**, *i.e.* making the worker proficient through specialized instruction and practice. It is becoming increasingly apparent, however, that those exposed to chemicals see safety as a function of **education**, *i.e.* developing knowledge and skill, or an awareness and understanding of the risks arising from the manufacture and use of chemicals through experience and study.

4 ON-SITE TRAINING: A MANAGEMENT RESPONSIBILITY

Management has a fundamental obligation to provide its employees with information about the substances liable to be present in the workplace, together with instruction and training on how to use those substances safely (see chapter by K. Kogi and S. Machida). Many countries have introduced legislation to this effect. Section 2(2)(c) of the UK Health and Safety at Work Act (1974) explicitly states that '...the employer is responsible for the provision of such information, instruction, training, and supervision as is necessary to ensure, so far as is reasonably practicable, the health and safety at work of his employees'. The US Occupational Safety and Health Administration's Hazard Communication Standard, promulgated in 1983, also requires employers to evaluate all chemicals used in their workplaces. For hazardous chemicals the employer must develop, implement, and maintain a written hazard communication programme covering container labelling, material safety data sheets, and employee training. It must include a list of hazardous chemicals in each work area, describe how the employer plans to meet the requirements of the Standard, and explain what methods will be used to communicate hazards to the employees.

4.1 Information: Labels and Data Sheets

In many countries the law stipulates that workers must be provided information via labelling of containers to show the identify of the chemical and appropriate hazard warnings. Chemical manufacturers, importers, and distributors are required to ensure that every container of hazardous chemicals they ship is appropriately labelled with such information and the name and address of the producer or other responsible party. In the USA employers are also required to have a Material Safety Data Sheet (MSDS), laid out in a standard format, for each hazardous chemical in use in their workplace. The MSDS is to be used to provide detailed information on each hazardous chemical, including its potential hazardous effects, its physical and chemical properties, and recommendations for appropriate protective measures. MSDSs should be readily available and accessible to employees in their work areas.

Unfortunately, however, this is not always the case. In 1984 a survey carried out by the American International Chemical Workers Union (ICWU)[9] indicated that 30% of ICWU members in the plants surveyed were either minimally informed or were totally uninformed. In only 25% of the workplaces surveyed was it found that hazard warning labels were used as a matter of routine, as opposed to being present only sporadically. It seems that the situation has hardly improved in recent years in the USA, as the Hazard Communication Standard ranks number one on the list of most frequently violated standards during 1990, accounting for 22% of all citations issued by the Occupational Safety and Health Administration.[10] Obviously, a more responsible attitude on the part of manufacturers, importers, and employers towards their obligations to downstream users is called for.

4.2 Training

If these violations of safety data sheet requirements are taking place among some of the biggest chemical companies in the USA, it follows that the situation can only be exacerbated in smaller enterprises, among the self-employed, and among workers in developing countries. To help address these weaknesses in national systems the International Labour Office (ILO) adopted a resolution in 1985[11] which calls for employers, *inter alia* to ensure that all workers, technicians, and managers who play a role in the safety control system of the enterprise be given adequate specialized training for this purpose, and to provide to all workers in the enterprise, **and in a language they can understand**, the necessary training, information, and instructions as well as equipment required for the protection of their individual and collective safety and health at the workplace (Author's emphasis). (See also chapter by K. Kogi and S. Machida.)

The importance of information and training being in a form and language available and understandable to the worker cannot be over-emphasized. It may seem an obvious right, and even unthinkable that it could be otherwise, but in reality, especially in importing countries or where the workforce may include minorities or immigrants, it is not always the case. In the case of the 1984 Bhopal disaster, for instance, language differences may have contributed to the lack of understanding about methyl isocyanate and other hazards. All signs and operating instructions were written in English, even though many of the workers spoke only Hindi. Workers stated that if they wrote in the log book in Hindi they were reprimanded.[12,13]

Information and training, then, in an understandable form and language is of utmost value to those directly engaged in the handling and use of chemicals. The emphasis at workplace level is thus often laid on the practical measures to be taken to work safely and avoid undue consequences. Each employee who may be exposed to hazardous chemicals when working must be provided with this information and be trained prior to initial assignment to work with a hazardous chemical, whenever perception of the risk changes, or whenever hazard information or operating instructions need updating. Whatever the means of instruction at the workplace – videos, hands-on training, classroom teaching, drills or exercises, safety posters, *etc.* – training must be as site-specific as possible. It will require a combination of reading material, videos, and other methods, including face-to-face presentations.

The absolute need for training in the particular technical risks of any one workplace or procedure has been highlighted by the same North American-based ICWU survey quoted above. In the survey, about 80% of workplace situations had associated written job procedures which required the use of personal protective equipment. In practice, fully 25% of the time these procedures were not followed. Almost half of the workers surveyed stated that **lack of training in the procedures was the primary reason for not following the procedures.** Furthermore, 40% reported that it would have been impossible to comply simultaneously with all the working procedures and meet production requirements. In such a situation the employer may be aware that the quota demands unsafe practices but may simply turn a blind eye. Clearly, no worker should be expected to take on an unfamiliar job without being instructed and shown how to do the job in question safely. It becomes apparent when a required job procedure is or is not feasible during hands-on management-instituted training. (See also chapters by T.A. Kletz, A. Mottershead, and D. Collington.)

4.3 Safety Culture

Even if management is committed to inspiring confidence in its workers, the problem still remains that the information and training given may be the bare minimum needed to perform a task, as well as being rather paternalistic. In the words of an employers' association communication[14] the employee requires clear, action-oriented instructions, **free of any need for interpretation** (Author's emphasis). This is **training** in its basic sense, and not **education**.

It may ensure that workers follow instructions under normal operating conditions, but does nothing to ensure that the worker can recognize when abnormal conditions arise, what their significance is, and how to respond correctly to restore safe conditions (accident management). While exercises and drills may exist in some workplaces in anticipation of responding to abnormal operating incidents, they are no substitute for developing an essential safety culture. This can be defined as a set of attitudes and qualities in individuals (and organizations) which ensures that safety issues receive the attention warranted by their significance. It refers to personal dedication and accountability of **all** individuals engaged in an activity which has a bearing on safety in the use of chemicals. It includes an all-pervading safety thinking which allows an **inherently questioning attitude**, the **prevention of complacency**, and a **committment to excellence**[15] (Author's emphasis).

5 EDUCATION: A WORKER'S RIGHT

While management training and instruction are essential at the workplace they put the emphasis on workers carrying out rules defined by people who are not (usually) exposed to the risks, or who may not know the realities of shopfloor work. A basic education in safety culture necessarily means that those exposed to chemical risks are in a position to influence the decisions taken to assess and manage the risks.

It is easy to gain the (false) impression that occupational hygiene is an objective, opinion-free, and neutral scientific endeavour, uninfluenced by the attitudes of companies or governments towards workers. Any workplace control of hazardous chemicals represents only an informed and negotiated compromise between a combination of technical experts, including occupational hygienists, and physicians, economics-driven employers, government agencies, and – more recently in some countries – workers and their trade unions. As those with decision-making powers on workplace chemicals rarely need to consider a personal threat of exposure day-in, day-out, their own interest in the stringency of controls rests only on professional ethics and goodwill. Both these qualities are open to influence in the powerful aura of industry, and this bias will not change until the whole system of control also becomes responsive to the experiences and demands of workers and their trade unions. In many cases workers probably know most about the practicalities of the day-to-day functioning or malfunctioning of a plant. Very often it is a worker who first notices a health problem at the workplace. (This has happened several times in the past and has helped avert personal tragedies.) Above all, it is workers themselves who ultimately stand to benefit - or lose - most from workplace policies on risk management of chemicals.

Therefore it is fundamental that workers, the users of chemicals, be consulted in any programme of workplace health and safety measures; that is, they should be actively involved in hazard assessment, risk assessment, and risk management.

5.1 Educating for Participation

In most industrialized countries worker and trade union participation in workplace health and safety issues is now an established feature of legislation. Indeed, it is no longer acceptable that workplace protection and regulation be separated from trade union activity at the factory floor where the realities of risk are actually encountered. The aim is for worker input into all decisions relating to workplace exposures, processes, and preventive methods mediated via trade unions within a joint union/ management framework. This implies that as many workers as possible are alert as a consequence of their being organized, educated, and informed to the potential dangers they may face at work with chemicals.

Informing and educating workers to become motivated to improve working conditions has become a major concern of trade unions throughout the world. Obviously, it is difficult for trade unions to reach every single worker on the shopfloor, but unions in most industrialized countries and in some developing countries are educating their members elected as safety representatives or to workplace health and safety committees to participate actively in decisions concerning safety at work, to monitor the workplace and identify hazards, to minimize the risks arising from those hazards, and to warn others of the potential dangers (actual risks).

Setting up and operating an effective workplace organization through the election and training of safety representatives and safety committee members is a key element in the trade union approach to reducing chemical risks at work. Substantial gains have been registered in the 1980s, and further advances depend on organizing workers into trade unions with a network of informed and motivated union safety representatives who can operate effectively in every workplace.[16]

That trade unions play a valid role in educating workers on the safe use of chemicals can be seen from a recent study by a US industrial relations expert[17] which shows a link between workers' involvement and reduced accident and illness rates in the USA. According to the study, trade union activity resulted in a dramatically increased enforcement of the US Occupational Safety and Health Act in workplaces. Workplaces that were unionized were more likely to receive health and safety inspections, faced greater scrutiny during these inspections, and paid higher fines for OSHA violations, than comparable non-union establishments. Workers who are members of trade unions were, according to the study, more likely to exercise their rights to a safe workplace than their non-union counterparts. It also confirmed that union health and safety programmes provide workers with up-to-date and detailed information on workplace risks, which made them better equipped to identify risks requiring attention.

This principle, that where there are unions, joint management/labour health and safety programmes are more likely to be established and worker education in safety will be increased, can be corroborated. The rate of fatal injuries over 17 years to 1990 has averaged about 1.4 per 100,000 members of the American United Auto Workers International Union, compared with about 4 fatalities per 100,000 unorganized workers in the US manufacturing industry in general.[18] Although not strictly related to the chemical industry, this may be taken to indicate that unions can greatly influence workplace health and safety.

5.2 The Role of the Health and Safety Representative

A primary role in managing chemical risks is that of the worker health and safety representative. If one aim of improving workplace health and safety is prevention through negotiation, then the safety representative needs to be familiar with the basics of risk assessment, risk evaluation, and risk management. These principles can be translated into the terminology of the safety representative as, respectively, the right-to-know, the right-to-act, and the right-to-refuse dangerous work. In many countries, including the UK, these are legally constituted rights. These rights are not, however, enough on their own to make sure that workers' health and safety is protected. Good organization, training, and technical support are necessary to make sure the rights are used to best advantage.

The health and safety representatives do not have to be technical experts. Their basic tools are common sense and their own eyes, ears, and noses. Those best qualified to identify dangerous situations have to face them every day in the workplace. The representative may not know all the details of the thousands of chemicals with hazardous properties, but he or she can find out about the few that are used in any particular working situation. The aim of trade union education for the safety representative then is to impart the basic skills to be able to spot dangers and to initiate action. Three basic skills – investigation, inspection, and representation – are necessary for the carrying out of the safety representative's functions.

5.2.1 Investigation. The investigation of risks and of dangerous occurrences at work includes examining the causes of work-related accidents and illnesses. The representative should be able to investigate complaints raised by any worker he or she represents relating to that worker's health, safety, or welfare at work. In other words, the representative should be trained as a detective to identify possible risks and then organize appropriate measures to investigate the risk more fully, and, if required, to control the risk.

5.2.2 Inspection. This involves monitoring the workplace on a general, regular basis, often using a checklist: *e.g.* are hazardous gases, liquids, or other materials properly stored, labelled, safe from fire, or access restricted to authorized personnel? Is exhaust ventilation adequate, and protective clothing and equipment available and maintained? In the UK, for instance, the law proposes that a maximum of one inspection may be requested by the safety representative every three months, unless the employer agrees to more frequent inspections, and inspection may be needed if there has been a substantial change in the conditions of work, or new information has been published relevant to the risks of the work. Workplace monitoring by the health and safety representative leads to better implementation of existing occupational health and safety regulations. This is especially important when government factory inspection is infrequent. In many industrialized countries, government cut-backs in public expenditure have meant that both the numbers and resources of factory inspectors are being diminished. In many developing countries factory inspectors are virtually non-existent or vastly undertrained. In India, for example, at the time of the Bhopal accident, the state of Madhya Pradesh employed just two factory inspectors for the whole Bhopal area. Both these inspectors were mechanical engineers with neither the training nor the equipment to assess potential risks posed by chemicals. India has now - since, and because of the Bhopal accident - one of the most stringent and far-reaching laws on health and safety at work, but of what use will it be without a similar stringency in enforcement?

A 100% safety regulation compliance rate would only occur if there were adequate numbers of full-time independent inspection officers with powers of enforcement at the workplace. This is obviously unfeasible. In the meantime, workers' health and safety representatives must monitor the workplace for their own safety and for that of their fellow workers.

5.2.3 Representation. Representation of workers is necessary. Worker safety representatives should be consulted by both management and government inspectors and should receive information from them. In many countries, the representative has the right to call in a government inspector, to accompany the inspector during any visit he or she may make to the workplace, and to receive a copy of any reports made. The employer or an inspector on a workplace visit should also contact the safety representative before the start of an inspection.

5.3 The Rights of the Health and Safety Representative

Obviously, the safety representative cannot be expected to fulfil all of these functions without adequate information, training, and technical support.

5.3.1 Access to Information. Employers are obliged by law in certain countries to provide the safety representative with information on:

i) The technical nature of hazards present in the workplace, and the precautions deemed necessary to eliminate or minimize risks. This includes information provided by consultants, designers, manufacturers, importers, and suppliers of any substance or material used at work;

ii) The operational plans and performance (against nominal capacity) of their undertaking, and any changes proposed insofar as they affect the health and safety at work of their employees;

iii) The occurrence of any accident, near-miss, or notifiable industrial disease, and any statistical records relating to such events;

iv) The results of environmental monitoring and personal health monitoring, provided that information is obtained only with the consent of the people involved, and shall not be used to the detriment of any individual.

In addition to the lax compliance in providing information, there are often a number of restrictions on the safety representative's right of access. Information which would be against the interests of national security or information of a commercially sensitive nature is often not available, the latter on grounds of trade secrecy. In practice, the effect is to weaken greatly the ability of the safety representative to obtain information he or she may need, such as the identity of chemical constituents in a product. He or she cannot then identify possible health effects. As the law stands in many countries, employers are in compliance with regulations if chemicals are labelled simply with a code or trade name, so that if a worker needs to know its chemical identity he or she must first approach the employer and run the risk of being labelled a troublemaker.

On an international level, the workers' right-to-know the substances they are working with, and the rights of safety representatives in this area, should receive added impetus under the provisions of a 1990 ILO Convention concerning Safety in the Use of Chemicals at Work,[19] negotiated between worker, employer, and government representatives from around the world. This Convention requires that all chemicals must be marked so as to indicate their identity. In addition, chemicals classified as hazardous are to be labelled in such a way that the classification, hazard, and safety precautions are easily understandable to workers. Workers and their representatives have the explicit right to receive all information required under the Convention. Effectively, the Convention places a responsibility on employers to ensure that workers and their representatives receive chemical safety data sheets and other information; that they maintain records of hazardous chemicals used and they ensure that workers and their representatives have access to these records; that they provide full information and training on chemical safety; and that they co-operate as closely as possible with workers and their representatives with respect to safety in the use of chemicals at work. As regards trade secrets, the Convention still allows national regulatory bodies to introduce a special provision to protect confidential information whose disclosure to a competitor would be liable to cause harm to an employer's business - but only so long as the safety and health of workers are not compromised thereby.

ILO Conventions are significant in that they can be used as a basis for health and safety policy in countries that have yet to implement any significant policy of their own. Once a country ratifies an ILO Convention, its own laws must comply with the Convention. The ILO can investigate and rule on complaints (including those from workers) that a Convention has been breached. Although, like all United Nations bodies, the ILO cannot ultimately enforce its rulings, they are made public and carry considerable moral weight. Most governments are keen to avoid criticism from this UN agency. (See also chapter by K. Kogi and S. Machida.)

5.3.2 Specialized Training. Safety representatives in those countries that have legally provided for safety representatives are entitled to time off during working hours with pay in order to perform their functions, and in order to undertake appropriate training. This provision has been critical to the success of the safety representative movement, for it has ensured that safety representatives are trained and have time on the job to take initiatives. This is seen as central to the successful functioning of a preventive strategy aimed at bringing management and workers closer together over health and safety issues.

The special training needs on the rights and functions of the safety representative, on the need for worker-oriented policies on occupational health and safety, on how to use the law, on how to organize at the workplace, and on how to encourage the employer to set up appropriate health and safety systems, are often carried out by the trade union to which the representative belongs. Initial basic trade union training for safety representatives is to be encouraged, although follow-up training can be provided by joint industry or employer courses, by educational institutions, or by national safety councils. (It must be pointed out, however, that any trade union training programme is an addition to, and not an alternative for training undertaken by the employer.)

The work of the safety representative is inevitably technical. In addition, technical terms are the currency of negotiation and these have to be learnt and understood. While he or she cannot, and need not, be an expert in the traditional sense in all the complex issues pertaining to health and safety at work, the representative's basic training should enable him or her to know when and where to go for expert opinion or information, and how to evaluate that option. Technical support and advice for trained safety representatives is of crucial importance – representatives who lack such support will be ineffectual and will rapidly become disheartened. Support may take the form of telephone and mail inquiry services on hazards; field inspection services; promotion of awareness through issuing hazard alerts, bulletins, and pamphlets; representation of workers on technical committees; and promotion of worker involvement in health and safety issues in public forums.

5.3.3 Right to Refuse Dangerous Work. If access to information and training allow the health and safety representative to participate in hazard and risk assessment and evaluation, it follows that he or she should be capable of participating in risk management. If, however, in spite of combined preventive measures, he or she believes that a significant risk still exists, then the safety representative should be able to call a halt to work in that situation. Stopping unsafe work should be seen as a **last resort**, to be used only in extreme cases. However, it is important to emphasize that this is a simple extension of the fundamental right that every worker has to refuse to perform work with chemicals that he or she believes can damage his or her health. The safety representative should be able to make use of this right.

Article 13 of the ILO Convention 155 of 1981[20] states that ...'A worker who has removed himself from a work situation which he has reasonable justification to believe presents an imminent and serious danger to his life and health shall be protected from undue consequences in accordance with national conditions and practices'. In this case, it is the individual worker only who may refuse to work in a given dangerous situation. What may be more relevant is legislation to enable the health and safety representative, with special training in workplace hazards as outlined above, to be able to call a halt to work which may endanger someone else's health or life rather than merely advise the worker to use his or her right to refuse to perform the work. The Nordic countries have fully developed the right of the safety representative to instruct colleagues to cease dangerous work pending an investigation by management, or by the labour inspectorate. Every year, the Swedish National Board of Occupational Safety and Health receives about 100 reports of incidents where safety representatives have ordered the suspension of work. Investigations by the Swedish National Board of Occupational Safety and Health have not pointed to any abuse of the right. This suggests that safety representatives have used knowledge gained through training to good advantage.

6 CHEMICAL USE IN DEVELOPING COUNTRIES: SPECIAL NEEDS

The rights of the safety representative outlined above may not be uniformly accepted in workplaces within an individual nation, and are not applied fully everywhere. Users of chemicals in developing countries in particular are at greater risk than their counterparts in industrialized countries where health and safety improvements have allowed for greater participation and knowledge. Safety and health problems associated with the use of chemicals in developing countries are exacerbated by such adverse factors as:

i) A lack of enforcement machinery leading to possible indiscriminate use of very toxic substances that are banned or whose use is severely restricted in industrialized nations.

ii) Low levels of functional literacy among workers using chemicals.

iii) The high comparative cost of personal protective equipment compared with income; its scarcity; lack of training in its use, repair, and maintenance; and scarcity of replacements for disposable components such as filters.

iv) Tropical climatic conditions that place added stress on workers, and also greatly hinder working with personal protective equipment.

v) Poor health care facilities and remoteness of health centres from workplaces.

All these factors tend to expose workers in developing countries to greater health hazards from the use of chemicals. Inducements such as risk allowances encourage workers to take greater risks. Their functional illiteracy may also be exploited. Equally, warning labels and data sheets will be of no use to illiterate workers. In some cases the training of illiterate workers is facilitated by the use of pictograms, or symbols to convey a message without the use of words.[21] Pictograms can be used for training purposes or by workers who have been trained in their meaning and use. Some systems of pictograms need full training courses in their own right, and cultural differences in interpretation may lead to ambiguities.

In addition to these problems, the very nature of much of the work in developing countries – that is generally smaller workshops, self-employed, agricultural workers – means that there is less potential for workers to organize and develop an awareness of health and safety problems. In these cases, it is especially important to minimize risks at the source. Judicious choices of less toxic substances and formulations become extremely important, especially with agrochemicals as those account for many thousands of deaths in developing countries each year.

All those who are responsible for the production, import, storage, sale, and use of chemicals in developing countries have a role to play in ensuring their safety–in–use. Multinational corporations, for instance, should not set up operations or export products which are considered too dangerous for use in their home countries. The standard of safety, levels of training, and worker rights in their foreign operations should be at least as high as in their home countries.[22]

A particularly challenging feature when examining the needs of developing countries is informing and educating an often very wide geographical spread of workers. Simple, ready–to–use information on safety in the use of chemicals must in some way reach a variety of target groups, often involving language and literacy problems.[23] Information about chemicals and how they may be used safely is produced by manufacturers, suppliers, professional associations, consultants, governments, and international agencies. Some of the information is presented in an easy–to–read form and is free of charge. An increasing amount of information is now available in the form of videos, and workers with reading difficulties may find it easier to assimilate in this form.

But...the distribution of this information poses the biggest challenge in information dissemination. As the national agencies of many developing countries lack financial resources, international associations have increasingly taken on the challenge of informing, training, and educating workers in these countries.

6.1 The Role of International Organizations

6.1.1 The International Labour Office. The ILO has, as one of its actions, to promote the implementation of high standards of safety in the use of chemicals at work. It is preparing a training manual on the safe use of chemicals. It has already produced other general manuals on health and safety at the workplace, and specialized manuals such as a guide on safety and health in the use of agrochemicals.[24] All these are intended for use as training aids in ILO technical co–operation projects and to encourage action at the national level. Such training activities are carried out through existing infrastructures – government authorities, employers, and workers organizations – to ensure tripartite involvement in promoting health and safety at the workplace.

6.1.2 The International Programme on Chemical Safety. Another contribution to transmitting basic chemical information to those who use chemicals has been the International Programme on Chemical Safety (IPCS) project on International Chemical Safety Cards[25] and Health and Safety Guides.[26] The Safety Cards occupy a niche somewhere between the product label and the detailed Material Safety Data Sheet. It is intended that they will provide immediate information on safe handling to users of chemicals or those having responsibility for programmes on the safe use and disposal of chemicals, and emergency response. These cards, now appearing at the rate of 200 per year, represent peer–reviewed information, which, according to a survey of workers in the USA, is more understandable by workers than the usual MSDS. (See also chapter by K. Kogi and S. Machida.)

The Health and Safety Guides provide more detailed information, aimed at occupational health services, and those in government agencies, industry, and trade unions who are involved with the safe use of chemicals and the avoidance of environmental health hazards. A manual on chemical safety for workers is in preparation. It will be a companion to the Health and Safety Guides and is likely to be used for training courses for workers.

6.1.3 Trade Union Education Programmes. Education and training programmes in health and safety at work have taken on a high profile in the international trade union movement. Week-long courses for small groups (15–20) of shop stewards or other motivated activists are very successful. Those trained then go back to the workplace and train colleagues in the basic knowledge and awareness of chemical hazards, how to identify risks, and solve problems. Some of the material has been used since the late 1970s, and has been translated into many languages.[27,28]

As workers may not be too interested in merely being told what to do (a major failing in instruction at the worksite, without backing-up information with the reasons for following a particular task) the training packages produced by the international organizations have been based on the use of small discussion groups. Each participant brings his or her experiences to the group. Building on local experience and discussing realistic situations helps develop new knowledge, and part of the experience is learning by doing. It is logical that the more different methods are reinforced, the stronger and more effective learning becomes; or, in the words of the old Chinese saying:

If I hear it, I forget it,
If I see it, I remember
If I do it, I know

Often workers in developing countries can also benefit from the experiences – the achievements, strategies, and defeats – of workers in other countries who may have already tackled an issue relating to chemical safety. Through international organizations and bilateral aid programmes, trade unionists in industrialized countries are helping their colleagues in developing countries to become aware of chemical hazards and how to take charge of their own health and safety when training and information may not be forthcoming.

7 CONCLUSIONS

Recent years have seen a general movement to improve health and safety in chemical use in most countries. None of the issues involved are entirely new or unknown. A tripartite working group in Europe,[29] for instance, concluded that:

i) Awareness of chemical hazards is low, and health and safety issues still focus more on traditional areas;

ii) There is a need for training;

iii) There is a gap between legislation and the reality of health promotion at the workplace;

iv) Responsible participation in decision-making needs includes, as a minimum, adequate consultation and information; and

v) Evaluation of health promotion needs to be accompanied by mechanisms for the transfer of ideas and practices.

What is essential, therefore, is application of preventive methods and an insistence that management invest in a safe working environment as a first priority, backed up by education and training of workers in the safe use of chemicals. Often, a lack of incentive and economic constraints on the part of employers is compounded by inadequately funded government safety enforcement agencies. Under such conditions, worker education programmes become vital. The first line of health defence at the workplace is the individual worker, supported by the information and resources of his or her own trade union.

8 REFERENCES

1 J. Rantanen, 'A Global Overview on Prevention of Occupational Safety and Health Hazards', in 'The Role of Workers in Preventing Occupational Hazards', ed. A. Rice, International Federation of Chemical, Energy, and General Workers' Unions, Belgium, 1986, pp. 5–19.

2 Public Health Impact of Pesticides Used in Agriculture. Report of a WHO/UNEP Working Group, Geneva (in preparation).

3 J. Jeyaratnam, 'Acute Pesticide Poisoning: A Major Global Health Problem', *World Health Statistics Quarterly*, 1990, **43**, 139–143.

4 H. Smets, 'Le Risque Nucléaire dans les Sociétés Industrielles', Colloque Droit Nucléaire, Paris, June 1987.

5 H. Smets, 'Compensation for Exceptional Environmental Damage Caused by Industrial Activities', IIASA Seminar 1985, in 'Insuring and Managing Hazardous Risks', Springer Verlag, 1987, pp. 79–138.

6 J. Theys, 'La Société Vulnérable', Presses de l'Ecole Normale Supérieure, France, 1987.

7 J. Mathews, 'Health and Safety at Work', Pluto Press, Australia, 1985.

8 R. Batstone, 'World Bank Program to Ensure Safe Operations in Underdeveloped Nations', *Chemical Regulation Reporter: Current Affairs*, 28 October 1988, 1153–1154.

9 A. Marcus, D.B. Baker, J.R. Froines, E.R. Brown, T. McQuiston, and N.A. Herman, 'ICWU Cancer Control Education and Evaluation Program: Research Design and Needs Assessment', *J. Occup. Medicine*, 1986, **28** (3), 226–236.

10 'Does Your Employer Comply with the Hazard Communication Standard?', in *Lifelines: OCAW Health and Safety News,* May–June, 1991, **18** (3–4), Oil, Chemical and Atomic Workers Union, Denver.

11 Resolution Concerning the Promotion of Measures against Risks and Accidents Arising out of the Use of Dangerous Substances and Processes in Industry, International Labour Office, 1985.

12 'The Trade Union Report on Bhopal', International Confederation of Free Trade Unions and International Federation of Chemical, Energy and General Workers Unions, Geneva, July 1985.

13 C.R. Krishna Murti, 'A Systems Approach to the Control and Prevention of Chemical Disasters', in 'Risk Assessment of Chemicals in the Environment', ed. M.L. Richardson, The Royal Society of Chemistry, London, 1988, 114–149.

14 CEFIC (European Council of Chemical Industry Federations), 'Information on Hazards of Substances at the Individual Workplace: A CEFIC approach to the issue of how to inform employees on the specific hazards of substances they have to handle at their workplace', Brussels, 1987.

15 J. Dunster, J. Libman, A. Schaefer, and M.F. Versteeg, 'Fundamental Principles for the Safe Use of Nuclear Power', Issue paper No. 1, International conference on the safety of nuclear power: strategy for the future, Vienna, September 1991.

16 A. Rice, 'Organising for Safety', in 'Organising for Safety: An International Trade Union Programme for Health and Safety', ICEF, Brussels, July 1991.

17 D. Weil, 'Enforcing OSHA: The Role of the Labor Unions', 1991, reviewed in *Labor Relations Week*, Jan. 23 1991, Bureau of National Affairs, Washington DC.

18 'Health and Safety Conditions in UAW Represented Plants: Problems and Solutions', International Union, United Auto Workers, *Occupational Health and Safety*, 1991, No.2, UAW, Detroit.

19 Safety in the Use of Chemicals at Work, ILO Convention (No. 170) and Recommendation (No. 177), Geneva, 1990.

20 Occupational Safety and Health and the Working Environment, ILO Convention (No. 155), Geneva, 1981.

21 'Pictograms for Agrochemical Labels', GIFAP (International Group of National Associations of Manufacturers of Chemical Products) in Co-operation with the UN Food and Agricultural Organization, Brussels, 1988.

22 'Is there a Bhopal near you? Unions' drive to prevent chemical disasters worldwide', International Confederation of Free Trade Unions, Brussels, 1985.

23 E.M. Ambridge, 'Pesticides in the Tropics - Benefits and Hazards', in 'Chemistry, Agriculture and the Environment', ed. M.L. Richardson, The Royal Society of Chemistry, Cambridge, 1991, pp. 453-458.

24 'Safety and Health in the Use of Agrochemicals: A Guide', ILO, Geneva, 1991.

25 International Chemical Safety Cards, International Programme on Chemical Safety in co-operation with the Commission of the European Communities, first series, Luxembourg, 1990.

26 Health and Safety Guides, International Programme on Chemical Safety, Geneva, 1987 onwards.

27 'Better Working Environment', International Metalworkers Federation, in co-operation with the Joint Industrial Safety Council of Sweden, Geneva, 1979.

28 A. Taylor, 'A Self-help Guide to Health and Safety in the Workplace', International Metalworkers Federation, Geneva, 1987.

29 'Working for Health at Work', The European Foundation for the Improvement of Living and Working Conditions, Dublin, 1991.

Managing Risk in Manufacture

6

Risk Management of Chemicals throughout Their Life Cycle

Rune Lönngren

FORMER VICE-CHAIRMAN, NATIONAL CHEMICALS INSPECTORATE, PO BOX 1384, 171 27 SOLNA, SWEDEN

1 INTRODUCTION

Long before man-made chemicals were known, mankind utilized many naturally occurring chemicals. Early observations concerning powerful properties of some of these materials resulted in their use for warfare (sulphur, pitch, *etc.*), for hunting (poisoned arrows), and for murder and suicide (hemlock, arsenic). These uses were based on the ability of the materials to cause injury or death. Some hazardous materials, if used properly were found to have beneficial effects. Examples of this are opium, cannabis, and other vegetable materials which in small amounts could mitigate pain and other kinds of suffering but in greater amounts could damage and kill.

Until recently the ability of some chemicals to cause injury or death by acute poisoning was the primary concern. Corrosive, flammable, and explosive properties of some chemicals occasionally raised additional concern.

As long as the risk-picture was limited to a small number of recognized risk factors and the total number of chemicals in practical use was relatively modest the risk management of chemicals was an uncomplicated matter. The overall method was to keep hazardous chemicals out of reach of the general public and to apply simple precautionary measures in the transportation and other occupational handling of these chemicals. The situation has changed drastically for several reasons.

First of all the number of chemicals has increased many-fold during a short period of time. In 1942 the total number of chemicals known was estimated to be 600,000. The number had increased to 4 million in 1977 and is today more than 11 million. Of all these chemicals some 100,000 are considered to be in commercial use. Combinations of two or more of these commercially available chemicals appear in an unknown, but probably very high number of preparations.

Secondly, the growing awareness of environmental pollution included chemicals as a threat to the environment and placed a new dimension on the concept of chemical risk in the sixties. Nowadays risk management of chemicals must take into account not only human risks but also environmental risks.

Thirdly, it has slowly become obvious that adverse long-term effects of chemicals occasionally appear. Effects like mutagenicity, carcinogenicity, and adverse effects on the reproductive system and the immune system are examples of severe obstacles to the use of many chemicals. However, it is costly and often rather time-consuming to investigate whether there is a risk due to these effects.

It must be presumed that there is no such thing as a safe chemical, but that almost all chemicals can be handled safely. The more a sound management of chemical risks can be established, the greater it will be possible to use chemicals in the modern society.

Finally, it is often a matter of dispute to what extent a particular chemical risk may be accepted. It is a societal choice rather than a scientific matter. The perception of risk varies considerably from person to person. A complicating factor is that often those who enjoy the greatest benefit of a certain chemical are not the same people who are at the greatest risk.

2 THE LIFE CYCLE CONCEPT

No chain is stronger than its weakest link. This is certainly true as far as risk management of chemicals is concerned. Therefore the risks in all phases of the life cycle of a chemical have to be managed efficiently.

There may have been a greater or lesser firm feeling of this for quite a time, but only recently has the life cycle concept been widely recognized and practised. An important contribution to the break-through of this concept was a report published in 1986 by Environment Canada. The report[1] had been elaborated by a task force representing industry, governments, labour, environmental groups, and consumers. The title of the report is 'From Cradle to Grave: A Management Approach to Chemicals'.

In the Canadian report the need of a comprehensive approach is stressed. As salient features the following seven life stages of chemicals are mentioned: research and development, introduction to the market place, manufacture, transportation, distribution, use, and disposal. All these stages will be discussed more or less in detail elsewhere in this book. It would therefore be a duplication of effort and a waste of time if these seven phases were reviewed here. Instead, I would like to highlight some general apsects of importance for the risk management of chemicals throughout their life cycle.

3 SOUND MANAGEMENT OF CHEMICALS: A WORLD-WIDE NEED

At the outset some figures were mentioned which demonstrated the almost exponential growth of the number of new chemicals during a short period of time. This enormous increase of chemicals for a great variety of uses has given rise to the expression 'The Chemical Society', usually mentioned in a censorious way. However, as the Secretary-General of the United Nations Conference on Environment and Development put it in a report[2] comprising components of an international strategy for environmentally sound management of chemicals:

> 'The use of chemicals and of chemical processes is an essential element of the development process, and in the promotion of human well-being. They are extensively used by all societies, irrespective of their stage of development. The benefits of chemicals are inestimable'.

The Secretary-General then added that chemicals can result in adverse effects on human health, and can have harmful consequences for the environment.

This is not the place to discuss further in detail the UN Conference on Environment and Development. It must be said that a solid basis for continued and expanded work has been established by extensive international work during the past few decades. The activities have resulted in a great deal of valuable supporting material for risk management of chemicals.

There are some five major types of programme in the UN system and elsewhere:

> Those dealing with one aspect of the management of chemicals, such as the assessment of carcinogenic risk by the International Agency for Research on Cancer (IARC), and the development of recommendations on the transport of dangerous goods by the UN Committee of Experts on the transport of dangerous goods.
>
> Those dealing with the assessment or control of risk for chemicals that are pollutants in environmental media or contaminants in consumer goods, including food, *e.g.* the development and promotion by the International Maritime Organization (IMO) of marine pollution control conventions, and the Codex Alimentarius Commission.
>
> Those dealing with several aspects of management of specific chemicals or groups of chemicals, *e.g.* Food and Agriculture Organization (FAO) Code of Conduct on the Distribution and Use of Pesticides.
>
> Those dealing with many aspects of the management of all types of

chemicals, the two main programmes at the international level being the co-operative venture of UNEP/ILO/WHO in the International Programme for Chemical Safety (IPCS), and the OECD Special Programme for the Control of Chemicals.

Those dealing with activities concerned with technical co-operation and information dissemination, *e.g.* the International Register of Potentially Toxic Chemicals (IRPTC) of UNEP, and the training activities of the UN Specialized Agencies.

Some of the outputs of these activities and programmes of special importance for risk management of chemicals will be mentioned subsequently.

4 RISK ASSESSMENT - PREREQUISITE FOR RISK MANAGEMENT

An efficient risk management requires that an adequate risk assessment has been performed. This means that the hazard of the actual chemical should be known and the exposure situation should have been identified and assessed. A serious problem is that a complete hazard assessment has been carried out for only relatively few chemicals.

The most comprehensive hazard evaluation of chemicals takes place within the International Programme on Chemical Safety (IPCS). These evaluations are published in the Environmental Health Criteria series issued by WHO. In May 1991 118 volumes in total had been issued, comprising evaluations of 94 chemicals or groups of chemicals.

The International Register of Potentially Toxic Chemicals (IRPTC), established in response to one of the recommendations by the 1972 UN Conference on the Human Environment, has incorporated so called Data Profiles for some 600 chemicals in its computerized register.

The European Chemical Industry Ecology and Toxicology Centre (ECETOC) has issued a publication[3] tabulating chemicals reviewed or evaluated by one or more of eighteen organizations. The list of reviewed chemicals comprises more than 2000 items. However, the completeness of the reviews varies considerably between the different chemicals.

A serious deficiency is that data are totally absent or of poor quality even for many high production volume chemicals. This is one of the conclusions of a comprehensive international study carried out within the framework of OECD.[4] A list of high production volume chemicals, based on submissions from 11 Member countries and the EC Commission, comprised 1338 chemicals. The experts identified 521 chemicals as being 'chemicals of concern' in that there appeared to be significant data gaps that need filling before their potential risks could be satisfactorily assessed. It was found that 155 of these high production volume chemicals had little or no data available. Work is in progress to fill the gaps.

A consequence of this lack of hazard data is that it often can be extremely difficult to obtain a solid basis for an adequate risk management. This is not an excuse to set aside risk management.

In the absence of sufficient knowledge a 'worst case' strategy should be followed. Often it may be possible to modify such a severe scheme if at least partial testing is performed. If so it is advisable to follow the test methods given in OECD Guidelines[5] for testing of chemicals and observe the OECD principles[6] of good laboratory practice.

5 ALL CHEMICALS SHOULD BE CONSIDERED AS HAZARDOUS CHEMICALS

Originally chemicals were considered as hazardous chemicals only if it had been proven that they caused injury or death to humans. During the past few decades the potential for damage to the environment has been added to the picture. More recently, there has been a growing tendency to regard exposure to a chemical as a risk factor until it has been proven that the chemical is safe.

The modern legislative view is to consider all chemicals as hazardous and to mitigate the precautionary requirements for those where it is beyond reasonable doubt that they are safe. This is the concept of the recently adopted ILO Convention[7] concerning safety in the use of chemicals at work just to mention one example.

Two matters should be mentioned in this context.

It is not possible to prove scientifically that a chemical is safe. What can be done, and what is done is to find out whether certain criteria, based on experience, are exceeded or not.

The risk of a chemical to health and environment is a function of both the intrinsic properties of the chemical and the exposure. This is the modern technical way of expressing what Paracelsus (1493–1541) almost 500 years ago postulated, saying that everything is a poison and nothing is a poison, the dose alone decides.

6 GOOD CHOICE OF CHEMICALS

In 1952, the ILO General Conference adopted a resolution expressing the desirability of applying the principle of substitution of harmless or less harmful substances for harmful substances. The observation of this principle has been stressed in several subsequent ILO Acts. One example is the ILO Convention[8] concerning Prevention and Control of Occupational Hazards caused by Carcinogenic Substances and Agents. According to the

Convention every effort shall be made for carcinogenic substances and agents to which workers may be exposed in the course of their work to be replaced by non-carcinogenic substances or agents, or by less harmful substances or agents. The choice of substitute substances or agents shall take into account the carcinogenic, toxic, and other properties of the substance.

Indeed, the careful choice of a chemical, taking safety aspects into account, is the most efficient means to protect human health and environment from adverse chemical effects. It simplifies risk management in all phases of the chemical's life cycle.

Applied in environmental contexts the substitution method has great advantages. When chemicals have been emitted to the environment, be it to air, to water, or as solid wastes, pollution problems of various degrees of severity arise. Persistency, toxicity, and other properties of the emitted products indicate how harmful the pollution may be, and how costly and time-consuming it may be to restore the environment.

The more degradable, the less toxic the emitted chemical is, the less complicated the pollution will be. It may even be negligible. As always, prevention is better than cure.

Under the label of Risk Reduction, OECD has recently introduced a programme to promote the choice of good chemicals. The Council[9] has decided that Member countries shall establish or strengthen national programmes aimed at the reduction of risks from existing chemicals to the environment and/or the health of the general public or workers.

Among other things it has been recommended that, where appropriate, Member countries undertake concerted activities to reduce the entire life cycle risks of selected chemicals. These activities could encompass both regulatory and non-regulatory measures including: the promotion of the use of cleaner products and technologies; emission inventories; product labelling; use limitations; economic incentives; and the phasing out or banning of chemicals. An example of this strategy is an initiative[10] taken recently in USA. A list of 17 chemicals has been published and EPA has called for 1988 releases of these chemicals to be reduced by 33% by 1992 and halved by 1995.

Clearly, risk management has to take place in every phase of the life cycle of the chemicals. Experience suggests that the weakest link of the chain is often downstream, *i.e.* due to its use in small businesses and/or by the general public. The awareness of risk and the resources to meet the risk often are less developed than those of the manufacturer. This fact puts a supplementary duty on the manufacturer to produce chemicals as harmless as possible with regard to human health and to the environment.

7 THE LIFE CYCLE BEGINS AT THE CONCEPTION

The Toxic Substances Control Act (TOSCA)[11] effective in USA on 1st January 1976 introduced the mechanism of an inventory listing all chemicals on the market (existing chemicals) and a premanufacturing notification system for all other chemicals (new chemicals). A Directive describing a similar system[12] prior to marketing was passed in the European Community in 1979 and was introduced in 1981/82. Several other countries have introduced similar legislation. Although details differ, in principle all these schemes for the notification of new chemicals include a procedure for a preliminary hazard assessment and allow the relevant authority to restrict the use of the chemical, if warranted.

These schemes were the implementation of 1974 OECD Council recommendation[13] that, prior to marketing of a chemical, its potential effects on man and his environment should be assessed. In 1982 the OECD decided that sufficient information on the properties of new chemicals should be available before they are marketed in Member countries to ensure that a meaningful assessment of hazard to man and the environment can be carried out. This decision, binding for Member countries, was supplemented by a recommendation that a specified set of data (MPD) should serve as a basis for a meaningful first assessment of the potential hazard of a chemical to health and the environment.[14]

The MPD consist of chemical identification data, production/use/disposal data, recommended precautions and emergency measures, analytical methods, physical/chemical data, acute toxicity data, repeated dose toxicity data, mutagenicity data, ecotoxicity data, and degradation/accumulation data. Moreover, the following provisions for flexible application of the OECD MPD were introduced: (i) Due regard may be given, on a case–by–case basis, to the scientific and economic factors that may influence the need for and the scope of testing; (ii) Member countries may omit or substitute certain tests or ask for them in a later stage of initial assessment, as long as they can justify their course of action.

The examples of national legislation and international action[5] given above focus on the need to examine the potential hazards posed by a chemical at an early stage in the research and development. However, the life cycle of a chemical does not begin when it is put on the market. It begins when the decision is made to produce it. Ideally part of the background material for such a decision should be a hazard assessment based on MPD or similar data.

The design of a new chemical should be accompanied by chemical and biological testing and studies of the structure–activity relationships (SAR). This is a time–consuming and expensive undertaking. It has recently been estimated[15] that the cost of a 3 month repeat dose study is about $600,000 and the cost of a chronic toxicity and carcinogenicity study is more than double this sum. Inevitable, a manufacturer involved in such costs for safety already in the planning stage must have some protection of his intellectual property in order to be motivated to innovate. This will be considered later.

8 AWARENESS, INFORMATION, TRAINING

The first condition for successful risk management of chemicals throughout their life cycle is an awareness of the hazards in every phase. It is a shared duty of government, industry, national and international organizations, press, and other media to give rise to such awareness.

To be successful all campaigns to increase the awareness of chemical hazards must be open, frank, and informative. The risks should not be over-estimated. Recommended ways and means to protect human health and environment should be realistic.

Everybody active in any business dealing with chemicals should be educated to protect their health and the environment properly from the adverse effects of chemicals which appear at the workplace. The employer has the main responsibility for such education. Most international organizations, active in the field of chemical safety, contribute to develop training programmes and to make supporting educational material available. This is the case with IPCS and its sponsors UNEP, ILO, and WHO. Also FAO, UNIDO, and the World Bank give valuable contributions.

A special problem is to what extent details on chemicals must be confidential. The most thorough discussion on this has taken place within the framework of the OECD Chemicals Programme. The discussion has resulted in three Council Acts[16–18] with recommendations concerning protection of proprietary rights, exchange of confidential data, and a list of data which should be non-confidential.

The recommendation concerning proprietary rights is very much in line with conclusions drawn at the FAO Governmental Consultation on Pesticides held in 1982 in which industry participated actively. Among other things the following was concluded.[19]

> 'It is important to appreciate that the issue of concern to industry is the unauthorized use either directly or indirectly by a competitor, of data generated by another company, *i.e.* without either license or agreement to do so. Confidentially *per se* is not important except where it is related to accepted trade secrets, *e.g.* manufacturing or formulating know-how'.

The OECD list of non-confidential data includes:

Trade name(s) or name(s) commonly used;
General data on uses;
Safe handling precautions to be observed in the manufacture, storage, transport, and use of the chemical;

Recommended methods for disposal and elimination;
Safety measures in case of an accident;
Physical and chemical data with the exception of data revealing the chemical's identity (*e.g.* spectra);
Summaries of health, safety, and environmental data including precise figures and interpretations.

It should finally be clearly understood that adequate education and training is a prerequisite to avoid chemical accidents.

A great number of chemical accidents occur yearly, although only the great disasters are highlighted by the news media. Three such disasters during the past two decades have resulted in an extensive international work to prevent chemical accidents. They are the dioxin release in Seveso in 1976, the runaway reaction of methyl isocyanate in Bhopal in 1984,[20] and the warehouse fire in Basle 1986 discharging contaminated waters into the Rhine. The Bhopal disaster at which more than 2500 people were killed and some 200,000 were injured has had a special impact.

Two concrete results in particular of the intensified international work in this area should be mentioned. One is a manual[21] concerning major hazard control issued by ILO. The other one is APELL, a publication[22] issued by UNEP presenting a process for responding to technological accidents. APELL is an acronym for Awareness and Preparedness for Emergencies at Local Level. The programme involves publication of technical documents, training, and technical assistance to support demonstration activities.

It would be beyond the requirements of this chapter to elucidate any details regarding the various activities to prevent chemical accidents. But clearly adequate education and training is the key issue. Although a number of major accidents have occurred with disastrous consequences, the great number of small accidents occurring every day world-wide, overall give rise to even more suffering and losses.

9 DISTRIBUTION OF DUTIES

Everybody coming in contact with a chemical has to manage the situation in such a way that possible negative effects of the chemical be minimized. A prerequisite for this is an unbroken chain from the manufacturer to the final user through which essential information on the chemical can be transmitted. Sufficient information on significant properties of the chemical and relevant precautionary measures to be taken are minimum requirements.

In practice it is necessary to have balanced and up to date legislation and a well functioning enforcement system. The competent authority must have the power to implement and adopt the legislation in a flexible way.

The main responsibilities to provide accurate information on a chemical stays with the producer. Every supplier, whether manufacturer, importer, or distributor, should ensure that adequate information follows the chemical to everybody handling it.

All those who use the chemical must note and act on the information given.

An adequate labelling of chemicals is beyond doubt the most fundamental means of giving information, including precautionary measures, on the chemical. As the trade in chemicals is international it is crucial that common principles and harmonized rules are observed world-wide. A lot of work has been performed with this goal in mind but much remains to be done.

The first productive international work in this area was carried out during the 1950s within the framework of the West Pact and the Western European Union. This work was continued by the Council of Europe, who published in 1962 the famous yellow books comprising lists of some 500 chemicals with recommendations for their labelling. This work is the basis for the current rules for labelling of chemicals within the Member countries of the European Communities. The harmonized labelling system within the EC has had an impact also outside the EC area.

Recently an initiative[23] has been taken in ILO to consider the harmonization of classification and labelling of chemicals world-wide.

10 REFERENCES

1 'From Cradle to Grave: A Management Approach to Chemicals', 56 pp., Environment Canada, Ottowa, 1986.

2 UN Document A/CONF.151/PC/35, 1991.

3 ECETOC, Existing Chemicals, Technical Report No. 30(3), 210 pp., Brussels, 1990.

4 R. Lönngren, 'International Approaches to Chemicals Control', Stockholm, 1991, in press.

5 OECD Guidelines for testing of chemicals, looseleaf binder, Paris, 1981, with later supplements.

6 OECD Document C(81)30(Final), Paris, 1981.

7 ILO Convention No. 170, Geneva, 1990.

8 ILO Convention No. 139, Geneva, 1974.

9 OECD Document C(90)163(Final), Paris, 1991.

10 US Environmental Protection Agency, 'Voluntary Program Aimed at Reducing Toxic Releases', Chemicals-in-Progress Bulletin 1991, 12 (1).

11 US Public Law 94-469, Washington, DC, 1976.

12 EC Council Directive 79/831/EEC, *Off. J. Eur. Commun.*, Brussels, 1979.

13 OECD Document C(74)215, Paris, 1974.

14 OECD Document C(82)196(Final), Paris, 1982.

15 IPCS/IRPTC, 'International Registry of Chemicals Being Tested for Toxic Effects, CCTTE', p. I/2, Geneva, 1990.

16 OECD Document C(83)96(Final), Paris, 1983.

17 OECD Document C(83)97(Final), Paris, 1983.

18 OECD Document C(83)98(Final), Paris, 1983.

19 FAO Report AGP: 1982/M/5, p. 18, Rome, 1982.

20 C.R. Krishna Murti, 'A Systems Approach to the Control and Prevention of Chemical Disasters', in 'Risk Assessment of Chemicals in the Environment', ed. M.L. Richardson, The Royal Society of Chemistry, London, 1988, pp. 114-149.

21 Major Hazard Control, A practical manual, 296 pp., ILO, Geneva, 1988.

22 APELL, 61 pp. UNEP, Industry and Environment Office, Paris, 1988.

23 Resolution by the International Labour Conference, ILO, Geneva, 1989.

7

An Introduction to Managing Risk in Chemical Manufacture

David M. Collington

DYESTUFFS AND CHEMICALS DIVISION, CIBA-GEIGY AG, BASLE, SWITZERLAND

1 GENERAL APPROACH

Chemical production is the basis of our industry and involves a wide spectrum of risks. Both the raw materials and finished products can be toxic, corrosive, flammable, explosive, and undergo decomposition. They are stored, processed, and transported with the aid of sophisticated equipment which has been designed, operated, and maintained by people who may make mistakes.

Since there have been repeated reminders that the risk in chemical production cannot be zero, a systematic approach to the management of these risks is required. By this means the residual risk can be reduced to an acceptable level.

2 ESTABLISHMENT OF BASIC DATA ON THE CHEMICALS, EQUIPMENT, AND PROCESSES

In the chemical industry, managers must know the properties and behaviour of the substances being handled in their area of responsibility. They must also be aware of the possibilities and limitations inherent in their equipment. Within CIBA-GEIGY a formal risk analysis method has been used for more than 10 years. Its aim is:

To identify hazards
To evaluate the risks
To take appropriate measures

In order to identify the hazards, the first step is to assemble basic data on the chemical process and equipment. To facilitate this process, special test methods for the characterization of key physical properties were developed in the 1960s. These continue to be a very important tool in current risk analysis.

The basic data are compiled into tables, which illustrate, for example, the interaction between raw materials and the materials of construction used for the equipment. It is important to realize that such data must include information on both human toxicity and environmental impact, as well the more typical chemical and thermal data.

3 EVALUATION OF RISK

When the decision on a particular process and the equipment to be used has been taken, it is then necessary to search for hazards:

What could go wrong?
What could be the consequences?

Our risk analysis procedure is based on techniques from the space and aircraft industry, specifically adapted to our needs. It is vital to proceed systematically at this stage. It requires that these two questions above be asked for each step of the process in relation to the physical, toxicological, and environmental impacts.

The hazards of risk scenarios identified are then evaluated with respect to severity and probability. Such evaluations are not normally quantified since this is not possible for such factors as maintenance, training, and leadership.

It is important that these evaluations are carried out by a team composed of the relevant disciplines and that the conclusions are carefully and properly documented.

Of the two factors, the severity of a possible incident is clearly the one which determines the preventive measures to be taken in the fields of chemistry/technology organization and personnel. In this context, we must be aware that all decisions made as to the necessity for and the extent of additional measures are a matter of judgement. This is particularly difficult in the case of high severity and low probability. This is why the relatively few operations in CIBA-GEIGY with a high risk potential have been repeatedly evaluated, using different teams and new methods, sometimes involving members of the corporate management.

The quality of this risk analysis is critical in reducing the residual risk to an 'acceptable' level. The translation of this analysis into practical safety measures will only be as complete and effective as the information on which it is based.

No measures will be taken against risks not recognized in the risk analysis. Inefficient or excessive measures are the result of incomplete basic data or inaccurate or incomplete assessment of the risks.

It is essential that the safety measures are practical. It should not be forgotten that the means of prevention must be applicable to the situation or country concerned. For example, it is wrong to introduce sophisticated electronic controls in a third world country where the availability of proper maintenance or spare parts is limited.[1] In addition, the use of such equipment transfers the residual risk from direct process error to that from failure of the control equipment. This often leads to the assumption by the process operator that the process cannot fail. Where complicated maintenance from very well trained personnel is required this is often not the case.

From the writer's personal experience, two examples can be cited of this problem. In one case, a physical loss of product from an automated plant was immediately disputed by the operator since the computer panel showed everything to be in order. In the second case, random actuation of the control valves occurred when a power pack partly failed giving a low operating voltage. Such risk must also be taken into consideration.

4 REDUCTION IN CONSEQUENCES

There are two principal ways in which the risk can be reduced: by reduction in either the severity or the probability of an incident. There is a principal difference. Reduction in the possible consequences (severity), usually leads to an inherent and permanent improvement in the safety of a process or installation. However, since the probability can never be zero, the improvement here often only remains effective if properly managed and maintained.

If for example, a solvent is replaced with one of sufficiently greater flash point, the risk of an explosion in case of leaks can be eliminated. The normal alternative of using solvents in explosion proof equipment still gives a residual risk despite elimination of sources of ignition by technical and organizational means.

This example also serves to illustrate the fact that the reduction or elimination of a risk cannot always be a matter of choice. The options for a suitable solvent for a process may not permit use of one with a high flash point. This trade–off should be made only when all the factors have been assessed. Again, the location of the unit and its sophistication play a key role. An extreme example is the manufacture of explosives in small remote units where a runaway reaction causes no significant damage (see chapter by T.A. Kletz).

The use of small, often continuous units is a good example of a reduction in severity. The use of a small plug flow reactor for the diazotization of aromatic nitroanilines reduces the volume and hence the consequences of a thermal decomposition by 50–100 times when compared with a conventional batch reactor.

The complete elimination of the storage of a hazardous intermediate can even be achieved. CIBA–GEIGY have developed a process for the continuous and variable generation of phosgene from chlorine and carbon monoxide. With this new technology it is no longer necessary to store 15 tonnes of liquid phosgene.

A study of material flow or supply logistics can also help to reduce the hazard. In one case it was decided to transfer ethylene oxide reactions to the location of an ethylene oxide producer thus eliminating storage and transport of this explosive toxic chemical in a densely populated area. In another case, it was recognized only after accidentally overfilling a chlorosulphonic acid tank that 4 month's supply was stored close to a reliable supplier.

5 CONTINUOUS REVIEW AND RESEARCH

The types of measure discussed so far are of a fundamental nature. The choice of the synthesis route, of the process technology, are all conceptual decisions which must be taken early in process development. For this reason, the risk analysis must begin as soon as process development commences. Thereafter, the major hazards should be documented and their reduction be made a prime goal of the development work. This forms the basis of the iterative process of risk analysis with review and research.

This is so important to the chemical industry that a proper appreciation should form part of our University curricula. It must be recognized that only if we sharpen the chemistry student's awareness and skills in 'risk reduction by means of process development' can the general culture be changed. At present attempts are often made to reduce the risks too late when the process and plant can no longer be changed.

From CIBA-GEIGY, two examples of research for safety can be cited:

Chemical engineers have developed a special reaction calorimeter which enables the development chemist to measure heat evolution throughout the reaction. In the past 20 years, this analytical instrument has made a major contribution to process safety. It is now available commercially from Mettler.

In the field of gas and dust explosions, full-scale experiments in the Jura mountains have enabled us to develop economical safety technologies. This knowledge is also relevant for other industries such as the food industry.[2,3]

In order to make the process of review cost-effective and efficient, good contact between the plant, engineering, and development is essential. A well trained development chemist can only be effective if he knows the possible constraints and opportunities available on the full scale.

6 CLEAR ASSIGNMENT OF RESPONSIBILITY

The responsibility for safety and also for the residual risk and possible incidents lies with the line manager, the plant manager, who must finally accept the results of the risk analysis. Within CIBA–GEIGY regulation on process safety the plant manager has the specific right to refuse to implement a process. In this way the operating responsibility is allocated to a person not a function.

In order to manage this responsibility, the manager must know what to expect of his team in terms of knowledge, abilities, and reliability. He must also know what measures in organization and technology awareness are required to ensure safety. Both of these areas demand systematic training, good organization, clear operating instructions, proper supervision, and a sensible system to check the operation. If these measures are properly introduced, they form the basis of safety–conscious plant management and contribute equally to the minimization of risks as technical safeguards.

To achieve the above, the safety expert must help the plant manager carry the risks. He must propose proper measures which are well understood and make realistic demands, not excessive ones in the hope that part will be fulfilled. He must help with the training process and give advice and clarification as needed.

A classic example of over specification is to demand that all operators wear safety visors when only spectacles are required. This is a certain way to ensure that the regulation falls into disrepute. A further example is the choice of special safety clothing. This must be acceptable in hot weather. Otherwise the operator may ignore the instruction unless the use is exceptional. Wherever possible, exposure risks should be engineered out so as to minimize the use of such clothing.

In order to establish a sensible standard, CIBA–GEIGY have developed their own safety and environmental regulations. To ensure that these are practicable, they are implemented only when they have been accepted unanimously by a committee of higher line managers representing all the manufacturing sites and divisions in the parent company. This is the Central Safety Committee. This group of about 15 people study the drafts in depth, consult with their various collaborators, and, following suggestions and amendments, bear the responsibility for ensuring that the safety policies and regulations are practical and appropriate. To date some 25 regulations have been compiled on which risk evaluations are based.

7 PROCESS AND EQUIPMENT MONITORING

This can be carried out both periodically and continuously. In CIBA–GEIGY, a properly recorded system of safety and maintenance checks has been introduced to ensure that the equipment is functioning correctly. In the case of automated processes, the appropriate process parameters are then built into the process control system to minimize the risk of dangerous process deviations. These range from simple pneumatic interlocks to sophisticated computer models. Again, it must be stressed that the systems themselves can be subject to human error or mechanical failure. Thus, the manager must apply local knowledge to obtain the right level of control sophistication.

8 RELATIONSHIP WITH THE AUTHORITIES

As a result of the systematic approach, a safety and ecology culture has been developed over many years and is deeply rooted in all levels of our organization. This is true of many chemical companies. Despite this, the atmosphere of mutual trust has been severely shaken by catastrophes such as those of Seveso, Bhopal,[4] and, most drastically for CIBA-GEIGY, the Sandoz fire in Schweizerhalle.

A possible explanation is that our safety tools, including risk analysis, have always been used internally according to our own standards. However, today not only industry but also the authorities are challenged by the public at large concerning their overall technical competence.

The current question is therefore, what goals must be set to address these challenges and reduce the tension between the chemical industry and society? If we take the state seriously, then it must be accepted that it will become active, through the various authorities, to ensure the protection of man and the environment against damage.

Support must therefore be provided by giving technical knowledge and experience. However, the state must also exercise its control function in a credible way. Here the chemical companies must help by making their safety endeavours, their risks, and their own control mechanism more transparent and understandable. Their risk analysis will play an increasingly important role in this communication.

In the same way, the analytical data on emissions will be a key tool in communication with the authorities concerning environmental protection. However, such data cannot replace judgement when evaluating the relevance of safety risks or emissions in discussion with the authorities.

9 RELATIONSHIP WITH OUR NEIGHBOURS AND EMPLOYEES

It is not enough to have the goal of a successful relationship with the authorities. The chemical industry must also take the fears of a sensitized public very seriously. One way to achieve this is to minimize the possible consequences of an incident. The consequences of accidental environmental pollution must be taken more seriously. Where the consequences of a leak could lead to a damaging effect, the chemical industry must be more ready to apply the principle of the 'second barrier'. This can be achieved by building a containment for a potentially hazardous installation, using double walled pipelines and so on. In this way a simple human error cannot lead to a catastrophic result. Evaluation of the need for this barrier will involve the risk analysis procedure to establish when the consequences are unacceptably high, however low the residual risk.

However, it is not sufficient to evaluate and minimize these risks; the chemical industry must also communicate with its neighbours to explain what its business is all about, the benefits it brings and the problems that are faced. This requires courage and the industry must be ready to challenge its own position. One problem is to explain to a layman the complex technology involved in chemical manufacture. Fear of the unknown often brings a desire for 'zero risk'.[5]

Another difficulty is that the risk and benefit are often experienced by different people. One's immediate neighbours are concerned by the risk from manufacture, whilst one's customers, employees, and shareholders experience the benefits of the enterprise.

To further the communication process, the chemical industry should be able to use its own employees as ambassadors. This brings a double benefit. Firstly, in explaining and sharing the knowledge involved in the operations, the process of risk reduction can be assisted. Secondly, confidence is created from the openness that spreads to one's neighbours.

10 CURRENT TECHNICAL DEVELOPMENT

Increasingly, anlaytical chemistry plays a key role in our technical endeavours. A few examples of this are:

The use of on-line analytical tools to control continuous processes, which reduce the product inventory and hence the risk potential.

The use of fast acting warning devices to activate the second barrier. For example, the switching of waste water to a separate emergency retention tank.

Detailed evaluation of the kinetics and reaction mechanism of complex chemical syntheses.

The continuous monitoring of waste emissions to improve the control on the environmental impact of one's activities.

From these examples it can be seen that modern analytical techniques are of great value in the evaluation and control of chemical processes.

These techniques are applied to all aspects of the supply chain: from the purchase of the raw materials to the delivery of the end product.

11 CONCLUSIONS

From a true understanding of chemical processes, a cost-effective, safe, and environmentally friendly process and product can be developed. A proper analysis of the risks involved can support this development and assist in informing and enlightening the public at large. In this way society can obtain the benefits from the chemical industry in a mutually acceptable manner.

12 REFERENCES

1 A. Amadi, 'The Effects of Nitrogen Fertilizer Plant Effluent Discharges in Soil', in 'Chemistry, Agriculture and the Environment', ed. M.L. Richardson, The Royal Society of Chemistry, Cambridge, 1991, pp. 221-231.

2 W. Bartknecht, 'Explosions, Cause, Prevention, Protection', Springer-Verlag, Berlin, 1981.

3 W. Bartknecht, 'Dust Explosions, Cause, Prevention, Protection', Springer-Verlag, Berlin, 1989.

4 C.K. Krishna Murti, 'A Systems Approach to the Control and Prevention of Chemical Disasters', in 'Risk Assessment of Chemicals in the Environment', ed. M.L. Richardson, The Royal Society of Chemistry, London, 1988, pp. 114-149.

5 M. Mercier, 'Risk Assessment of Chemicals: A Global Approach', in 'Risk Assessment of Chemicals in the Environment', ed. M.L. Richardson, The Royal Society of Chemistry, London, 1988, pp. 73-91.

8

Managing Risk in Chemical Manufacture

Trevor A. Kletz

DEPARTMENT OF CHEMICAL ENGINEERING, UNIVERSITY OF TECHNOLOGY, LOUGHBOROUGH, LEICESTER LE11 3TU, UK

'... a concern for safety which is sincerely held and repeatedly expressed but, nevertheless, is not carried through into action is as much protection from danger as no concern at all.' – Official report.[1]

1 INTRODUCTION

The UK chemical industry has a good safety record despite the fact that many of the materials handled are flammable, toxic, or explosive, and sometimes all three. Nevertheless, the record could be better and the standard of many smaller companies is below that of the best. This is realized by directors and senior managers who continually exhort their staff to do better, as a glance at any company newspaper will show, and, on the whole, they do not grudge the resources necessary. There is often, however, a significant difference between their approach to safety and their approach to other problems.

If cost, output, efficiency, or product quality cause concern, senior managers identify the problems, agree actions, and receive regular reports on progress but all too often safety problems do not get this same detailed attention. Instead there are merely exhortations to do better. Unfortunately, there is no substitute for personal attention to the questions discussed in the following pages. The procedures described are based on the practice of the best companies.

2 INHERENTLY SAFER DESIGN

The traditional way of making a plant safer has been to *add on* protective equipment to control the hazards, often a great deal of equipment. In an inherently safer design the hazards are avoided by the use of so little hazardous material that it does not matter if it all leaks out (intensification), use of a safer material (substitution), use of hazardous materials in the least hazardous forms (attenuation), or elimination of hazardous equipment or operations.[2]

As an example, consider the manufacture of nitroglycerine (NG). At one time it was made batchwise in large stirred pots containing about a tonne of material. The operator had to watch the temperature closely as if the reactor got too hot it blew up. To make sure he did not fall asleep he sat on a one-legged stool; if he fell asleep he fell off. This process continued in use until the 1950s.

If we were asked to make this process safer, most of us would add on to the reactor instruments for measuring temperature, pressure, flows, rate of temperature rise, and so on, and use these measurements to operate valves which stopped flows, increased cooling, opened vents and drains, *etc*. By the time we had finished, the reactor would hardly be visible beneath the added-on protective equipment.

However, when the NG engineers were asked to improve the process they did not proceed in this way. They asked why the reactor had to contain so much material. The obvious answer was because the reaction is slow. But the chemical reaction is not slow. Once the molecules come together they react quickly. It is the chemical engineering - the mixing - that is slow. They therefore designed a small well-mixed reactor, holding only about a kilogram of material, which achieved about the same output as the batch reactor. The new reactor resembled a laboratory water pump. The rapid flow of acid through it created a partial vacuum which sucked in the glycerine through a side-arm. Very rapid mixing occurred and by the time the mixture left the reactor the reaction was complete. The residence time in the reactor was reduced from 120 minutes to 2 minutes and the operator could now be protected by a blast wall of reasonable size. Similar changes were made to the later stages of the plant where the NG is washed and separated.[3]

Safety studies usually take place late in design. To design inherently safer plants it is necessary to carry out safety studies much earlier than in the past, at the conceptual or business analysis stage when we decide which product to make, by what route, and where to locate the plant, and at the flowsheet stage when we decide which type of equipment to use. Often there is not enough time for the necessary development work, as the new plant is required quickly. To break out of this impasse we should be thinking of the plant after next as well as the next plant. While designing a plant we are conscious of all the changes in design that we would like to make but do not have time to make. We should start work on them now, ready for the plant after next.[4]

Inherently safer plants will not occur unless senior managers are involved. Only they can ensure that safety studies are carried out early in design, that resources are made available for 'plant after next' studies, and that inherently safer design is accepted as one of the aims of the organization. They should ask for regular reports on progress in reducing inventories. When lecturing to design engineers and operating managers on this subject the writer is often told that he is speaking to the wrong audience. He should be speaking, his audience say, to their senior managers.

3 HUMAN ERROR

At one time it was usual (in some companies it still is) to blame most accidents on human failing, meaning by that a failure by the injured man or a fellow worker. There is little that managers can do, it was said, except tell people to take more care. Today it is widely recognized that people will inevitably make mistakes from time to time and therefore we should design user-friendly plants that can withstand human error (and also equipment failure) without serious effects on safety, output, or efficiency. Plants should be designed which, as far as possible, are free from opportunities for human error (or, if that is not possible, which provide opportunities for recovery or protect people from the consequences). We should accept people as we find them and design accordingly; we should not expect people to change to match the deficiencies in our designs. We can get better pumps, valves, reactors, *etc.* but we are left with Mark 1 man. Every accident is therefore a management failure.[5]

A common mistake is to regard all human errors as examples of the same phenomenon. In fact, errors occur for different reasons which require quite different actions and they can be classified as follows.

3.1 Errors that Could be Prevented by Better Training or Instructions

Someone does not know what to do or, worse still, thinks he (or she) knows but does not. These errors are sometimes called mistakes as the intention was wrong. Sometimes there is a lack of elementary skills or knowledge, sometimes of sophisticated skills or knowledge, such as methods of diagnosing faults and deciding on the action required. (No matter how many instructions we write, we cannot foresee every eventuality and people should be trained to handle unforeseen situations.) Sometimes there are no instructions; sometimes they are not clear or are out-of-date. The action required seems obvious but before training people to carry out complex or difficult tasks we should ask whether it is possible to redesign the tasks so that they are no longer so complex or difficult.

Sometimes people are given contradictory instructions. Operators may be asked to complete a batch or a repair by a certain time. It may be difficult to do so unless the normal safety procedures are relaxed. What should they do? The manager may prefer not to know! In such cases the unfortunate operators are in a 'Heads I win, tails you lose' situation. If there is an accident they are in trouble for breaking the rules; if they keep to the rules they are in trouble for not completing the job in time. A manager should never put his staff in such a position. If he thinks that a relaxation of the rules is justified – sometimes it is – he should say so clearly, preferably in writing, and accept the responsibility. If he believes that the usual safety rules should be followed, he should say so when asking for extra output or urgent repairs. What we do not say is as important as what we do say.

3.2 Errors Due to a Lack of Motivation

Someone knows what to do but decides not to do it. An operator may decide not to wear the correct protective clothing; a manager may decide to keep the plant running to complete an urgent order, despite a serious leak. These errors are sometimes called violations but often the person involved genuinely believes that a departure from the rules, or the usual practice, is justified.

To prevent these errors we should make sure that the correct procedure is not unduly difficult or inconvenient to follow, explain the need for the rules, check from time to time to see that they are followed, and not turn a blind eye when we observe a rule being broken.

3.3 Errors Due to a Lack of Physical or Mental Ability

Someone is asked to do a job beyond his (or her) ability or, more often, beyond anyone's ability. The equipment or method of working should be redesigned.

3.4 A Slip or Momentary Lapse of Attention

Someone knows what to do, intends to do it, and is able to do it but forgets to do it or does it wrongly. These slips are similar to those of everyday life and are impossible to prevent, though they can be made less probable by reducing stress, distraction, and boredom. Routine tasks are delegated to the lower levels of the brain and are not continuously monitored by the conscious mind. If everything we did required our full attention we would be exhausted soon after we got up. So we put ourselves on autopilot. If anything disturbs the smooth running of the programme a slip occurs. We cannot prevent these failings but we can remove opportunities for error by changing the work situation, that is, the design or method of working. Alternatively we can guard against the consequences of errors or provide opportunities for recovery.

To illustrate these four types of human failing consider a common situation on process plants: an alarm sounds and an operator has to select the right valve and close it within, say, 10 minutes. If he fails to do so, it may be for one of several reasons.

He may not know that he was supposed to close the valve, or which valve.

He may decide not to close the valve although he knows that he should. He may consider it unnecessary in the circumstances prevailing at the time.

He may be unable to close the valve as it is too stiff or out of reach or he may be so overloaded that he cannot cope with the workload.

He may be busy on other jobs and forget to close the valve or be distracted and close the wrong valve.

We cannot prevent these errors by telling him to be more careful. We have either to accept the occasional error or install an automatic system. (Note that this will prevent errors of all four types but it will not remove our dependence on men. We now depend on the men who design, instal, test, and maintain the automatic equipment. They may also fail but they probably work under conditions of less stress and have more opportunities to check their work).

Often more than one type of error is involved.

Estimates of human error rates apply only to the last sort of error. It is hardly possible to estimate the probability that someone, or the whole workforce, will be poorly trained, unwilling, or incapable but it can be assumed that these errors will continue in a company at much the same rate as in the past.

4 LEARNING THE LESSONS OF THE PAST

Organizations have no memory; only people have memories and they move. A serious accident is followed by an investigation, recommendations are made,' and action is taken: new equipment and/or new procedures are installed. After about ten years staff have changed, the reasons for the new equipment and procedures have been forgotten, and someone in a hurry to increase output or efficiency, both desirable things to do, says, 'Why are we using this time-consuming procedure or cumbersome equipment?' Nobody knows, the equipment and procedures are abandoned, and the accident recurs.[6] For example, in 1928 some pipework was isolated by a closed valve while it was modified. The valve leaked, the escaping gas exploded and a man was killed. The report said that no one should rely on a closed valve. A slip-plate (blind), it said, is easy to insert and is absolutely reliable. This instruction was repeated in a safety handbook given to every employee.

In 1967, in the successor company, a large oil pump was being dismantled for repair. No blinds were inserted as it was no longer custom and practice to use them. The suction valve was left open and hot oil escaped and caught fire. Three men were killed.[7]

In 1988, close to the site of the first incident, equipment was prepared for welding by isolating it and then purging it with nitrogen. No blinds were inserted, one of the isolation valves was leaking, and again the leaking gas exploded. A welder was injured.

Minor accidents are forgotten and repeated after about two years.[8]

To prevent people forgetting the lessons of the past we should:

Describe past accidents as well as recent ones in safety bulletins or, better, discuss them and their relevance to the present day at regular safety training sessions.

Not allow standards to slip. The first step along the road to the second and third accidents occurred when someone turned a blind eye to a missing blind.

Design plants so that the recommendations made after accidents can be carried out. Blinding may have lapsed because pipework was designed with insufficient flexibility.

Standards and instructions should inform everyone why we do something and not just what we should do.

Never remove equipment or procedures unless we know their purpose.

If transfer of knowledge within a company is often poor, transfer between companies is worse. Although some companies publicize their mistakes, so that others can learn, others are reluctant to do so, often (particularly in the USA) on the advice of their lawyers. In many companies people intend to write up their accidents for publication, but pressure of work prevents them doing so; jobs which do not have to be completed by a particular time get repeatedly postponed.

5 WHICH RISKS SHOULD BE REMOVED, WHICH TOLERATED?

We cannot remove all hazards, however trivial or unlikely to occur, and the law in the UK does not ask us to do so. We have to do only what is 'reasonably practicable', weighing in the balance the size of a risk and the cost, in money, time, and trouble, of removing it. We have to decide which hazards should be removed, which tolerated, at least for the time being. The decision is often qualitative but in the process industries quantitative methods [often called hazard analysis (or hazan), risk analysis, quantitative risk assessment (QRA), or probabilistic risk assessment (PRA)] are now widely used to help decide which risks are so large that they should be reduced, which so small that they can be ignored.[9]

In applying hazard analysis three questions need to be answered:

How often will an accident occur? Experience can sometimes answer the question but often the equipment is new or has never failed, and we have to estimate an answer from the known failure rates of the components of the system, using fault tree analysis.

How big will the consequences be, to employees, to members of the public, and to the plant? Again, whenever possible experience should be our guide but often there is no experience and we have to use synthetic methods.

Finally, we have to compare the answers to the first two questions with a target or criterion. Various criteria, usually based on the risk to life, have been proposed and the Health and Safety Executive have recently published proposals.[10]

The answers to the first two questions are matters for estimation by experts but the third question is a matter for everyone, and especially for those exposed to the risk.

There has been much debate about the legitimacy of setting criteria for tolerable (or acceptable) risks to life but unless we do so we do not know where to stop. We can always spend more and make an accident less likely. Absolute safety is often approached asymptotically. Most of the debate has concerned risks to the public and the use of hazard analysis for quantifying risks to employees is much more widely accepted.

6 IDENTIFICATION OF RISKS

The biggest source of error in hazard analysis is failure to identify all the hazards or all the ways in which they can occur. As a result effort is wasted quantifying some hazards, with ever greater accuracy, while greater hazards are ignored. Identification of hazards is therefore more important than their assessment.

The traditional method of identifying hazards was to build a plant and then see what happened. If an accident occurred, the design or method of operation was changed to prevent it happening again (or it was blamed on human error and someone was told to be more careful; see Section 3 above). This 'dog is allowed one bite' philosophy was defensible when the size of an accident was limited but is no longer acceptable now that we keep dogs as big as Bhopal (over 2000 killed)[11] or even Flixborough (28 killed). Hazards need to be identified before accidents occur. Hazard and operability studies (hazops) are used widely in the chemical industry for this purpose, particularly for studying new designs or plants that have been extensively modified.

In a hazard and operability study each line on a line diagram is examined in turn by a team of designers and the commissioning manager under an independent chairman. The team asks whether no flow or reverse flow could occur in the line under examination. If it could, they ask whether it would be hazardous or prevent efficient operation and, if so, what changes in design or method of operation would overcome the hazard operating problem. The team then applies similar questioning to more and less flow, temperature, pressure, and any other important parameters. Finally, they ask about the effect of changes in concentration or the presence of additional materials or phases. These questions should be asked about all lines, including steam, water, drains, and other service lines, for all modes of operation, start–up, shut–down, and preparation for maintenance, as well as normal operation. On batch processes the instructions should be examined in the same way as well as the lines.[9] For example, if an instruction says 'Add 1 tonne of A'. the team should query the effects of adding more or less A, something else as well as or instead of A, part of A (if A is a mixture), reverse flow of A (that is, flow from the reactor to the A container), and adding A too quickly or slowly or at the wrong time. If the plant is computer–controlled the instructions to the computer should be examined as well as the instructions to the operators.[12]

Hazop is a powerful technique for identifying potential problems but at the line diagram stage it is too late to *avoid* the hazards (as discussed in Section 2 above) and all one can do is *control* these hazards by *adding on* protective equipment. To avoid hazards one should carry out similar studies at the flowsheet stage and even earlier at the conceptual stage when we decide which product to make, by what route and where. Unfortunately many companies who carry out hazops do not carry out these earlier studies. Their safety professionals do not get involved, their safety studies do not take place until late in design, and safety then becomes an expensive (though necessary) addition to capital cost. If they carried out the two earlier studies they would be able, in many cases, to design plants that are both cheaper and safer.

All existing plants should have safety audits carried out from time to time. Technical hazards and procedures (for example, the way in which equipment is tested or prepared for maintenance) as well as physical hazards should be considered. There has been increased interest recently in methods for quantifying the results of audits using a system such as the International Safety Rating System.[13]

7 TESTING PROTECTIVE EQUIPMENT

All protective equipment should be tested regularly, otherwise it may not work when required. The frequency of testing should depend on the failure rate. Relief valves are very reliable, they fail only about once per hundred years on average, so testing every one or two years is adequate. In contrast, protective systems based on instruments, such as trips and alarms, fail much more often, about once every couple of years on average, so more frequent testing is necessary, about once per month.

The following types of protective equipment should also be tested or inspected regularly, though they are often overlooked:

Drain holes in relief valve tailpipes. Are they clear?
Drain valves in tank bunds. Are they closed?

Emergency equipment such as diesel–driven fire water pumps and generators.

Earth connections, especially the portable connections used for earthing road tankers.

Fire and smoke detectors and fire–fighting equipment.

Flame arrestors.

Hired equipment. Who will test it, the owner or the hirer?

Labels are a sort of protective equipment. They vanish with remarkable speed and regular checks should be made to make sure that they are still in existence and legible.

Lifting gear.

Mechanical protective equipment such as overspeed trips.

Nitrogen blanketing (on tanks, stacks, and centrifuges).

Non–return valves, if relief valves have been sized on the assumption that they will operate or if their failure can affect the safety and operability of the plant in other ways.

Open vents. These are in effect relief devices, the simplest possible sort of relief device, and should be treated with the same respect.

Passive protective equipment such as fire insulation.

Spare pumps, especially those fitted with auto–starts, and the clearances on throttle bushes.

Steam traps.

Trace heating (steam or electrical).

Valves, remotely operated and hand–operated, which have to be used in an emergency.

Ventilation equipment

Water sprays and steam curtains.

All protective equipment should be designed so that it is possible to test or inspect it and designers should say how often it should be tested or inspected. They should note the points to be looked for. Audits should include a check that the tests are carried out and the results acted on.

Tests should be like 'real life'. A high-temperature trip failed to work despite regular testing. It was removed from its case before testing so the test did not disclose the fact that the pointer rubbed against the case and prevented it indicating a high temperature.

Test results should be displayed in the control room. A good practice is to list all the protective equipment on a board, showing the dates on which tests are due, and the test results. Everyone can then see when testing is overdue.

Operators sometimes regard testing as a nuisance and may say, 'Why are the instrument section always wanting to test their trips?' These operators fail to realize that the trips are there for *their* protection and that they should accept responsibility for seeing that they are kept in working order.

It is easy to buy protective equipment. All we need is money and if we make enough fuss we usually get it in the end. It is much more difficult to make sure that it is tested regularly and kept in good reliable working order.

8 OTHER PROCEDURES

The last section stressed the importance of having a system for testing protective equipment of every type and ensuring that the system is followed. Other systems of comparable importance are those for preparing equipment for maintenance, controlling contractors, and controlling modifications.

Many accidents have occurred because equipment which had to be repaired or overhauled was not prepared adequately. On many occasions maintenance workers have opened up equipment and found that it was full of flammable or toxic gas or liquid, often under pressure. Sometimes the equipment had not been isolated adequately and gas or liquid had entered through open or leaking valves (as in the incidents described in Section 4); sometimes it had not been freed from gas or liquid already present; sometimes the wrong equipment was opened up; sometimes insufficient protective clothing was worn. Sometimes the instructions were poor; at other times they are not followed.[14]

To prevent these accidents a three-pronged approach is needed:

We need good methods of working, explained in language that people can understand, not in legal language. Some instructions give the impression that they have been written to protect the author rather than help the user.

We need to explain the instructions to those who have to use them, not just send them through the post. We do not live in a society in which people will always follow a procedure just because they have been told to do so. They have to be convinced that the procedure is necessary. One way of doing this is to describe and discuss accidents that have occurred because the procedure was not followed.

Finally, we have to check to make sure that the procedure is being followed and we should never turn a blind eye when the procedure is broken. Of course, there are times when exceptions are justified and there should be a system for authorizing them. (See the discussion of contradictory instructions in Section 3.)

Maintenance and construction are often carried out by contractors and particular attention is needed in these cases. It is not sufficient to give the contractor's foreman a copy of the plant instructions. The relevant instructions should be explained to him, and if possible to the whole workforce. Supervision of contractors should not be left to a busy process foreman who is primarily responsible for operating the plant. A special supervisor should be appointed.

Many accidents have occurred because changes to the plant or the method of working, often quite trivial ones, had unforeseen effects. No change should be made unless it has been authorized by the manager and engineer concerned who should first carry out a critical examination of the proposal by hazop (see Section 6) or a similar technique. Afterwards they should make sure that their intentions have been followed.

As with the preparation of equipment for maintenance, we have to divide a procedure, convince people that it is necessary, and check that it is being followed.[15]

9 SENIOR MANAGEMENT, THE WEAK LINK?

To prevent accidents we need good equipment, we need to understand the causes of human error and design accordingly, and we need good management. The left-hand side of Figure 1 shows the effort devoted to these three aspects and the right-hand side shows the effort that ought now to be devoted.[16]

The effort devoted to equipment design is now enormous. Not only is basic equipment (reactors, tanks, heat exchangers, pumps, *etc.*) designed to carry out its duty safely but a wide variety of protective equipment warns us of leaks and other untoward events and allows us to control them.

For many years human error as usually understood, that is, errors by those at the bottom of the organization who cannot pass on the blame to someone below them, was neglected in comparison with equipment design but today there is a much better understanding of the reasons for human error and ways of reducing its effects (see Section 3).

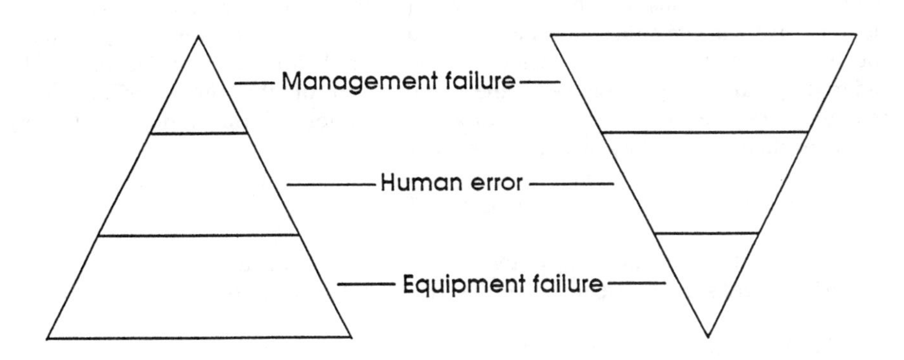

Figure 1 *The effort expended on various causes of accidents and their relative importance*

Comparatively little attention is still paid to management failings at the senior level. Many senior managers hardly realize that there is anything they can do to prevent accidents other than exhort their staff to do better and hand out the necessary cash. I shall therefore conclude by describing an accident that at first sight was due to equipment failure and human error but was primarily due to weaknesses in the senior management.

A chemical company, part of a large group, made all its products by batch processes. They had a reputation for safety and efficiency. Sales of one product grew so much that a large continuous plant was necessary. No one in the company had much experience of such plants so they engaged a contractor to whom they left the design. If they had consulted the other companies in the group they would have been told to observe the contractor closely and be informed of some of the points to watch.

The contractor sold them a 'pup'. The process design was good but the layout was very congested, the drains were open channels, and the plant was a mixture of series and parallel operation.

When some of the parallel sections of the plant were shut down for extensive overhaul and repair other sections were kept on line, isolated by blinds (slip-plates). One of the blinds was overlooked. Four tonnes of a hot flammable hydrocarbon leaked out and was ignited by a diesel engine being used by the maintenance team. Two men were killed and the plant was damaged. The congested design increased the damage and the open drainage channels allowed the fire to spread rapidly.

The immediate causes of the fire were the presence of the diesel engine and the missing blind. The positions of the blinds were not planned well in advance, as they should have been, at the design stage, but left for a foreman to decide a few days before the shutdown. The underlying cause, however, was the amateurism of the senior management and their failure to consult the other companies in the group. If they had consulted them they would have been told:

To observe the contractor closely.

Not to let the plant get too congested but to divide it into blocks with breaks in between, like the fire breaks in a forest.

To have underground drains.

If possible, not to maintain half the plant while the other half is on line. If it is essential to do so then build the two halves well apart and plan well in advance, at the design stage, where blinds will be inserted to isolate the two sections. Mark these places clearly. Do not leave it for a foreman to sort out a few days before the shutdown.

I doubt if it ever occurred to the senior managers that their actions (or lack of them) had led to the accident though they fully accepted that they were responsible for everything that went on. However, a few years afterwards the group was re-organized and responsibility for the continuous plant was transferred to another part of the group.[17]

10 REFERENCES

1 A. Hidden, 'Investigation into the Clapham Junction Railway Accident', HMSO, London, 1989, p. 163.

2 T.A. Kletz, 'Plant Design for Safety – A User-Friendly Approach', Hemisphere, New York, 1991.

3 N.A.R. Bell, 'Loss Prevention in the Manufacture of Nitroglycerine', in 'Loss Prevention in the Process Industries', Institution of Chemical Engineers Symposium Series, No. 34, 1971, pp. 50–53.

4 R. Malpas, 'The Business of Stimulating Process Innovation', in 'Research and Innovation for the 1990s', ed. B. Atkinson, Institution of Chemical Engineers, Rugby, 1986, pp. 28–47.

5 T.A. Kletz, 'An Engineer's View of Human Error', 2nd Edn., Institution of Chemical Engineers, Rugby, 1991.

6 K. Asher, *Chemical Engineer*, 1983, **395**, 25.

7 T.A. Kletz, *Loss Prevention*, 1990, **13**, 1–6.

8 T.A. Kletz, *Loss Prevention*, 1976, **10**, 151–154.

9 T.A. Kletz, 'Hazop and Hazan – Notes on the Identification and Assessment of Hazards', Institution of Chemical Engineers, Rugby, 2nd Edn., 1986, 3rd Edn., 1992.

10 Health and Safety Executive, 'Risk Criteria for Land-use Planning in the Vicinity of Major Industrial Hazards', HMSO, London, 1989.

11 C.R. Krishna Murti, 'A Systems Approach to the Control and Prevention of Chemical Disasters', in 'Risk Assessment of Chemicals in the Environment', ed. M.L. Richardson, The Royal Society of Chemistry, London, 1988, pp. 114–149.

12 T.A. Kletz, *Plant/Operations Progress*, 1991, **101**, 17–21.

13 J. Bond, *Loss Prevention Bulletin*, 1988, **080**, 23–29.

14 T.A. Kletz, 'What Went Wrong? – Case Histories of Process Plant Disasters', 2nd Edn., Gulf, Houston, Texas, 1988, Chapter 1.

15 T.A. Kletz, *Chemical Engineering Progress*, 1976, **72**, 48–55.

16 R.J. Batstone, International Symposium on Preventing Major Chemical Accidents, Washington DC, Feb. 1987 (not included in published proceedings).

17 T.A. Kletz, 'Learning from Accidents in Industry', Butterworths, 1988, Chapter 5.

9

Managing Risk on a CIMAH Site – A Case History

A. Mottershead

IMPERIAL CHEMICAL INDUSTRIES PLC, C & P LTD., NORTH WEST OPERATIONS, RUNCORN, CHESHIRE WA7 4JE, UK

1 INTRODUCTION

Chlorine has been manufactured in Runcorn, Cheshire for almost one hundred years. A chemical site was first established at Weston Point in 1896 to exploit a new method for the production of high-purity caustic soda by the electrolysis of brine using a mercury cathode and graphite anode. The key factors which influenced the directors of the then named 'Aluminium Company' to select the Runcorn site included access to brine from the Cheshire Salt Area, good supplies of coal and water, and good transport facilities. Sufficient land was also available to allow for future development. The forty-six acre site was purchased for the princely sum of £5650.

For an electrolytic process, a power plant for the generation of electrical energy was of fundamental importance. The initial plant consisted of four hand-fired boilers supplying three steam engines of 200 hp each. The original chlorine cell room contained 270 individual rocking cells with a target output of 20 tonnes day^{-1} of caustic soda liquor. Equivalent amounts of chlorine and hydrogen were also produced. In the early days caustic soda was the most saleable product since it was used in the manufacture of sodium. The demand for chlorine depended solely on its use as a bleaching agent.

Through the decades the production of chlorine at Runcorn steadily increased from the initial 10 tonnes day^{-1} to 2000 tonnes day^{-1} to satisfy an ever expanding market. The manufacture of sodium hypochlorite and hydrochloric acid was developed around 1910. Subsequently, the chlorine was converted into chlorine derivatives of acetylene and later ethylene, for use as degreasing agents. Production of vinyl chloride (for making PVC) started in 1941, with production now at tens of thousands of tonnes year^{-1}.

The majority of the chlorine produced is consumed on site with some distributed by road, rail, and sea to other plants in cylinders, drums, and transport containers. In the early 1980s chlorine stock capacity totalled up to 15,000 tonnes.

The area of the site has expanded three-fold over the years and the site now employs over 3000 people. The landscape has changed from rural at the turn of the century to an urban district. The local community has grown to a population of 10,000 who live or work within 1.5 km of the site. Consequently, the need to manage the risks has become increasingly important.

2 PROPERTIES OF CHLORINE

Chlorine is a greenish yellow gas at ambient temperature and pressure. The liquid is about one and a half times heavier than water and the gas two and a half times heavier than air. Liquid chlorine has a boiling point of −34 °C. Any liquid leaks are likely to vaporize quickly unless the spillage can be contained. The gas has an 8 h Occupational Exposure Standard (OES) of 0.5 p.p.m., which coincidentally corresponds to the threshold of smell. The extent of any injury from exposure to chlorine gas is dependent on many factors – concentration and duration of exposure being most significant. Irritation of the mucous membranes of the eyes and nose, and especially the throat and lungs, is caused by exposure to chlorine at levels of around 15 p.p.m. Above 15 p.p.m. a choking sensation is generally experienced and above 40 p.p.m. pulmonary oedema can develop depending on the dose and health of the individual affected.[1–3] Exposure to liquid can produce burns. Traces of moisture in chlorine lead to rapid corrosion of steel and the gas reacts vigorously with many organic compounds.

The storage, handling, and use of both liquid and gaseous chlorine thus require very high standards in the design, construction, and operation of the plants. The following sections of this chapter describe the development of the systems and procedures which promote and sustain such standards.

3 LESSONS LEARNED FROM ACCIDENTS

Until the 1960s one of the principal driving forces for reducing the probability of incidents came from learning from the mistakes of others.

Worldwide production rates for chlorine exceed 20 million tonnes year^{-1} and, despite the acute toxic properties of the chemical, casualties from accidental releases during manufacture have been relatively low over the past century. V.C. Marshall[4] reported 152 fatalities from 30 incidents worldwide involving the release of a total of 271 tonnes over a period of 60 years. Six of the above incidents involved loss of containment from bulk liquid storage. They occurred prior to 1953 and accounted for 115 fatalities. Although chlorine production has increased 10-fold since then, there have been no further catastrophic storage tank failures causing fatalities. Tank failures have been reduced dramatically and this can be attributed to improved knowledge of metallurgy, instrumentation, operation, *etc*.

Generally, fatalities have not occurred beyond 400 m from any incident - more often within 250 m. The exception is an incident at Zarnesti, Romania, where the furthest fatality was at half a mile. Consequences are related to wind/weather conditions prevailing at the time and the population density, location, and vulnerability of people in the vicinity.

Records on how the loss of containment occurred, and how much was released in a given time, enable us to improve our understanding of the hazards and their causes, and also enable us to check our preventative, corrective, and emergency measures.

It has been established practice at Runcorn to carry out data searches and incident reviews which continually feature in management discussions on the operational viability of our plants and designs for new systems. However, in the 1960s the growth of the solvents market, plastics, and water treatment plants had led to a significant increase in chlorine production and storage requirements at Runcorn. With this increased inventory there was a growing realization of the potential for a major incident. Also a gas release in 1970 affected children in a local playground. This resulted in demands from senior management that 'this must never happen again'. Reliance on learning from accidents was no longer enough.

4 TOXIC GAS EMISSION STUDIES

Following the 1970 gas release, ICI at Runcorn embarked on a series of studies aimed at examining the chlorine handling plants systematically to identify potential and realistic hazard scenarios which could lead to consequences affecting people most at risk beyond the site boundary. Targets were set based on the consequences and frequency of incidents. The consequences were assessed using simple gas dispersion calculations to assess the down-wind ground level centre-line concentrations (expressed as the average 3 minute mean in p.p.m.). Wind and weather probability data were used to calculate a spectrum of centre-line concentrations at defined distances from the sources of emission. The effect on the community at these distances was evaluated generally from a knowledge of the duration of the release and the centre-line concentration. This 'dose' was compared with physiological data to judge the likely average effect in three hazard categories, *viz.* detectable/irritating, mild casualties, severe casualties. Judgements were made on the expected public reaction to these three categories of 'effect' to set site targets for event frequency. Studies were then carried out to assess the size, likelihood, and consequences of foreseeable plant incidents and to examine whether the events were within the defined targets. These were some of the first quantitative risk assessments produced in the UK or worldwide in the chemical industry.[5,6] They were referred to as 'toxic gas emission studies'.

Over a period of six years 34 reports were produced on the Runcorn complex which described the potential events and assessed the risks against the three defined hazard targets. Since these studies more has been published about chlorine toxicity.[7-9] Quantitative risk assessment techniques[10] and gas dispersion theory have developed considerably, but nevertheless the early studies have helped our management to make more objective decisions on safety issues. This initiative was not imposed by external authorities, but about 300 recommendations were made at a cost of several million pounds in present day terms to improve the plants to meet the defined targets.

Management were made much more aware of the hazards on their plants, safety became a 'No. 1 priority', and better equipment was installed to ensure more effective control of the hazards.

5 MODIFICATION CONTROL

Even prior to these gas emission studies, management had appreciated the risk of hazards originating from ill-conceived changes to equipment and procedures. In the mid-1960s protocols referred to as 'site instructions' were developed to promote consistent standards and safe working practices. Control procedures for modifications to plants were amongst the first instructions to be written, and they have been revised a number of times to incorporate improvements. Modifications are only authorized on the site after careful assessment by at least two people using a list of guide words and prompts to stimulate lateral thinking on factors which could be affected by the proposed changes. Checks to ensure that the modifications are inspected before use confirm that the design intent has been achieved and that appropriate documentation has been updated and training provided.

The original procedure was first aimed at controlling costs, but it is now promoted with hazard avoidance as the prime motivator. No change is allowed unless the senior site business manager is convinced of the safety and process benefits.

6 HAZARD STUDIES

The Runcorn site was one of the first installations to use the ICI Hazard Study Procedure[11] in the design of new plant and extensions to existing plants.

The original single-stage procedure has been extended to consist of a series of six distinct studies which are carried out at defined stages throughout a project – from initial project development through to operation to ensure that the hazards of the process are understood and adequately controlled.[1 2]

The first of the six studies aims to provide a clear definition and understanding of the proposed new system or process. It ensures that the hazards of the chemicals and intermediates are appreciated and contributes to key policy decisions on siting. It also prompts reference to codes/regulations and requires appropriate contact with functional experts and external authorities.

The second study aims to identify the significant risks and provides the opportunity for elimination or minimization by examining both event likelihood and consequence. It is a very important study and provides the fundamental reasoning and justification for Safety, Health, and Environment (SHE) control measures characteristic of the requirements in the Control of Industrial Major Accident Hazards (CIMAH) Safety Reports (Regulation 7, Schedule 6, Item 5).

Study 3 is a detailed examination of the design/procedures to identify unexpected deviations from design intent. Intended originally for the study of engineering line diagrams, it has been developed over the years to apply to batch and mechanical handling systems and more recently to construction and computer activities. The study relies on a multi-disciplinary team systematically examining the design, prompted by guidewords designed to encourage lateral thinking.

Unlike the first three studies which necessitate a series of meetings, Studies 4, 5, and 6 rely on the commissioning manager or site representative to check progressively that the design intent has been achieved and that standards and systems will be sustained. Study 4 is conducted before hazardous chemicals are introduced to the process. It checks that all documentation is accurate and adequate and that people are trained to operate the process safely and can respond effectively to any emergency. Routine test/inspection procedures are checked at this stage together with start-up and shut-down procedures.

Study 5 is an examination of the equipment and systems during one or more plant inspections and sets the standards for future audits. The study ensures that the process is safe to operate and meets Company and legislative requirements. The final Study 6 is usually scheduled to review plant performance and experience several months after start-up and aims to feed back lessons learned in order to improve future designs. It confirms that early operation is consistent with design intent and that the documentation is complete and adequate to sustain continued safe operation.

One opportunity to apply this systematic hazard study procedure at Runcorn was provided, for example, in the early 1980s when the chlorine handling plant at Runcorn was subjected to a rationalization programme. Four key objectives were identified, *viz.*

To make the process:

Simpler

Easier to maintain

Safer to operate

With reduced inventory

Among other improvements, over a period of several years, the maximum liquid chlorine inventory has been reduced by more than a factor of 5 and the number of stock tanks and pumps in use has been halved. In addition, pipework has been replaced to improved standards and remote-operated shut-down valves have been introduced at appropriate locations.

Like the modification control procedure, the hazard study procedure has been reviewed and revised in the light of experience. Most notable in a new complete revision of the procedures is the emphasis placed on health and environmental protection as well as safety.[13] The procedure now used will ensure consistency of approach in ICI worldwide.

7 QUANTIFIED RISK ASSESSMENTS

Although our first objective is always the elimination of hazards, there are cases where this is not possible and the risk must be reduced to tolerable levels. Since the early 1970s, where the risk was felt to be significant, 'quantified hazard analysis' has been used, as an adjunct to the hazard study procedure. It consists of the identification and analysis of undesired hazardous events together with basic estimates of magnitude and frequency. Sensitivity studies can reveal the band of statistical uncertainty attributed to varying assumptions and degree of confidence in failure data. Estimates of risk can generally be accepted to be no more accurate than ± one order of magnitude, but such differences between independent quantitative risk assessments will usually lead to useful dialogue between the hazard analysts and hence lead to an improved mutual understanding. Quantified risk assessment extends the procedure, at the expense of some complexity, to include the probability of harm being caused.

Quantified risk assessment (QRA) was used in a modified form in the early toxic gas emission studies. A number of fuller studies were carried out in the 1970s in connection with the transport of chlorine and other hazardous substances.

With the advent of stricter planning controls[14] on, and in the vicinity of, major hazard sites in the late 1970s and early 1980s, QRA has become an essential tool:

i) To ensure that the risk from new major hazard developments is kept within strict criteria;

ii) To assess and rank the risk from existing installations in order to target areas for improvement if necessary;

iii) To allow the Company to assess the advisability of proposed developments in the vicinity of its major hazard installations whether initiated by the Company or even by outside developers.

In the case of the Runcorn complex there are several vessels containing bulk quantities of liquid chlorine and significant lengths of liquid/gas pipework.

Use of generic failure data based on past experience will give rise to risk levels, in a range broadly acceptable for older plant. What we have to do is to show ourselves, our neighbours, and the authorities that current plant can be operated more safely than this (retrospective) average and that new plant fall into a risk range which will meet the more stringent criteria which the Company sets as a contribution to an overall policy of reducing risk.

Computerized risk assessment packages[15] can give plant management an improved understanding of the size and potential consequences of a range of hazardous event scenarios.

It is this improved understanding that raises awareness on the importance of conforming with standards and working to best practice.

8 SAFETY ASSURANCE

Avoidance of unnecessary hazards, particularly major hazards, has long been part of our site culture. We now review our operations to ensure that:

i) Hazards are recognized and their consequences are understood;

ii) Appropriate equipment/facilities are provided and they are fit for the purpose;

iii) Systems and procedures exist to ensure that we operate to design intent and maintain system integrity;

iv) Appropriate staff are given information, instruction, and training;

v) Emergency procedures exist and are practised;

vi) Effective arrangements exist for promoting and progressing safety issues and for auditing.

HAZARD SUMMARY:

		MINIMIZE LIKELIHOOD ←		*MINIMIZE CONSEQUENCE* →
Hazard	*Cause*	*Preventative measures*	*Corrective measures*	*Emergency measures*
Scenario (inc. consequences)	**Mechanism**	**Means for ensuring safe operation under under normal operating conditions**	**Means to avoid an unprevented disturbance leading to loss of containment**	**Means adopted for reducing the scale of consequences**
		i.e. Prevents the initiation of a sequence of events that could lead to a hazardous event	*i.e.* The means for intervention	*i.e.* Reduce quantity discharged or effects following discharge
e.g.	*e.g.*	*e.g.*	*e.g.*	*e.g.*
Acute toxic exposure (on/off–site)	Loss of containment – damage – overpressure – overfill – corrosion – error *etc.*	Control loop Procedure Design detail Inspection Analysis *etc.*	Alarm/operator action Trip system Dump tank Relief system *etc.*	Bunding Water curtains Remote SDVs Restrictor plates Gas alarms Emergency procedure *etc.*

Figure 1

In order to describe systematically and effectively the key hazards and control measures, plant managers on the Runcorn site have produced summary tables of the hazards in a format similar, but in more specific detail, to that shown in Figure 1. This method of tabulating the key features which minimize the likelihood and consequences of potentially hazardous events is useful in training, handover, and auditing. In addition to examining the safety aspects we also now include health and environmental event scenarios.

9 STANDARDS

Our Company is implementing a process (called stewardship) by which safety, health, and environmental (SHE) performance can be constantly evaluated and regularly improved by Business Groups and Sites. Within the Company, businesses are aligned with specific products and are responsible for commercial success, together with preservation of standards and the corporate reputation. Site management is responsible for ensuring that these standards are complied with on the operating plants. The result of this process is a committment to 'stewardship'. This is the term now widely used to describe the totality of the responsibility for our manufacturing activities and products.

Nominated managers within our Business Group, Product Stewards, have been given a formal responsibility for ensuring that the products within their care meet company requirements, which are often in excess of the legal requirements relating to SHE standards in development, promotion, distribution, marketing, and use. These stewards are also responsible for maintaining up-to-date records of relevant legal standards and product performance.

Similarly, the process technologies associated with our products are the responsibility of Process Stewards. The same systematic approach that applies to our products characterizes the SHE assessment of new and existing processes, from the raw materials they use to any waste they may create.

In short, the Businesses define the standards to be met and in compliance with these standards more detailed site-wide instructions are produced and used in the generation of local process and maintenance procedures and instructions.

These vital links between executive management and the workforce are characteristic of a quality approach to reducing our risks and managing our operations safely and responsibly.

10 SYSTEMS AND PROCEDURES

High–quality maintenance and safe working practices can only be achieved if standards of workmanship and equipment integrity are maintained and the instructions are documented, applied, and audited.

Considerable resource has been devoted on the site to updating site instructions, updating operating procedures, reviewing test procedures, updating engineering line diagrams, and, more recently, to formalizing inspections of critical equipment (*e.g.* pipelines, pipebridges, machines, *etc.*). Independent formal vessel inspections and relief vessel inspections have been a long–standing feature of our maintenance strategy.

During the life of a plant valuable documentation is generated by hazard studies, operating experience, and inspections and audits. This key information must survive. Safety, Health, and Environment Dossiers have been created on all our Runcorn plants and these will be used to sustain the corporate memory.

However, effective implementation of the procedures requires that regular and appropriate training is given to our plant personnel.

11 TRAINING

All new employees on our site are given induction training during their first week to introduce them to Site Instructions and Procedures. The depth of training for all levels on the site is related to previous knowledge and experience. It is divided into induction, engineer/management, supervisor, and operative training. The induction training is basic 'need to know material'. It includes formal training on: plant hazards, site emergency procedures, reporting of accidents, protective equipment, factory law/site rules, and individual responsibilities.

The majority of managers and engineers are technically qualified to graduate level. Plant familiarization and experience in ICI systems accrues from selective development projects and structured training spanning several years. Involvement in working parties, hazard studies, plant meetings, incident investigations, and familiarity with Codes of Practice are just some of the many features of a plant manager's training. Selective refresher training and emergency training are routinely injected into the job, with opportunities to pursue additional specialized training from many other internal and external safety functions. 'Risk assessment' is one such example.

It is our corporate policy that all managers are responsible for the training and development of the people who report to them. In the case of supervisors this would be the plant manager or engineer. Procedures are becoming more formalized with improved documentation and validation. The training is both systematic and recorded. It reinforces familiarity of the plant with consolidation of key principles of operation and emergency response.

New supervisors are normally promoted from senior process operators or experienced craftsmen who have progressed through various jobs on other plants. They are given on-the-job training under the direction of their management, working alongside experienced supervisors until they are considered competent to carry out the job. They are also given specific off-the-job instruction and attend an introductory supervisor training course which includes a large safety element. This initial training is further developed in time with specific functional training modules which also contain safety as a key element.

Process operators and tradesmen follow a general induction course as new starters. A period of on-the-job training by supervisors is followed by working side-by-side with competent experienced local supervision. They are also given specific off-the-job instruction and are subsequently tested and approved by process/engineering supervisors before being allowed to work on their own.

All employees on the Runcorn complex are regularly familiarized with Site Instructions and safety requirements by communicating down the line prepared modules on key safety themes.

12 AUDITING

The requirement to conduct regular audits of plants and systems is well understood and captured in a number of Company and site documents.

However, at Runcorn, improvements in auditing feature in the Improvement Plans of many of our units. The aim is to provide a continuous process with different aspects receiving attention daily, weekly, monthly, *etc.* depending on the activity being audited.

Three levels of audit are required:

Level 1 Regular pro-active checks to examine compliance with local instructions.

Level 2 Regular in-depth audits to check compliance with site and Company standards.

Level 3 Overall performance assessments to confirm that the audit plan functions effectively.

The functional requirements of process and engineering safety management are sufficiently distinct to require separate dedicated auditing processes:

Process	*Engineering*
Equipment	Pressure systems and mechanical handling
Occupational health	Civil structures and portable equipment
Emergency procedures	Electrical/Instrument systems
Site procedures	
Local procedures	

Such processes are under continual review to benefit from experience in conducting the audits.

13 CIMAH SAFETY REPORTS

The ICI site at Runcorn is a top tier major hazard site which, because of its toxic and flammable plants, requires a number of Safety Reports to comply with the UK CIMAH regulations. Compliance with such regulations is a minimum standard as far as ICI is concerned.

The report for the chlorine handling plant took many man months of effort to prepare and contained over 100 pages of text, drawings, and references and requires updating every three years. Despite initial reservations we have found the preparation of the Safety Report a valuable exercise. It contributes to the corporate memory with a concise summary of critical safety features.

The preparation of the Safety Report proved to be a fairly rigorous self-examination, not only of the adequacy of the procedures, software, and hardware laid down for the control of major accident hazards, but also of the standards of upkeep. Although this was time consuming, it was nevertheless a valuable contribution towards the control of major accident hazards, not merely a regulatory imposition. A particular benefit was the improved awareness and understanding of the major accident control measures associated with the process. The exercise also identified some areas where technical/procedural improvements were needed (typically the need for updating instructions, procedures, *etc.*)

Many of the topics addressed in this chapter feature in the Report because they describe the many facets to managing the risks on a CIMAH plant. For all new management trainees the Safety Reports can provide very useful information and safety site induction training requires early reference to appropriate reports.

The Reports describe the measures in place to avoid the hazards and also the emergency procedures and equipment necessary to minimize the consequences if an accident were to occur. In the case of chlorine, the three basic steps required if containment were to be lost are **isolate, depressurize,** and **drain** the pipework or vessel. Measures subsequently taken by the external emergency services would relate to rescue and treatment of casualties.

Continuing management effort to reduce the number of potential sources and the size of release is essential and, as Trevor Kletz[16] has frequently quoted, 'What you don't have, can't leak'.

14 CONCLUSIONS

ICI supports and contributes to many national and international associations, in particular the UK Chemical Industries Association – Chlorine Group, Euro Chlor (formally known as the Bureau International Technique du Chlor – BITC), and the American Chlorine Institute. All these bodies have a common aim which is to serve the chlor-alkali industry, the authorities, and the public in matters of safety, health, and the environment connected with the production, transportation, handling, and use of chlorine. These links enable us to share ideas in our constant drive to minimize the risks.

We make chlorine because it is highly reactive and versatile as a reagent. There are huge benefits to society from chlorine and related alkalis. However, as with life, making chemicals cannot be divorced from risks, but nonetheless we are determined to minimize these risks by proper controls.

15 REFERENCES

1 'WHO/IPCS Environmental Health Criterion 21. Chlorine and Hydrogen Chloride', World Health Organization, 1984.

2 'Loss Prevention Bulletin Chlorine Toxicity Monograph', Institution of Chemical Engineers, Rugby, 1987.

3 R.M. Turner and S. Fairhurst, 'Toxicity of Substances in Relation to Major Hazards. Chlorine', HMSO, London, 1990.

4 V.C. Marshall, *The Chemical Engineer*, August 1977, p. 573.

5 I.G. Sellers, 'Quantification of Toxic Gas Emission Hazards', Process Industry Hazards Symposium, Institution of Chemical Engineers, Symposium Series 47, 1976.

6 A.N.A. Dicken, 'The Quantitative Assessment of Chlorine Emission Hazards', Electrochemical Society, San Fransisco, May 1974.

7 ECETOC Technical Report No. 43, 'Emergency Exposure Indices for Industrial Chemicals', ECETOC, Brussels, 1991.

8 H.P.A. Illing, 'Assessment of Toxicity for Major Hazards: Some Concepts and Problems', *Human Toxicol.*, 1989, **8**, 369.

9 'Major Chemical Disasters – Medical Aspects of Management', ed. V. Murray, Royal Society of Medicine, International Congress and Symposia Series No. 155, London, 1990.

10 J. Withers, 'Major Industrial Hazards. Their appraisal and Control', Gower Technical Press, Aldershot, 1988.

11 S.B. Gibson, 'Hazard Analysis in the Design of New Chemical Plants', Institution of Chemical Engineers, Symposium Series No. 40, 1976.

12 R.D. Turney, 'Techniques for the Analysis and Assessment of Accident Hazards', Institution of Engineers and Factories and Machinery Dept., Malaysia, 1987.

13 R.D. Turney, 'Hazard Studies for Safety, Health & Environmental Protection: Application to New and Existing Plants and Processes'. Safety in Refining and Petrochemical Plant, The First European Conference, 6 March 1991, London.

14 HSE Risk Criteria for Land Use Planning in the Vicinity of Major Hazards, HMSO, London, 1989; HSE Quantified Risk Assessment – Its Input into Decision Making, HMSO, London, 1989.

15 OCPS & Technica, 'Guidelines for Chemical Process Quantitative Risk Analysis', American Institute of Chemical Engineers, 1989.

16 T.A. Kletz, 'Cheaper, Safer Plants or Health and Safety at Work?', in 'Loss Prevention', Institution of Chemical Engineers, Rugby, 1984.

10
Risk Management Case History – Detergents

George Calvin

PROCTER & GAMBLE EUROPEAN TECHNICAL CENTER,
TEMSELAAN 100, 1853 STROMBEEK-BEVER, BELGIUM

1 INTRODUCTION

Detergents are an example of a widely used consumer product. The application of the general principles of risk management to this product category will be illustrated by reference to a specific case history. This involved the incorporation of a new chemical, sodium isononanoyloxy-benzenesulphonate, into a detergent product. An examination of the toxicology of the new chemical showed that it had the potential to induce skin sensitization, and to produce a specific lesion in the male rat kidney. The risks presented by both these effects had to be assessed and managed before a new product which contained the chemical could be considered as safe for the consumer.

2 THE PHILOSOPHY OF RISK MANAGEMENT

The concept of risk has been with man from earliest times, but its articulation into a set of universal principles has only occurred during the last decade or so, at least as those principles are applied to chemical exposure.

This situation has largely been corrected with the appearance of several quality publications on risk assessment,[1-4] risk analysis,[5,6] and risk management.[7] In some instances, the principles of risk management have been developed for very specific toxic and points, such as carcinogenesis.[8]

It is now generally accepted that regulators and business managers, who are involved with toxicological risks, should base their actions on the distinct elements of risk assessment and risk management. Within these elements, there is a clear sequential approach to the management of risk.[9] For a particular chemical to be introduced into commerce, the following steps would be taken:

i) A determination of the chemical's **toxic potential**, which will give an understanding of:

- its target organs;
- the dose–response relationship between the magnitude of the effect on the target organ and dose;
- the no observable adverse–effect–level;
- the relevance to humans of the toxic response observed in animals.

ii) A determination of the **exposure** likely to be encountered in practice. At least two distinct groups of people are recognized:

- those exposed in a work environment;
- the general population who are 'the consumers' and in the latter case we recognize exposure resulting from:

  normal use
  foreseeable misuse
  accidental exposure.

When this process has been completed it will be possible to identify the **hazard**. An integrated view of all the evidence gathered from hazard identification, dose–response relationships, and exposure, which is often referred to as risk characterization[4,8] allows one better to define the nature of the risk, *i.e.* by taking into account each exposure situation.

All these factors allow for a **risk assessment** to be made. Such an assessment is based on objective, scientific information which is largely quantifiable. This differentiates it from the final step of **risk management**, which carries with it decisions and judgements which may be societal and economic in nature. Political factors may also be important in risk management.

3 RISK ASSESSMENT AND RISK MANAGEMENT IN THE DETERGENT INDUSTRY

All the components of detergent products are subjected to rigorous toxicological evaluation. When the component is 'new' under the terms of regulations within the European Community, it must conform to the Directive on Dangerous Substances before it can be placed on the market.[10–12] This legislation lays down a framework for the conduct of a series of toxicological studies of escalating complexity. The number and type of studies carried out are determined by the so called 'trigger tonnages' of 1 tonne $year^{-1}$ for a Base Set notification, 10 tonne $year^{-1}$ for Level 1 and 1000 tonne $year^{-1}$ for Level 2. The studies which are mandated are those that would normally be considered in drawing up any toxicological programme on a new chemical. They range from studies of acute toxicity, sensitization, sub–chronic toxicity, and genotoxicity at the Base Set level, through teratology, fertility, and reproduction with options for further mutagenicity tests and sub–chronic studies at Level 1. Level 2 deals with chronic toxicity and carcinogenic potential while leaving the option open for additional studies on any of the toxic end points previously examined and for studies in a second species.

The regulatory framework also covers the determination of physico-chemical properties and the establishment of ecotoxicological potential. These issues are outside the scope of this chapter and they will not be discussed further.

It is interesting to note that even within this regulatory framework, which is primarily a scheme for CLASSIFICATION, the concept of RISK is included. The preamble to the Level 2 testing programme states: 'The latter (*i.e.* **the Competent Authority**) shall then draw up a programme of tests to be carried out by the notifier in order to enable the Competent Authority **to evaluate the risks of the substance** for man and the environment'. (The emphasis is the author's). (See also reference 11).

So, even within the regulatory framework the concept of risk assessment is considered to be important. As mentioned earlier, this process involves a definition of toxic potential, an estimation of exposure, and an assessment of the risk of harm occurring for any given exposure situation.

The determination of exposure is a complex process but has recently been described in an authoritative manner for consumer products in general[13] and for laundry and cleaning products in particular.[14] Through the way in which such products are used, it is clear that the skin is a major route of exposure, but oral exposure and inhalation exposure must not be forgotten. The former can occur when, for example, residues are consumed from dishes. The latter could occur when product is poured from its container or a measuring device and when dryer exhausts are vented indoors. It is sufficient to say that methods exist for determining such human exposure.

When that is known, the information can be integrated with the toxicological data, including critically the specific information on target organs or systems, and an estimation made of the risk posed by a particular chemical in a particular use situation.

Such an assessment will be greatly aided if information on the metabolic fate of a chemical is known, and if there are data available from clinical studies.

The principles outlined above have been elaborated further. Beck *et al.* outlined the risk assessment process (nitrilotriacetic acid, sodium aluminosilicate, and linear alkyl benzenesulphonate).[15] Robinson *et al.* detailed a risk assessment process in a specialized situation involving the assessment of allergic contact sensitization.[16]

In the approach recommended by Robinson *et al.* the animal skin sensitization test (Buehler method) was seen as a means of assessing sensitization potential. An ultimate consumer risk assessment can only be developed through a series of controlled human tests such as a human repeat insult patch test, provocative use tests, and extended product use tests. The post-marketing surveillance of skin-related consumer comments was seen as a further source of valuable information which helps in the overall risk assessment process. This scheme is described in detail later.

Decisions on the management of the risks presented by laundry detergent products do not differ substantially from those taken for other consumer products. Any individual decision process developed for a particular product category must be compatible with the recently promulgated Dangerous Preparations Directive.[17] In the scheme defined in the Directive, the risks inherent in any particular product are managed through the provision on the product package of appropriate warning symbols, and 'Risk Phrases'[12] which are intended to warn the consumer of a potential threat. These are backed up by a series of 'Safety Advice Phrases' which are intended to minimize any risk. Information from human experience may override the provision in the Directive for using either the results of tests in animals or the calculation ('conventional') method in determining appropriate labelling information.

The approach taken by the European Commission is a specialized and highly formalized scheme which is within a general framework. Such a framework allows two strategies to be followed:

i) A default position where a decision may be taken not to use a particular chemical in a particular product;

ii) An informed decision may be taken after the risk assessment process has been completed and other options, listed below, examined:

- control of ingredient levels
- risk–benefit ratios
- package labels } This is the approach taken in the Dangerous Preparations Directive
- warning statements }
- advice statements }
- special packaging, *e.g.* child resistant closures
- special product form
- use instructions
- education of specialist groups, *e.g.* Poison Information Centres, dermatologists
- consumer complaint follow-up
- economic considerations.

4 A CASE HISTORY: THE INCORPORATION OF SODIUM ISONONANOYLOXYBENZENE-SULPHONATE INTO A DETERGENT PRODUCT

The principles outlined above of managing the risks inherent in the use of chemicals in detergent products will be illustrated by reference to one particular chemical. This is sodium isononanoyloxybenzenesulphonate (isoNOBS), which is more correctly identified as sodium 3,5,5-trimethylhexanoyloxybenzenesulphonate (Figure 1).

A risk assessment for the incorporation of this chemical into a laundry detergent product was made possible by obtaining a thorough understanding of its solution chemistry in use, its toxicology including skin sensitization potential and the major toxic end points, and importantly likely human exposure.

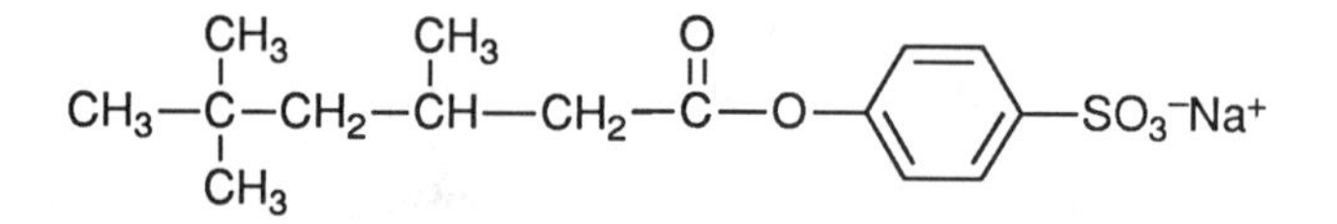

Figure 1 *Isononanoyloxybenzenesulphonate (isoNOBS)*

4.1 The Chemistry of isoNOBS In a Detergent Context

For many years the bleaching component of detergent powders has been either sodium perborate or sodium percarbonate, both of which release hydrogen peroxide in the wash solution. This bleaching system gives optimal performance at 80–90 °C and has met consumer needs until wash temperatures began to be reduced, wash times shortened, and more coloured and synthetic fabrics were included in the wash.

Detergent product manufacturers have therefore identified the need to find safe, practical routes to cost-effective, colour-safe bleaching at wash temperatures between 40 and 60 °C. The most commonly used route has been to develop chemicals that form the peroxy group *in situ* through perhydrolysis.[18,19] In the case of isoNOBS this is shown in Figure 2.

To be effective, the peracid precursor must be highly water soluble, undergo rapid and complete perhydrolysis, and produce a peracid bleach with a degree of surface activity; isoNOBS fulfils these requirements. It is designed so that the ester perhydrolysis, the preferred reaction, dominates over straightforward hydrolysis and diacyl peroxide formation, as shown in Figure 3.

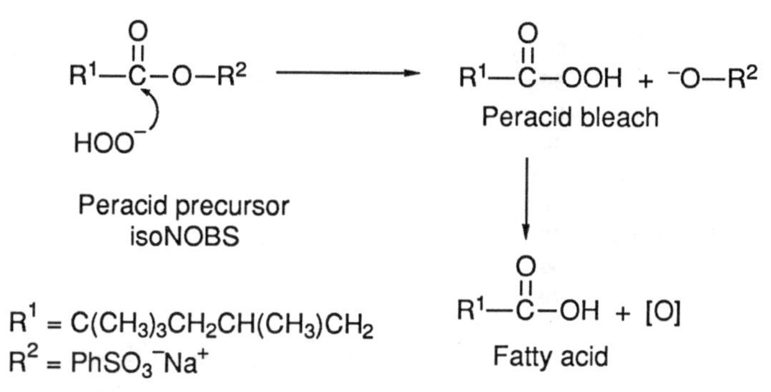

Figure 2 *The perhydrolysis of isoNOBS*

The hydrolysis process

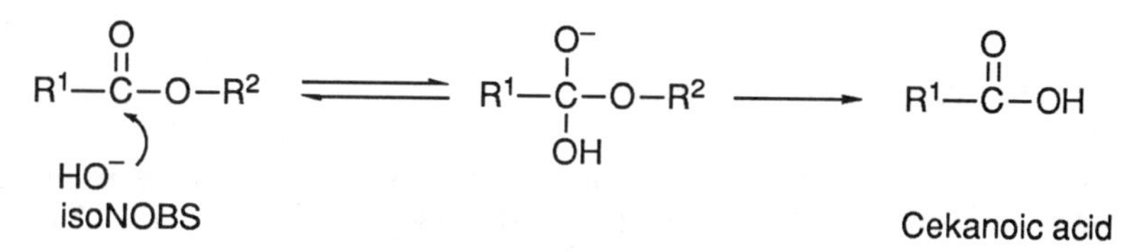

The formation of diacyl peroxide

R^1—C(=O)—OOH
Peracid bleach (see Figure 2)

⇅

R^1—C(=O)—OO⁻ + R^1—C(=O)—O—R^2 ⟶ R^1—C(=O)—O—O—C(=O)—R^1

Peranion   isoNOBS   Diacyl peroxide

Figure 3 *The hydrolysis of isoNOBS and the route to diacyl peroxide formation, R^1 and R^2 as in Figure 2*

The success of the design of the isoNOBS molecule is illustrated in Figure 4. This shows the rate of loss of isoNOBS from a solution of a granular laundry detergent product (100 g in 8 l of water) at 40 °C, pH 10.5, and with an initial isoNOBS solution concentration of 0.04%. By the time the first measurement was made at 30 seconds, essentially all the isoNOBS had disappeared. In other studies, it was shown that the chemical species formed was the peracid.*

*Acknowledgement is given to the assistance provided by Dr. A.J. Pretty in the preparation of the section on the solution chemistry of isoNOBS.

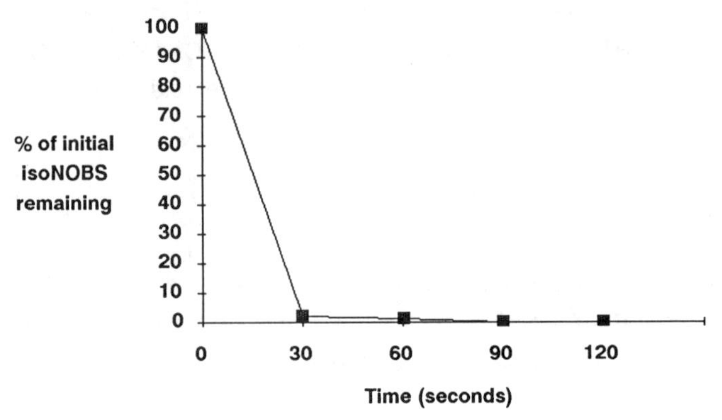

Figure 4 *Percentage isoNOBS remaining in a handwash solution*

4.2 The Toxicology of isoNOBS

The toxicology of isoNOBS is well understood. It has been evaluated in studies addressing all the major toxic end points. These have already been detailed in Section 3 above.

As the chemical is 'new' under the terms of the European Communities legislation, it has undergone all the prescribed toxicology studies at Level 1 to the satisfaction of the Competent Authority in the country of manufacture. As part of the overall programme, three metabolic studies were carried out with the chemical labelled in three different positions. From this we were able to confirm that the immediate metabolites of isoNOBS are sodium *p*-hydroxybenzenesulphonate and 3,5,5-trimethylhexanoic acid (cekanoic acid). Further metabolic work will be described in a later section.

The toxicology of isoNOBS is unremarkable in all but two areas. It is a skin sensitizer in the guinea-pig, and a nephrotoxic response is induced in male rats in 91 day sub-chronic feeding studies. These two issues will be dealt with separately in the following sections.

4.2.1 Sensitization

4.2.1.1 Guinea-pig tests. It was established at an early stage that isoNOBS is a skin sensitizer in that it gave positive responses in a Buehler test[20] at concentrations in the range 1–10% in water. A comparison was made between responses observed in these tests (variation of skin grades with concentration) and the responses observed with another commonly used detergent ingredient, sodium linear alkylbenzenesulphonate (LAS) in the same test. The surfactant was known to give comparable responses in the Buehler test at comparable concentrations, and it was known to have a long history of safe use in detergent products. This comparison allowed the conclusion to be drawn that isoNOBS possessed weak skin sensitization potential in animals at higher concentrations than would be encountered in use, and that human exposure could proceed cautiously.

4.2.1.2 Human tests. In controlling the human exposure that occurred in the early stages of the development of isoNOBS, recognition was given to the exposure that the consumer and the volunteer panellists might receive in relation to established no-effect concentrations in the guinea-pig skin sensitization test.

The concentration of isoNOBS in the detergent products was *ca.* 4% w/w. As the highest exposure was foreseen in the handwashing of laundry, where the concentration of the product is *ca.* 1% w/v, an initial isoNOBS concentration in the wash solution of *ca.* 0.04% w/v needs to be considered. Subsequently in the wash solution, perhydrolysis would rapidly reduce this concentration, as has been explained in the preceding section.

The first experimental exposure of people to isoNOBS was in human repeat insult patch tests[21] using either the product containing isoNOBS, or isoNOBS itself. The experimental procedure received Ethical Review Board approval. Five such tests were eventually completed on the product which contained isoNOBS in 436 volunteer panellists with uniformly negative results. These tests were carried out in aqueous solution at a product concentration of 0.15% w/v. The initial isoNOBS solution concentration was 0.006% w/v.

The second form of human exposure was in a human repeat insult patch test on isoNOBS itself to confirm that low concentrations under exaggerated exposure conditions would give negative results. The concentration selected in the first study was 0.1% w/v in water, which is below the no-effect level for sensitization in the guinea-pig (1% w/v) but above the highest expected human exposure (0.04% w/v). The test produced a negative response in 92 volunteers.

A second test with a concentration of 0.2% w/v in water produced an unequivocal positive response in two volunteer panellists out of 87 tested. This result was not anticipated as the concentration tested is below the lowest concentration giving positive responses in the guinea-pig test but is an event foreseen as a possibility in the design of such a study. It is not possible to predict that in this type of test there will never be a case of induced sensitization. The design of the test is such that a positive response is seen as an extremely remote possibility.

A series of studies were undertaken on human volunteers. These included:

i) Investigative clinical studies involving a home use test on two sensitized panellists who were patch test positive to isoNOBS[22] (no adverse reactions were observed); clinical studies (on 600 volunteers) namely T-shirt wear test (200 male panellists); clinical mildness study (200 female panellists); and extended use tests (200 female volunteers). The general conclusions were that exposure to isoNOBS under normal use conditions does not induce skin sensitization;

ii) Exaggerated exposure tests were also undertaken as it is well known that many persons regularly use a paste of granular products for special tasks such as collar and cuff cleaning. This involved two further studies: the two sensitized persons mentioned above rubbed their hands for 60 seconds in a 50:50 paste containing *ca.* 4% w/w isoNOBS and water three times on alternative days – no sensitization or change in hand condition was noted; a panel of 100 unexposed volunteers also experienced no adverse reactions when subjected to a similar exposure routine over an eight week period;

iii) A diagnostic patch test programme in an exposed population of 117 consumers which were compared with 115 controls indicated no evidence of a sensitization response;

iv) A follow-up of consumer complaints – this was of particular significance in view of a potential market of 300 million consumers in Europe alone. There were consumer comments including several which were health related, but after consideration by a consultant physician (who was also a dermatologist) there was no evidence of any complaint being related to exposure to isoNOBS.

From the foregoing investigations it was deduced that exposure to isoNOBS in a detergent formulation under real life exposure conditions does not lead to induced skin sensitization.

4.2.1.3 Discussion. The incorporation of a recognized skin sensitizer, isoNOBS, into a detergent product neatly illustrates the principles outlined by Robinson *et al.*[16] The data generated in the guinea-pig define the potential of isoNOBS to induce skin sensitization while at the same time giving an indication of the dose-response relationship. This is a key element in risk assessment as it allows effect and no-effect concentrations to be defined. The clinical work with human volunteers confirmed both the potential and the dose-response relationship.

At this stage there were only limited data available on the real-life situation where consumers are exposed to isoNOBS. A risk assessment for the exposure of larger populations of consumers was possible as a result of the various clinical studies which were designed to confirm that a potential for sensitization did not lead to induced reaction in practice. The number of people involved in these various studies, taken together with the different exposures examined, gives a good basis for making a risk assessment.

The other key pieces of information that helped in that process were: firstly, an understanding of the chemistry of isoNOBS in a solution of the product; secondly, the relationship between the potential of isoNOBS to induce sensitization and that of other commonly used detergent chemicals;

and thirdly, the lack of any evidence of induced sensitization from real life exposure.

In this situation, the management of the risk involved is relatively straightforward. Clearly, it is important always to maintain the isoNOBS level at about the level evaluated in product. At the same time, any use of the chemical at anything other than a low level would have to be in products where perhydrolysis will occur in solution.

4.2.2 Nephrotoxicity

4.2.2.1 The toxic response. A 91 day dietary feeding study was carried out in rats. Groups of 15 male and 15 female CrL:CD (SD) BR rats were fed a standard diet containing 2, 10, or 100 mg isoNOBS kg^{-1} b.w. day^{-1} continuously. A control group of similar size received only normal diet. Two responses related to treatment were observed in the liver of males and females, and in the kidney of male rats only.

The liver lesion of minor periportal vacuolation was seen in three male and five female rats in the high-dose group. The effect was of questionable significance as reviewing pathologists questioned whether or not the effect was real. The no-effect level was 10 mg kg^{-1} day^{-1}.

The histopathological examination of the tissues revealed that the kidney of male rats showed significant toxicity. This will be the subject discussed in the remainder of this section.

A thorough investigation of the kidney lesion has been carried out by Lehman–McKeeman and co-workers[23] and the remainder of this section draws freely upon her work with acknowledgement.

4.2.2.2 Structure–activity relationships. A male rat specific nephrotoxicity, which is referred to as hyaline droplet nephropathy, has been induced by a variety of chemicals, including volatile hydrocarbons.[24] This nephropathy is characterized by three changes in the kidney: the appearance of increased levels of hyaline droplet formation, the appearance of granular casts in the outer medulla, and exacerbated changes that cannot be distinguished from progressive necrosis.[25] Gasoline,[26] decalin,[27] 2,4,4-trimethylpentane,[28] *d*-limonene,[29] and 1,4-dichlorobenzene[30] are established male rat-specific nephrotoxins. The determinant of the sex and species specificity of this toxicity is the presence of α2u-globulin.[25] This protein is synthesized exclusively by adult male rats[31] in which it is the major urinary protein.

When rats are exposed to the chemicals listed above, α2u-globulin accumulates in the renal cortex within the secondary lysosomes. This protein accumulation can be identified histologically as hyaline droplets.[32–36] It appears that the accumulation of α2u-globulin in the kidney results from the binding of the chemical, or a metabolite, to α2u-globulin[28–30] and that this disrupts the lysosomal degradation of the protein.[23] Evidence for the key role of α2u-globulin in the development of this lesion comes from a knowledge that female rats and male NCI Black Reiter rats, which do not synthesize α2u-globulin,[37] do not develop hyaline droplet nephropathy when exposed to these chemicals.

The hydrolysis product of isoNOBS, cekanoic acid (see Figure 3) is structurally similar to 2,4,4-trimethylpentane (see Figure 5) which is a known inducer of the nephropathy.

$$CH_3-C(CH_3)_2-CH_2-CH(CH_3)-CH_2-C(=O)-OH$$

Cekanoic acid

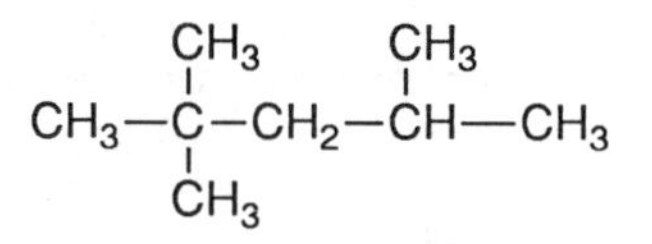

2,4,4-Trimethylpentane

Figure 5 *The chemical structures of cekanoic acid and 2,4,4-trimethyl-pentane*

4.2.2.3 isoNOBS nephrotoxicity investigative programme. The principal finding in the 91 day dietary feeding study on isoNOBS was that of the kidney lesion observed only in the male rat. The lesion met the criteria for hyaline droplet nephropathy established by Swenberg *et al.*,[25] of increased levels of hyaline droplet formation, the appearance of granular casts, and progressive necrosis. It has been established with other chemicals that induce this lesion that the chemicals themselves, or a metabolite, bind reversibly with α2u-globulin.[37] It was therefore important to conduct the necessary experiments to determine whether or not isoNOBS, or a metabolite, binds to α2u-globulin. At the same time, it would be important to show that similar interactions did not occur in the mouse kidney.

To this end, Lehman-McKeeman and co-workers[23] showed that acute oral dosing of isoNOBS to adult male and female rats causes the male rat-specific hyaline droplet nephropathy. Experiments were carried out to determine whether or not the accumulation of protein in the male rat kidney cortex was caused by the binding of radiolabelled isoNOBS equivalents to α2u-globulin. Several key items of evidence showed that this was the case:

i) Two–dimensional gel electrophoresis showed that α2u–globulin levels in the male rat kidney increased 24 hours after a single oral dose of isoNOBS (500 mg kg^{-1}).

ii) The sex–dependent retention of isoNOBS was approximately 10 times higher in the male rat kidney than in the female rat kidney.

iii) Equilibrium dialysis experiments showed that 40% of radiolabelled isoNOBS equivalents bind reversibly to male rat kidney proteins.

iv) No interaction was observed between isoNOBS and female rat kidney proteins.

Several techniques were used to establish the specific binding of isoNOBS to α2u–globulin. These included anion–exchange HPLC followed by identification of metabolites in the α2u–globulin fraction by gas chromatography with parallel radioactivity–mass spectrometry and mass spectrometry–matrix isolation Fourier–transform infrared analysis. Of the four metabolites of isoNOBS identified in these procedures, the major component (approximately 70%) was identified as the *cis*–γ–lactone of 3,5,5–trimethylhexanoic acid (TMHA; Figure 6).

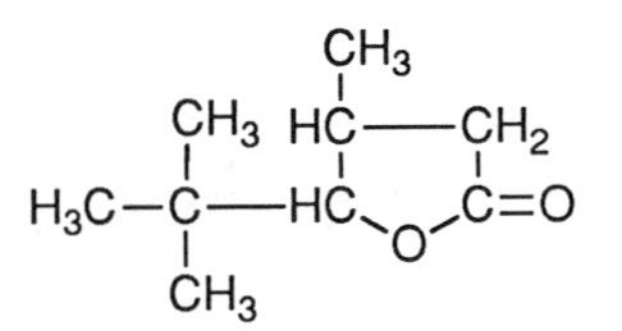

Figure 6 *The* cis–γ–*lactone of 3,5,5–trimethylhexanoic acid*

Studies carried out in mice showed that there was no interaction between isoNOBS and mouse urinary protein, a protein that shares significant homology with α2u–globulin.

These comparative exposure studies show that isoNOBS exacerbates hyaline droplet formation in male rat kidneys by binding to α2u–globulin. This causes the protein to accumulate in the renal cortex. The absence of an interaction between isoNOBS and a protein similar to α2u–globulin found in the mouse indicates that the interactions found in the male rat are unique.

This work gives a very clear demonstration of the fact that isoNOBS is an inducer of classical hyaline droplet nephropathy and that it can therefore be related to the history of chemicals previously identified with this syndrome.[26] The central fact is, of course, the ability of these chemicals, or a metabolite, to bind reversibly with the male rat specific α2u–globulin. No such protein is produced by man, and as Lehman–McKeeman and her group showed,[23] the key metabolite in the case of isoNOBS does not bind with a structurally related protein in the mouse, mouse urinary protein, MUP.

4.2.2.4 Discussion. There are parallels between the approach used in assessing the risk posed by the kidney lesion induced by isoNOBS, and the approach taken in the previous section on sensitization. In neither case did the evaluation of risk stop when either the effect (sensitization) or the lesion (nephrotoxicity) had been established. In both cases a set of investigations was initiated which was aimed at obtaining a better understanding of the effects produced. Programmes on both fronts attempted to place the effects into a wider context, and they succeeded.

In the present case of the nephrotoxic response, a risk assessment is relatively straightforward. The lesion observed is directly related to the ability of the lactone metabolite of isoNOBS to bind reversibly with the male rat specific protein, α2u-globulin. As man, let alone female rats and male and female mice, does not produce this protein, the nephrotoxic response can be discounted. The understanding obtained from a thorough examination of the toxicokinetics of isoNOBS allows one to conclude that this lesion is inappropriate for risk assessment purposes.

Risk management of this toxic response, which is the final stage in providing adequate protection for exposed people, is now straightforward. Consumers and factory workers who might be exposed to isoNOBS, and hence its metabolite, are simply not at risk.

5 CAN CHEMICALS BE SAFELY USED?

The answer has to be 'YES'. The same answer would, of course, be given to the question 'Is it possible for chemicals to cause harm?'. The same answer would also be given to the question 'Do chemicals confer significant benefits to society?'.

The situation is complex. Society has developed a complex chemical industry supplying producer companies with an almost unbelievable range of chemicals including food additives, sweeteners, pesticides, solvents, preservatives, dyestuffs, plastics, aerosols, and general chemicals of such diversity as to be almost limitless.

It is perhaps not surprising to find that the science of toxicology has grown in parallel with the growth of the chemical industry. This growth has been both at academic level and in the growth of professional groups and learned societies in toxicology. This growth has stemmed from the need to know what exposure to this myriad of chemicals implies for the health and well being of society.

The partnership between toxicologists and society has not been an easy one. Many painful lessons had to be learned before the concept began to be accepted that the determination that a particular chemical has a certain toxic potential was in and of itself no basis for establishing whether or not the chemical presented a significant risk to man. A deeper understanding of such matters as dose–response relationships, no–effect levels, safety factors, mechanisms of action, metabolic comparability, and exposure are essential before a risk assessment can be made.

As indicated previously for isoNOBS, such identification of a toxic potential is but one step along the road to risk management. Exposure calculations are equally important and enable the risk to be estimated of an effect being observed at any particular level of exposure. The management of that risk is a societal question as it can no longer be answered by scientists alone. The option not to use a chemical must always be available, but more constructively a range of quality toxicology studies made in conjunction with exposure calculations in normal circumstances allows use to be made of chemicals with adequate safety margins.

6 ACKNOWLEDGEMENT

I would like to acknowledge the help provided by the late Dr. R.A. D'Amato, Dr. R.D. Laurie, Dr. L.D. Lehman–McKeeman, and Dr. P.M. McNamee in reviewing the manuscript. My appreciation of Dr. Lehman–McKeeman's permission to quote freely from her work[23] has been stated previously. I also appreciate Dr. A.J. Pretty's permission to quote from his work on isoNOBS solution chemistry.

7 REFERENCES

1 'Risk Assessment, A Study Group Report', The Royal Society, London, 1983.

2 'Risk Assessment in the Federal Government: Managing the Process', Commission on Life Sciences, National Research Council, National Academy Press, Washington, DC, 1983.

3 'Risk Assessment of Chemicals in the Environment', ed. M.L. Richardson, The Royal Society of Chemistry, London, 1988.

4 'Presentation of Risk Assessments of Carcinogens', American Industrial Health Council, Washington, DC, 1989.

5 J.J. Cohrssen and V.T. Covello, 'Risk Analysis: A Guide to Principles and Methods for Analyzing Health and Environmental Risks', Executive Office of the President of the United States, The National Technical Information Service, 1989.

6 'The Risk Assessment of Environmental Hazards. A Textbook of Case Studies', ed. D.J. Paustenbach, John Wiley & Sons, Inc., 1989.

7 'Risk Assessment and Management of Industrial and Environmental Chemicals', ed. C.R. Cothern, M.A. Mehlman, and W.L. Marcus, in 'Advances in Modern Environmental Toxicology', Series Editor M.A. Mehlman, Princeton Scientific Publishing Co. Inc., Princeton, NJ, 1988.

8 'Management of Assessed Risk for Carcinogens', ed. W.J. Nicholson, *Ann. N.Y. Acad. Sci.*, 1981, **363**, 1–299.

9 'Risk Assessment in the Federal Government: Managing the Process', Commission on Life Sciences, National Research Council, National Academy Press, Washington, DC, 1983, p. 2.

10 'Legislation on Dangerous Substances, Classification and Labelling in the European Communities', Volume 1, Office for Official Publications of the European Communities, Graham and Trotman Ltd., 1987.

11 J.L. Vosser, 'The European Community Chemicals Notification Scheme and Environmental Hazard Assessment', in 'Toxic Hazard Assessment of Chemicals', ed. M.L. Richardson, The Royal Society of Chemistry, London, 1986, pp. 117–132.

12 H.P.A. Illing, 'The European Community Chemicals Notification Scheme: Human Health Aspects', in 'Toxic Hazard Assessment of Chemicals', The Royal Society of Chemistry, London, 1986, pp. 133–149.

13 P.J. Hakkinen, C.K. Kelling, and J.C. Callender, *Vet. Human Toxicol.*, 1991, **33**, 61–65.

14 P.J. Hakkinen, 'Handbook of Hazardous Materials, Cleaning and Laundry Products, Human Exposure Assessments', Academic Press, in press, 1992.

15 L.W. Beck, A.W. Maki, N.R. Artman, and E.R. Wilson, *Regulatory Toxicol. Pharmacol.*, 1981, **1**, 19–58.

16 M.K. Robinson, J. Stotts, P.J. Danneman, and T.L. Nusair, *Food Chem. Toxicol.*, 1989, **27**, 479–489.

17 'European Communities Directive on the Classification, Packaging and Labelling of Dangerous Preparations', in 88/379/EEC, *Offic. J. Eur. Commun.*, June 1988.

18 K. Grime and A. Clauss, *Chem. Ind. (London)*, 1990, 15 October, 647–653.

19 B.D. Sully and P.L. Williams, *Analyst (London)*, 1962, **67**, 653–657.

20 H.L. Ritz and E.V. Buehler, 'Planning, Conduct and Interpretation of Guinea Pig Sensitization Patch Tests', in 'Current Concepts in Cutaneous Toxicity', ed. V.A. Drill and P. Lazar, Academic Press, 1980.

21 J. Stotts, 'Planning, Conduct and Interpretation Human Predictive Sensitization Patch Tests' in 'Current Concepts in Cutaneous Toxicity', ed. V.A. Drill and P. Lazar, Academic Press, 1980.

22 S. Fregert, 'Manual of Contact Dermatitis', 2nd Edn., Year Book Medical Publisher, 1981, pp. 74–76.

23 L.D. Lehman–McKeeman, P.A. Rodriguez, D. Caudill, M.L. Fey, C.L. Eddy, and T.N. Asquith, *Toxicol. Appl. Pharmacol.*, 1991, in press.

24 C.V.R. Murti, M.J. Olson, B.D. Garg, and A.K. Roy, *Toxicol. Appl. Pharmacol.*, 1988, **96**, 380–392.

25 J.A. Swenberg, B. Short, S. Borghoff, J. Strasser, and M. Charbonneau, *Toxicol. Appl. Pharmacol.*, 1989, **97**, 35–46.

26 C.A. Halder, T.M. Warne, and N.S. Hatoum, 'Renal Toxicity of Gasoline and Related Petroleum Naphthas in Male Rats', in 'Renal Effects of Petroleum Hydrocarbons', ed. M.A. Mehlman, C.P. Hemstreet, J.J. Thorpe, and N.K. Weaver, 'Advances in Modern Environmental Toxicology', Princeton Scientific Publishing Co. Inc., Princeton, NJ, 1984, Vol. II, pp. 73–88.

27 C.L. Alden, R.L. Kanerva, G. Ridder, and L.C. Stone, 'The Pathogenesis of the Nephrotoxicity of Volatile Hydrocarbons in the Male Rat', in 'Renal Effects of Petroleum Hydrocarbons', ed. M.A. Mehlman, C.P. Hemstreet, J.J. Thorpe, and N.K. Weaver, 'Advances in Modern Environmental Toxicology', Princeton Scientific Publishing Co. Inc., Princeton, NJ, 1984, Vol. II, pp. 107–120.

28 E.A. Lock, M. Charbonneau, J. Strasser, J.A. Swenberg, and J.S. Bus, *Toxicol. Appl. Pharmacol.*, 1987, **91**, 182–192.

29 L.D. Lehman–McKeeman, P.A. Rodriguez, R. Takigiku, D. Caudill, and M.L. Fey, *Toxicol. Appl. Pharmacol.*, 1989, **99**, 250–259.

30 M. Charbonneau, J. Strasser, E.A. Lock, M.J. Turner, and J.A. Swenberg, *Toxicol. Appl. Pharmacol.*, 1989, **99**, 122–132.

31 A.K. Roy, O.W. Neuhaus, and C.R. Harmison, *Biochim. Biophys. Acta.*, 1966, **127**, 72–81.

32 M.D. Stonard, P.G.N. Phillips, J.R. Foster, M.G. Simpson, and E.A. Lock, *Toxicology*, 1986, **41**, 161–168.

33 M. Charbonneau, E.A. Lock, J. Strasser, M.G. Cox, M.J. Turner, and J.S. Bus, *Toxicol. Appl. Pharmacol.*, 1987, **91**, 171–181.

34 R.L. Kanerva, M.S. McCracken, C.L. Alden, and L.C. Stone, *Food Chem. Toxicol.*, 1987, **25**, 53–61.

35 M.J. Olson, B.D. Garg, C.V.R. Murty, and A.K. Roy, *Toxicol. Appl. Pharmacol.*, 1987, **90**, 43–51.

36 N.G. Read, P.J. Astbury, R.J.I. Morgan, D.N. Parsons, and C.J. Port, *Toxicology*, 1988, **52**, 81–101.

37 B. Chatterjee, W.F. Demyan, C.S. Song, B.D. Garg, and A.K. Roy, *Endocrinology*, 1989, **125**, 1385–1388.

11

Priority Setting and Risk Assessment of Chemicals: Human Health Effects

R. O. Shillaker

HEALTH AND SAFETY EXECUTIVE, MAGDALEN HOUSE, STANLEY PRECINCT, BOOTLE, MERSEYSIDE L20 3QZ, UK

1 INTRODUCTION

Commercial chemical substances that were declared as having been supplied on the European Community (EC) market between 1 January 1971 and 18 September 1981 are listed on the European Inventory of Existing Commercial Chemical Substances.[1] For legal purposes, the *ca.* 100,000 listed substances are referred to within the EC as 'existing' substances, whilst substances placed on the market after 18 September 1981 are regarded as 'new'.

Regulatory appraisal of new substances supplied on the EC market is currently achieved through the 6th Amendment (79/831/EEC)[2] to the Dangerous Substances Directive. In the UK the requirements of the Directive are implemented by the Notification of New Substances Regulations.[3,4] Before a new substance can be marketed a notification must be made, with some limited exceptions, to the relevant national competent authority. For substances supplied at one tonne year^{-1}, a 'Base Set' dossier of specified technical information (including physico-chemical, toxicological, and ecotoxicological data) is required in order to evaluate the foreseeable risks, whether immediate or delayed, which a substance may pose for man or the environment. The toxicological data requirement consists of information on acute and subacute toxicity, skin and eye irritancy, skin sensitization, and mutagenicity. The dossier of information is studied by the competent authority in order to assess the intrinsic hazardous properties of a substance and, using the limited information normally supplied about exposure, to conduct a preliminary evaluation of risk. For more details of the new substances scheme see accounts by Vosser[5] and Illing.[6]

Under the proposed 7th Amendment[7] to the Dangerous Substances Directive greater emphasis is placed on risk assessment.* More exposure data are required at Base Set than under the 6th Amendment so that a better evaluation of risk should be possible. Risk management, in the proposed 7th Amendment, is addressed in terms of: classification, packaging, and labelling; recommended methods and precautions relating to handling, storage, transport, fire hazards, and other dangers (particularly reaction with water); emergency measures in case of accidental spillage or injury to persons; the possibility of rendering the substance harmless, *e.g.* by incineration; the provision of safety data sheets; and recommendations for legislative control measures associated with marketing the substance. Any restrictions on marketing and use will be implemented via Directive 76/769/EEC[8] or other appropriate Community legislation. It should be noted that there is a growing view in the EC that new substances should be safer than those with equivalent uses already on the market. In practice, however, such judgements may be hindered by the lack of toxicological and ecotoxicological information available about existing substances.

For existing substances, regulatory emphasis at Community level has, until recently, also been on assessing information on intrinsic hazardous properties for purposes of classification and labelling. However, an EC Council Regulation[9] was proposed in 1990 which requires that information be submitted to the EC Commission by manufacturers/importers of existing substances in order to evaluate and control the risks to man and the environment. The proposed Regulation applies to any EC manufacturer/importer who produces or imports a substance in quantities greater than 10 tonnes year^{-1}. Risk evaluation may lead to workplace controls or, where appropriate, action under the Marketing and Use Directive (76/769EEC)[8] or other Community legislation.

With greater regulatory emphasis within the EC on risk assessment of new and existing substances, there is a clear need for a harmonized approach throughout the Community. This chapter describes on-going work in the UK to meet this need, primarily with respect to setting priorities for human health risk assessment of existing substances.

2 SELECTION OF SUBSTANCES FOR RISK ASSESSMENT

2.1 New Substances

Selection of new substances for more detailed risk assessment is not regarded as posing a particular problem because a standardized package of technical information will be available for all substances at Base Set (and

*At the time of writing this chapter, the 7th Amendment had not been adopted but the enacting terms of an amended version of the published proposal[7] had been approved by the EC Environment Council.

at pre-Base Set under the proposed 7th Amendment). In addition, at least in the case of Base Set notifications, the numbers of substances are relatively small. In 1990, 138 substances were the subject of Base Set notifications to the EC, for the first time; of these, 44 were to the UK. Substances classified as dangerous to human health would be obvious candidates for more detailed risk assessment. However, owing to the limited testing at Base Set, the absence of a need to classify would not automatically make more detailed risk assessment unnecessary.

2.2 Existing Substances

In the case of existing substances, a formalized priority setting procedure is essential because of the large number of candidate substances and consequent data sets anticipated to be submitted to the EC Commission within a short period of time. The total number of substances likely to fall within the scope of the draft Regulation has been estimated to be about 10,500. Annex 1 of the draft provides a pragmatic list of *ca.* 2000 substances produced or imported into the Community in quantities exceeding 1000 tonnes year^{-1} (high volume substances): these Annex 1 substances should be covered by the first phase of a three-phase data collection programme.

Priority setting for existing substances is also important in order to maximize the benefit from the limited resources available for risk assessment, including the acquisition and evaluation of additional toxicity and/or exposure data.

A number of priority setting schemes[10-13] have been proposed for selecting substances for further assessment in the context of European and OECD existing chemicals initiatives. These and other priority setting schemes have been reviewed, and limitations indicated, by the Department of the Environment.[14] One scheme, that of Sampaolo and Binetti[13,15,16] is of particular interest because it was developed at the request of the EC Commission. In this scheme numerical scores are allocated to hazard and exposure data, with overall priority scores being calculated separately for human health effects and for environmental effects.

The Sampaolo and Binetti scheme has both strengths and weaknesses. A wide variety of endpoints are scored but some of the physico-chemical parameters are of no obvious relevance to human health risk assessment. Simple and objective criteria, mostly based on EC classification and labelling requirements[17] are used for scoring toxicological effects, but in the absence of data an expert assessment of structure-activity relationships (SARs) is required. This reliance on SARs is seen as a serious limitation because the current state-of-the-art is not sufficiently well advanced; the implication for introducing a high degree of expert judgement (subjectivity) into the scheme is also disadvantageous.

Mutagenicity, carcinogenicity, and effects on reproduction are identified by Sampaolo and Binetti as areas of special concern by the high scores that can be obtained, primarily when there is convincing human data. However, the scores attributed to positive animal data are often relatively low. In particular an *in vivo* somatic cell mutagen (category 3 mutagen), which many regard as a potential carcinogen, is only scored the same as an eye irritant. The absence of any scoring of *in vitro* mutagenicity data has the advantage of simplicity but it reduces the sensitivity of the scheme.

Given the perceived difficulties with the Sampaolo and Binetti approach it was considered appropriate to develop an alternative scheme in order to contribute to the debate on this important subject.

3 PRIORITY SETTING FOR HIGH VOLUME EXISTING SUBSTANCES

3.1 Introduction

A simple, numerically based scoring system for setting priorities has been developed to meet the needs of the draft Existing Substances Regulation. The scheme, which provides a pragmatic and objective initial screen, requires minimal direct expert input and is suitable for computerization. By taking account of both hazard and exposure data, the scheme is of relevance to priority setting for risk assessment.

The scheme is designed specifically for high-volume substances and may require slight adaptation for substances of lower tonnages. The parameters scored are those for which information is requested, at the initial reporting stage, on the Data Set forms for high-volume substances. This information includes tonnage and classification/labelling details, with brief data, 'if available or easily obtainable', on use, toxicity, and physico-chemical properties. Since information may be submitted by different manufacturers/importers for the same substance, it is proposed that tonnages are summed to obtain a value for the EC as a whole; for other data, worst case values are taken where there are differences in the information submitted.

It is anticipated that the quality and quantity of information supplied on the Data Set form will vary considerably. However, because of the large numbers of substances involved, a priority setting scheme will be necessary that does not require a detailed appraisal of the quality and validity of the information provided. This appraisal, together with consideration of possible structure-activity relationships, is considered to be a function best performed at the subsequent risk assessment stage.

The overall priority score allocated to a substance is intended solely to give an approximate idea of the priority to accord to a substance. It is essential that a list of priority scores is not seen as a rigorous ordering of substances but rather as a means of highlighting substances, or a group of substances, for risk assessment. The scheme presented here is not necessarily regarded as being in its final form, as it is recognized that further refinement may be necessary following more experience of its use in practice.

The scheme, which has been developed in parallel with a comparable priority setting scheme for environmental risk assessment,[18,19] uses four types of data: toxicity (a measure of hazard), physico-chemical properties, use pattern, and tonnage (exposure indicators).

3.2 Toxicity Data

3.2.1 Test Data Required.* Although most of the tests required for priority setting purposes are outlined in Annex V[20,21] of Directive 79/831/EEC, a test will be acceptable for these purposes irrespective of whether it has been conducted fully in accordance with Annex V protocols. If the required data for an area of toxicity are not supplied and no labelling for that area is proposed, a default score is necessary. Data requirements for priority setting are as follows:

Acute toxicity. Data for one route of exposure (oral, dermal, or inhalation) are sufficient.

Irritation. Data for either skin or eye irritation are sufficient.

Sensitization. Data for one skin sensitization test are sufficient.

Repeated exposure effects. Data for repeated exposure over at least a 28 day period are sufficient. Oral, inhalation, or dermal data are acceptable.

Mutagenicity. The basic data requirement is one *in vitro* test for gene mutation and one somatic cell test (*in vitro* or *in vivo*) for chromosome aberrations. These tests correspond to categories (a) and (b) or (c), respectively in Table 1.

Table 1 *Categorization of mutagenicity tests*

(a) *In vitro* tests for gene mutation

Reverse mutation test in *Salmonella typhimurium* or *Escherischia coli*
Mammalian cell gene mutation test

(b) *In vitro* somatic cell test for chromosome aberrations

Mammalian cytogenetics assay

*The terms 'required' and 'acceptable' tests are used solely in the context of this priority setting scheme and do not relate to any wording in the draft Existing Substances Regulation.

Table 1 cont.

(c) <u>*In vivo* somatic cell test for chromosome aberrations</u>

Mammalian bone marrow assay, cytogenetics, or micronucleus test

(d) <u>Other *in vivo* somatic cell tests</u>

Mouse spot test
Liver unscheduled DNA synthesis assay

(e) <u>*In vivo* germ cell tests</u>

Mammalian germ cell cytogenetics test
Rodent dominant lethal test
Mouse heritable translocation test

This Table, which lists all the mutagenicity tests used for priority setting, is comprised of the best validated and most reliable Annex V tests for somatic cell and germ cell mutation. Data for other mutagenicity tests may be available for some substances but this information is not used in the priority setting scheme.

<u>Reproductive toxicity</u>. The requirement is for a reproductive toxicity (fertility) test in one or more generations and for a teratology study.

<u>3.2.2 Scoring Systems</u>. There are difficulties in justifying, scientifically, the relative weighting of priority scores for different toxicological effects. A pragmatic approach has therefore been adopted of, wherever possible, relating scores to the EC criteria for classification and labelling of substances.[17] The classification order of precedence (*i.e.* very toxic > toxic > corrosive > harmful > irritant) has been used as a guide for allocating scores, with extra weighting being given to effects of special concern.

<u>Toxicity excluding mutagenicity and reproductive toxicity</u>. The scoring system, including default scores, is shown in Table 2. When toxicity data are available, scoring is based on the proposed/agreed R ('risk') phrases applied to a substance.

Respiratory sensitization is an effect of special concern because it is potentially life threatening at low concentrations. Hence a substance is given a high score (9) if it is labelled with R42 (may cause sensitization by inhalation). There is no predictive test for respiratory sensitization, and hence data for this effect are not a requirement for priority setting. Labelling with R42 will be based on structure-activity relationships or human data.

In the absence of data, a low default score (1) is generally used. A repeated exposure study is, however, considered to be of importance for high-volume substances and so a default score of 3 is applied.

Table 2 *Toxicity scores (excluding mutagenicity and reproductive toxicity)*

Acute toxicity		
Very toxic	(R26, R27, R28)	3
Toxic	(R23, R24, R25)	2
Harmful	(R20, R21, R22)	1
No test conducted		1 (default)
Test conducted, no R phrase needed		0
Irritation/corrosion		
Corrosive	(R34, R35)	2
Serious damage to eye	(R41)	2
Eye irritant	(R36)	1
Skin irritant	(R38)	1
Respiratory irritant	(R37)	1
No test conducted		1 (default)
Test conducted, no R phrase needed		0
Sensitization		
Respiratory sensitizer	(R42)	9
Skin sensitizer	(R43)	2
No skin test conducted		1 (default)
Skin test conducted, no R phrase needed		0
Repeated exposure effects		
No test conducted		3 (default)
Serious damage to health	(R48)	3
Cumulative effects	(R33)	2
Test conducted, no R phrase needed		0

Mutagenicity. Scoring for mutagenicity is more complicated and is based on a number of possible combinations of test results (Table 3). The scoring system reflects the testing strategy recommended by the Department of Health[22] for investigating the potential for somatic cell mutation; in particular the need to conduct one or two *in vivo* tests if an *in vitro* positive result is obtained. It should be noted that for priority setting purposes *in vivo* germ cell mutagenicity test data are used only when positive (score 9) and a substance labelled for carcinogenicity is regarded as an *in vivo* mutagen (score 9).

Table 3 *Mutagenicity scores*

(a)	Labelled for carcinogenicity or mutagenicity (or positive in an *in vivo* somatic or germ cell test)	9
(b)	No test for gene mutation (*in vitro*), and no test for chromosome aberrations in somatic cells (*in vitro* or *in vivo*), conducted	9 (default)
(c)	Positive in one *in vitro* test **but no** *in vivo* somatic cell test conducted that was unequivocal	9
(d)	As (c) **but one** *in vivo* somatic cell test conducted and was negative	6
(e)	Test(s) for gene mutation (*in vitro*), **or** for chromosome aberrations in somatic cells (*in vitro* or *in vivo*), conducted (*i.e.* **not** for both end-points) and negative	3 (default)
(f)	As (c) **but two** or more *in vivo* somatic cell tests conducted which were all negative	0
(g)	Test(s) for gene mutation (*in vitro*) **and** for chromosome aberrations in somatic cells (*in vitro* or *in vivo*) conducted, and all were negative	0

Although it is undesirable to base risk assessment decisions on equivocal data, it is recognized that such data may have to be used for priority setting. Until such time as an equivocal result can be overridden by a clear negative or positive result in a repeat of the same test, the following rules are applied:

(a) An equivocal result in an *in vitro* mutagenicity test is regarded as positive;

(b) An equivocal result in an *in vivo* mutagenicity test is regarded as an *in vitro* positive.

If there is a mixture of *in vitro* test results, including results for the same type of test, the worst case is used for priority setting. The same rule applies to *in vivo* data.

Mutagenicity test data are considered crucial for priority setting and if there are none of the required tests the substance is given a high default score (9).

Reproductive toxicity. Scoring for reproductive toxicity is shown in Table 4. The term 'Toxic to reproduction' is introduced in the proposed 7th Amendment Directive to replace the more limited term 'Teratogenic'. The new term covers developmental toxicity (including teratogenicity) and impaired fertility.

Reproductive toxicity data are also regarded as crucial for priority setting of high-volume substances and if there are none of the required tests the substance is given a high default score (9).

A repeated dose toxicity study with histological examination of reproductive organs is **not** considered an adequate reproductive toxicity study. If, for a particular substance, this is the only investigation of reproductive effects, and no labelling for reproductive toxicity has been proposed as a consequence, the substance would be given a high default score (9).

Table 4 *Reproductive toxicity scores*

Labelled for toxicity to reproduction or positive in a reproductive toxicity test	9
No test for reproductive toxicity conducted	9 (default)
No fertility test conducted but negative in a teratogenicity test	3 (default)
No teratogenicity test conducted but negative in a fertility test	2 (default)
Negative in fertility and teratogenicity tests	0

Calculation of the total toxicity score. The total toxicity score is taken as the sum of the highest score for each area of toxicity, *i.e.* score for acute toxicity + irritation + sensitization + repeated exposure effects + mutagenicity + reproductive toxicity. A substance would have a total toxicity score of 0 if all the required toxicity data were present and no effects of concern were identified.

3.3 Physico–chemical Properties

3.3.1 Scoring System. Three physico–chemical properties are used for priority setting: boiling point, vapour pressure, and the n–octanol/water partition coefficient (Log *P*). The scoring system, including default values, is shown in Table 5. The system for scoring when data are available shows similarities to that proposed by Sampaolo and Binetti.[13]

The total physico–chemical score is calculated by adding the highest score for boiling point/vapour pressure to the Log *P* score. The lowest possible total physico–chemical score is 1.

Table 5 *Physico–chemical scores*

			Score
Boiling point:	≤60 °C at 700–800 mm Hg (933–1067 hPa)		3
Vapour pressure:	≥10 hPa at 25 °C (±5 °C)		
Boiling point:	>60 ≤200 °C at 700–800 mm Hg (933–1067 hPa)		2
Vapour pressure:	≥1 <10 hPa at 25 °C (±5 °C)		
Boiling point:	>200 °C at 700–800 mm Hg (933–1067 hPa)		1
Vapour pressure:	<1 hPa at 25 °C (±5 °C)		
n–Octanol/water partition coefficient (Log *P*)		≥3	1
		<3	0

Notes: (a) If neither boiling point nor vapour pressure data are available, for the specified pressure and temperature conditions, a default score of 2 is applied.

(b) If no Log *P* value is available a default score of 1 is applied.

3.3.2 Comments. Boiling point and vapour pressure are measures of the ability to form a vapour and thus are important with respect to the potential inhalability of a substance. Another index of inhalability, particle size, is not scored because this parameter is not listed on the Data Set form.

Log *P* is a measure of lipophilicity and gives an approximate indication of the ability of a substance to accumulate in fatty tissue (the indication is only approximate because no account is taken of the reduction in lipophilicity by metabolism). The cut–off value selected was based on the view that at Log *P* ≥ 3 accumulation in humans becomes more significant. It should be noted that it may be appropriate to set an upper limit for Log *P* to reflect the fact that at high Log *P* values absorption (and therefore accumulation) may decrease.

3.4 Use Pattern

The following categories of use are described in the draft Regulation:

Use in closed systems. Exposure is very limited. Emissions into the environment (workplace and external environment) are normally limited to losses during production and disposal of production residues or losses due to accidents, *e.g.* corrosion inhibitors in a hot water heating system.

Use resulting in inclusion into or onto a matrix. Substances are fixed into or onto matrices from which, under normal conditions, they cannot be removed. Emissions and exposure may occur during the application process and to a limited extent after disposal, *e.g.* antioxidants in rubber.

Non-dispersive use. Substances are emitted during application, and exposure may take place but only where there are trained personnel and under controlled conditions, *e.g.* paints in a special paint spraying area.

Wide dispersive use. Substances will be released into the environment to a large extent during use. There is also significant exposure to untrained personnel, *e.g.* paints applied in domestic premises.

Scores for the four use categories are allocated on a 1–5 scale (Table 6). Wide dispersive use is given the highest score because of the low level of control over human exposure in such circumstances. If a substance has more than one category of use, the category giving the highest score is taken for calculating the overall priority score; in the interests of simplicity no account is taken of the quantity of substance in each use category.

Table 6 *Use pattern scores*

Wide dispersive use	5
Non dispersive use	3
Use resulting in inclusion into or onto matrix	2
Use in a closed system	1

3.5 Tonnage

The annual production or importation tonnage for a substance is reported on the Data Set form as one of a series of fixed banded tonnages (*e.g.* 1000–5000 tonnes year^{-1}) up to one million tonnes, or as greater than one million tonnes. For priority setting purposes the tonnage score is taken as the $\log_{10}$ of the appropriate upper band value, *i.e.* a substance produced by a manufacturer at 2000 tonnes year^{-1} falls within the 1000–5000 tonnes band and would score 3.7. Substances in excess of one million tonnes score 6.

3.6 Calculation of the Overall Priority Score

The overall priority score is calculated as follows:

> Total toxicity score x total phys/chem score x use pattern score x tonnage score.

The overall priority score for a substance is not a fixed value but may change following re-assessment of submitted information or the provision of new information. It is anticipated that substances with high initial overall scores will be examined closely at the risk assessment stage to determine whether further testing is necessary.

3.7 Preliminary Trial of Priority Setting Scheme

A preliminary trial of the priority setting scheme was performed using data for 52 high-volume substances examined in the initial phase of the OECD Existing Chemicals Programme.[23] The substances, all organic chemicals, had been selected by the OECD primarily on the basis that no publicly available toxicity and ecotoxicity data could be found.

A certain amount of judgement was applied when scoring the limited information summarized in the OECD 'dossiers'. It was necessary to assess whether R ('risk') phrases were applicable. The EC system of use categories had not been used, and so it was necessary to judge which category(ies) applied. A large number of the substances were chemical intermediates and if there was no evidence of use solely in a closed system they were, in the main, allocated to the non-dispersive use category. A degree of discretion was also applied when interpreting the test data requirements described in Section 3.2.1. For instance, in the absence of a 28 day study, a 14-day repeated exposure study was deemed acceptable if no effects were reported at very high dose levels (*ca.* 1 g kg^{-1}, orally). It should be noted that tonnage data were not available on a standard geographical or temporal basis and so the highest per annum figure quoted for a substance was used for scoring purposes.

The results of the trial are shown in Table 7. No acceptable reproductive toxicity data were available for 50 substances; 14 substances also lacked any acceptable mutagenicity tests and 36 lacked an acceptable repeated exposure study. However, in spite of the high toxicity scores, resulting primarily from the absence of data, there is a good spread of overall priority scores. Use scores contributed to this spread, with substances at the top of the priority list having high use scores.

Table 7 *Results of preliminary trial of priority setting scheme with 52 high-volume existing substances*

Substance	*CAS RN*	*Toxicity score*	*Phys/chem score*	*Use score*	*Tonnage score*	*Overall priority score*
But-2-ene	107-01-7	24	3	5	4	1440
Dodecan-1-ol	112-53-8	22	2	5	5.7	1254
4-Methylbenzenesulphonamide	70-55-3	21	3	5	3.7	1166
Dipentaerythritol	126-58-9	24	3	5	3	1080
[29*H*,31*H*-Phthalocyaninato-(2-)]-copper	147-14-8	14	3	5	4.7	987
2,6-Bis(1,1-dimethylethyl)phenol	128-39-2	23	2	5	4	920
N-(1-Methylethyl)-*N*'-phenyl-1,4-benzenediamine	101-72-4	24	3	3	4	864
Potassium amylxanthate	2720-73-2	24	3	3	4	864
3-Nitrobenzenamine	99-09-2	24	3	3	3.7	799
4-Ethoxybenzenamine	158-43-4	24	3	3	3.7	799
2-[2-(2-Butoxyethoxy)ethoxy]-ethanol	143-22-6	20	2	5	3.7	740

Cyclododecane	294–62–2	17	3	3	4.7	719
1,1–Dimethylethyl hydroperoxide	75–91–2	25	3	2	4.7	705
Aminoiminomethanesulphinic acid	1758–73–2	20	3	3	3.7	666
3–Methylpyridine	108–99–6	19	2	3	5.7	650
4–Nitro–*N*–phenylbenzenamine	836–30–6	23	2	3	4.7	649
2,3,5,6–Tetrachloropyridine	2402–79–1	21	2	3	5	630
[Nitrilotris(methylene)]tris–phosphonic acid	6419–19–8	13	2	5	4.7	611
3–Oxobutanoic acid, ethyl ester	141–97–9	13	2	5	4	611
Propionaldehyde	123–38–6	22	3	3	3	594
4–Methylpyridine	108–89–4	26	2	3	3.7	577
Dichlorodimethylsilane	75–78–5	24	4	1	6	576
Phosphorous acid, triethyl ester	122–52–1	17	3	3	3.7	567
Octadecan–1–ol	112–92–5	9	3	5	4	540
Isobutyraldehyde	78–84–2	20	3	3	3	540
1,2–Dichloro–3–nitrobenzene	3209–22–1	24	2	3	3.7	533

3–Methoxybenzenamine	536–90–3	24	2	3	3.7	533
Prop–2–enoic acid, iso–octyl ester	29590–42–9	16	3	3	3.7	533
2–Methylpyridine	109–06–8	24	2	3	3.7	533
2–Methylpropanoic acid, monoester with 2,2,4–trimethyl–pentane–1,3–diol	25265–77–4	13	2	5	4	520
Octamethylcyclotetrasiloxane	556–67–2	10	2	5	5	500
2–Methylbut–3–enenitrile	16529–56–9	18	3	3	3	486
Trichloromethylsilane	75–79–6	24	4	1	5	480
Dichloromethylsilane	75–54–7	24	4	1	4.7	451
Penta–1,3–diene	504–60–9	18	3	2	3.7	400
Maleic acid, dibutyl ester	105–76–0	18	2	3	3.7	400
4–Aminobenzenesulphonic acid	121–57–3	17	2	3	3.7	377
Butane–1,2–diol	584–03–2	23	3	1	4.7	324
Hexafluoroprop–1–ene	116–15–4	23	4	1	3	276
Dec–1–ene	872–05–9	16	3	1	5.7	274

Chlorotrimethylsilane	75–77–4	16	4	1	3.7	237
2–Phosphonobutane–1,2,4–tricarboxylic acid	37971–36–1	6	3	3	4	216
2–Ethyl–2–(hydroxymethyl)propane–1,3–diol	77–99–8	23	1	2	4.7	216
2,2–Dimethylpropane–1,3–diol	126–30–7	23	1	2	4.7	216
Pyridine–3–carboxylic acid	59–67–6	9	1	5	4.7	212
2,3,4–Trichlorobut–1–ene	2431–50–7	25	2	1	3.7	185
3–Methylbutanal	590–86–3	16	3	1	3.7	178
3,7–Dimethyloct–6–en–1–yn–3–ol	29171–20–8	17	2	1	4.7	160
1–Methyl–2–nitrobenzene	88–72–2	15	1	3	3	135
3,5–Bis(1,1–dimethyl)benzene–propanoic acid	6386–38–5	13	2	1	4.7	122
2,2–Bis(prop–2–enyloxy)methylbutan–1–ol	682–09–7	16	1	2	3.7	118
Dodecanedioic acid	693–23–2	12	2	1	4.7	113

It should be noted that the objective of the trial was to assess the priority setting scheme rather than to produce a definitive priority listing of the 52 OECD chemicals. Refinements to the scheme were indeed deemed necessary to achieve the degree of resolution reported here. Further refinements may be judged necessary following current work to computerize the scheme and subsequent application to more existing substances. In particular, given the influence a high score (5) for wide dispersive use can have on the overall ordering of substances, it can be argued that substances with only a very small percentage of the total tonnage in this use category should be given a lower score. Introduction of a Harmonized Electronic Data Input Set form for use in both the EC and OECD existing chemicals programmes may also result in some changes to the priority setting scheme.

4 FUTURE DEVELOPMENTS: RISK ASSESSMENT

Work is now in progress on developing guiding principles for human health risk assessment of new and existing substances. Attention is first being paid to the preliminary qualitative stage of considering identified toxicological hazards in relation to anticipated human exposure. At this stage, numerical probabilities of adverse effects occurring are not calculated. The assessment, which is iterative and based on a considerable degree of expert judgement, involves evaluating the potential for a substance to cause harm to people under particular exposure conditions. It should be noted that the EC Commission has advocated use of the term 'hazard assessment' for this stage of the risk assessment process.*

The objectives of 'hazard assessment' (qualitative risk assessment) are to identify:

i) Substances which may be set aside as being of no immediate concern;

ii) Substances of concern for which information may be needed to refine the assessment;

iii) Substances for which special risk management measures should be taken.

With respect to the second objective, HSE, in consultation with UK industry and the Department of Health, is contributing to EC Commission initiatives to formulate mutagenicity, reproductive, and inhalation toxicity test strategies for new substances.

Human health 'hazard assessment' is being considered in terms of occupational exposure, the use of consumer products, and dispersal of a

*See Editor's Preview for definitions and the chapter by P.P. Koundakjian and H.P.A. Illing.

substance into the general environment (including incorporation into the food chain). A comparable initial 'hazard assessment' procedure has been developed for evaluating risks to the environment.

5 ACKNOWLEDGEMENTS

This chapter is based on the work of a UK government-industry working group on priority setting and risk assesssment. The author would also like to thank his colleagues at HSE for their valued advice and, particularly, Dr. J. Delic for his considerable input into the preliminary trial. The views expressed in this chapter are the author's own and are not necessarily those of the Health and Safety Executive, other government departments, or the UK chemical industry.

6 REFERENCES

1 Annex to *Offic. J. Eur. Commun.*, C 146A, Volume 33, 15.6.1990

2 Council Directive 79/831/EEC; *Offic. J. Eur. Commun.*, No. L259, 15.10.1979, p. 10.

3 Statutory Instrument SI 1982/1496, HMSO, London, 1982.

4 Statutory Instrument SI 1986/890, HMSO, London, 1986.

5 J.L. Vosser, 'The European Community Chemicals Notification Scheme and Environmental Hazard Assessment', in 'Toxic Hazard Assessment of Chemicals', ed. M.L. Richardson, The Royal Society of Chemistry, London, 1986, pp. 117–132.

6. H.P.A. Illing 'The European Community Chemicals Notification Scheme: Human Health Aspects', in 'Toxic Hazard Assessment of Chemicals', ed. M.L. Richardson, The Royal Society of Chemistry, London, 1986, pp. 133–149.

7 Proposal for a Council Directive COM (89) 575 final - SYN 227, *Offic. J. Eur. Commun.*, No. C33, 13.2.1990.

8 Council Directive 76/769/EEC, *Offic. J. Eur. Commun.*, No. L262, 27.09.1976.

9 Proposal for a Council Regulation COM (90) 227 final - SYN 276, *Offic. J. Eur. Commun.*, No. C276, 5.11.1990.

10 H. Schulze and W. Mückle, *Chemosphere*, 1986, **15**, 771.

11 M. Weiss, W. Kördel, D. Kuhnen-Clausen, A.W. Lange, and W. Klein, *Chemosphere,* 1988, **17**, 1419.

12 H. Könemann and R. Visser, *Chemosphere*, 1988, **17**, 1905.

13 A. Sampaolo and R. Binetti, *Reg. Tox. Pharmacol.*, 1986, **6**, 129.

14 M.J. Crookes and I.R. Nielsen, Building Research Establishment Note 99/89, Department of the Environment, 1989.

15 A. Sampaolo and R. Binetti, *Reg. Tox. Pharmacol.*, 1989, **10**, 183.

16 A. Sampaolo and R. Binetti, 'Risk Assessment of Chemical Substances', ed. M. Ragno, Rome, 1990.

17 Commission Directive 83/467/EEC, *Offic. J. Eur. Commun.*, No. L257, 16.09.1983.

18 J. Rea, R. Shillaker, and P. Smith, Toxic Substances Bulletin (published by the Health and Safety Executive), 1991, No. 15, 2.

19 M.J. Crookes and J.D. Rea, Draft for Comment, Directorate for Air, Climate and Toxic Substances, Department of the Environment, July 1991.

20 Commission Directive 84/449/EEC, *Offic. J. Eur. Commun.*, No. L251, 19.09.1984.

21 Commission Directive 87/302/EEC, *Offic. J. Eur. Commun.*, No. L133, 30.05.1988.

22 Department of Health, 'Guidelines for the Testing of Chemicals for Mutagenicity', Report on Health and Social Subjects 35, HMSO, London, 1989.

23 J.E. Brydon, V.H. Morgenroth, A.M. Smith, and R. Visser, International Environment Reporter, June 1990, 263.

12

The Management of Safety in Transport

Ian C. Canadine

IMPERIAL CHEMICAL INDUSTRIES PLC, 9 MILLBANK, LONDON SW1P 3JF, UK

1 INFLUENCES ON SAFETY

1.1 Management Performance

The safety performance of operations involving the transport of dangerous substances is implicitly dependent on the management performance of those controlling the operations. This statement is supported by numerous investigations of accidents and incidents[1,2] which have, in most cases, shown that they are due, in some measure, to factors which should or could have been under management control. In many cases there was legislation requiring certain precautions to be taken, but compliance with that legislation and sensible application of additional precautions was dependent upon the management in place at the time.

The quantification of risk and hence the prediction of future safety performance[3] is usually dependent both upon historical data and on various assumptions such as a level of future management performance equivalent to that in the historical data gathering period. It is, therefore, essential that this standard, at least, is achieved, otherwise benefits obtained from improvement in equipment and even from more stringent legislation are likely to be dissipated by inadequate management control.

Systems which assist and enable management to perform more effectively and consistently are likely, therefore, to result in improved safety performance.

1.2 Legislation

Balanced, practicable legislation defining the conditions and precautions under which dangerous substances may be transported is generally accepted as being desirable.[4] Demands for further legislation of greater stringency after any serious incident in which dangerous substances are in any way involved are, however, less supportable.

The *Herald of Free Enterprise* disaster was an example of an incident which was neither caused by dangerous substances, nor did those present cause injury. However, media headlines[5] at the time gave the impression that dangerous substances were a major feature of the incident and that they should not have been allowed on a passenger ferry.

One aspect of the above incident which was a cause for concern was the non–compliance with certain dangerous substances regulations which was revealed. Better enforcement of existing regulations would seem to be a more logical way forward than producing additional ones.

Although there may be cases where legislation is deliberately ignored for commercial reasons, it seems likely that in the main non–compliance is caused by ignorance or misunderstanding.

1.3 Technical and Engineering Factors

The safe transport of dangerous substances requires a thorough understanding of their properties and of the equipment used to transport them. The chemical, physical, and hazardous properties of the goods must be known, recorded, and thoroughly understood as a basis for the proper labelling, documentation, packaging, selection of transport, segregation, and provision of other precautions and emergency response procedures. Usually, particularly for goods of lower hazard and in small quantities, these requirements are adequately codified in various regulations[6–11] and may be applied by competent staff of relatively limited knowledge. The more hazardous the substances and the larger the quantities to be transported, then, in general, the more sophisticated the requirements are for equipment, procedures, and emergency response.

1.4 Human Factors

It is now widely accepted that the majority of accidents in industry (and this must include the transport of dangerous goods) are in some measure attributable to human as well as technical factors. In other words people initiated or contributed to the accidents, or might have acted better to avert them.[12] Although many accidents are caused by carelessness or the inattentiveness of those concerned there are many other reasons for human error. These include inadequate information, lack of understanding, mistaken priorities, and even stubborn willfulness.

Mistaken priorities can often result from lack of clarity in management objectives. For example, managers can take short–cuts or other risks to attempt to improve productivity only to find that when an accident results there is no support for what has been done. Clear communication is needed to ensure that such a misunderstanding of corporate objectives does not occur!

Senior managers have a clear role to play in the prevention of human error by ensuring the clarity of the organization and its objectives, the study and proper design of the job, and the careful selection, training, and management of the person doing the job.

2 RESPONSIBILITIES

2.1 The Role of Line Management

Perhaps the most important contribution of line management to the safe operation of any enterprise is to be seen to be utterly committed to safety as an integral part of that enterprise. This means that every level of management from the very top to the bottom not only accepts full responsibility for safety, but is prepared to commit both effort and, perhaps more importantly, time to ensuring that the highest levels of safety performance are achieved.

In order to achieve this, line management must ensure that there are both safety standards for the operations of the enterprise and safety objectives which can be measured in quantitative terms. Safety objectives must be clear to everyone in the enterprise, progress to achievement must be measured,[13] and safety performance must be included in individual performance assessments.

The safety standards of the enterprise must be clearly written down and understood by everyone and must be incorporated in all plans, operating instructions, and training. Conforming with the safety standards must be a way of life within the enterprise and it is particularly important that junior staff can see that senior managers are completely committed.

2.2 The Role of Technical and Engineering Professionals

This role is essentially to provide advice on technical matters in order to assist decision making. This can include the specification or design of equipment to meet specifications or levels of performance required by regulations, codes of practice, or company requirements. It is essential to avoid confusion between this role, however qualified those persons filling it, and the line management role.

The technical side of safety is often a specialist matter requiring highly competent professionals in order to produce the right answers. Line management, however, must be quite sure that in seeking the help of these professionals the jobs which they do are carefully defined with a clear understanding of what is being asked and any limitations to answers being supplied. It is very important that line management does not abdicate from this area thus allowing gaps in knowledge or misunderstanding to cause unsafe situations.

2.3 The Role of Safety Professionals

The role of safety professionals in the chemical industry[14] is well established as an advisory and monitoring resource for line management. There is still a tendency in some companies to delegate safety matters to the 'safety officer', but in general this is now seen to be an unacceptable and ineffective practice which marginalizes safety. The safety professional has an important role as an expert in safety legislation, specialist safety matters, safety training, the measurement and monitoring of safety performance, and many similar matters without being expected to take on the proper line management duty associated with safety.

However, the role of the safety professional has in the main, been traditionally seen to lie within the works boundary. Since transport operations are off-site and are often carried out by third party contractors, the safety professional has often not been involved in advising on or monitoring these operations. This situation was highlighted in the HSE report on the Peterborough explosion.[15] This comments 'the Safety Department was primarily concerned with manufacture, *i.e.* plant and processes, and had little involvement in product development and design, including correctness for transport'.

Some companies already have safety professionals within the transport or distribution function itself and doubtless others have included this duty in the remit of existing safety professionals. The proper monitoring and auditing of the transport function, however, would seem to require the attention of safety professionals.

3 A MODEL STRUCTURE FOR MANAGEMENT PRACTICES

3.1 Introduction

The duties and opportunities of management in the transport of dangerous goods can be formalized, defined, and explained in such a way that everyone in an organization is clear what is required and is assisted in achieving these requirements. Such clear definition is also the basis for auditing to ensure that the requirements are indeed being met.

The Canadian Chemical Producers Association (CCPA) pioneered such a system as part of their 'Responsible Care' initiative several years ago.[16] This initiative has since been taken up in the United Kingdom by the Chemical Industries Association (CIA)[17] as well as in the United States[18] and several countries in Continental Western Europe. In each case there has been developed, or is under development, a 'Code of Management Practices' which is in effect a broad statement of safety and environmental standards which are required by those companies involved in the initiative.

In essence the structure of what is required under 'Responsible Care' is that companies shall have:

Safety policies – specifying what is required overall in the safety field and the responsibilities of those involved;

Safety standards – stating what must be achieved;

Safety procedures - stating in detail how to do the job in such a way as to achieve these standards;

Safety guidance - explaining as necessary how to convert standards into procedures.

3.2 Policies

The safety policy of a company is a statement describing the overall results required. Such a policy in general terms is required in the UK under the Health and Safety at Work Act (1974)[19] and is a written statement of the general policy with respect to the 'health and safety at work' of employees, and the organization and arrangements for carrying out that policy. As applied under 'Responsible Care' such a policy should include the total transport operation and off-site activities of the company and be concerned not only with the safety of employees but also of the public and any other third parties. The policy for example might start with a statement such as 'It is this company's policy that in the manufacture, transport, storage, and use of its products, proper care be taken to protect the safety and health of all persons involved and to protect the environment'. It might then go on to discuss how the company would manage its activities, how it would conform with legal requirements and appropriate codes of practice, and possibly any additional measures it would wish to take.

Very importantly the policy should make it clear who, within the company, carries responsibility for all aspects of transport safety and who is responsible for the provision of safety standards, professional safety advice, and monitoring and reporting on performance.

3.3 Standards

The safety policy should be supported by safety standards which set down the basic requirements of behaviour or operation which underpin the safety policy and must be complied with. These standards should describe what is to be achieved and will reflect not only legal requirements but the best management practice within the industry. Such standards must be written down in such a form that subsequent auditing can clearly decide whether they are being complied with. For example, such a standard might read 'There shall be a system to ensure that suitable tanks and equipment, including any special requirements, are specified and used for each product, and that adequate product information is given to the haulier and to the driver'. This standard clearly specifies what is required in this particular instance but it does not state in detail how to achieve it. Procedures are required in order to achieve this, and it might well be that guidance is required to give assistance to line management in what sort of procedures would best fulfil requirements of the standard.

3.4 Procedures

At the operating level what is required to ensure that complicated jobs are carried out accurately, consistently, and safely are written procedures which can be referred to when required and used as a basis for training. These procedures can of course only be produced in the context of the individual job or operation and by those closely involved with it. They must be written in such a way that the procedure achieves what is required in the associated standard and also in such a way that the proper carrying out of the procedure can be audited at regular intervals. Although primarily associated in this case with the proper management of safety, this requirement is basically the same as that required in the Quality Registration process which is now generally accepted as necessary in order to ensure technical and administrative quality to the satisfaction of customers.

3.5 Guidance

In this context this is material describing what standards are required and how to produce procedures for achieving the standards. This guidance should reflect not only legal requirements but the 'best practice' in the industry and in the company concerned. Guidance is discretionary in that the best way to achieve some standards may already be available in company procedures or be self-evident whilst the achievement of other standards might require comprehensive guidance.

3.6 The Model in Brief

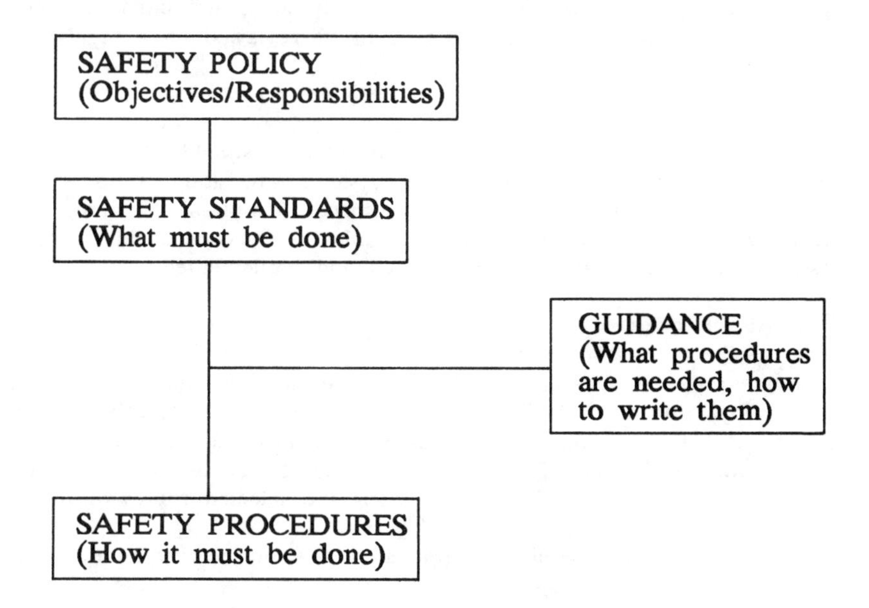

Figure 1 *Proposed structure of management practices*

Thus we have a proposed structure of management practices composed of general policy on safety matters, specifying responsibilities, of declared standards saying what must be achieved, and of procedures saying exactly how to achieve the standards (see Figure 1). This structure should be backed up by guidance where necessary on how to achieve the standards.

4 GUIDANCE ON STANDARDS

4.1 Introduction

In the model structure of management practices as outlined above, there is a requirement for operational standards for various critical aspects of the transport of dangerous substances. Items requiring standards, and what those standards might be, are suggested below.

4.2 Safety and Environmental Policy

The safety of employees, customers, and the public and the protection of the environment from the adverse effects of transport operations are paramount considerations. The prime responsibility for this lies with line management which is required to demonstrate a high level of commitment to safety and environment protection during transport.

The Company should have a safety and environmental policy which covers transport outlining the organization and arrangements for achieving the required results. Transport safety and environmental improvement programmes should be developed at each location aimed at improving performance, and these programmes shall be reviewed on a regular basis.

4.3 Auditing

Adequate local formal auditing procedures should be defined and implemented to ensure that written systems are adopted to meet these standards and are soundly established, maintained, and observed. Deficiencies identified during audit should be recorded, their implications assessed, and remedial action prioritized and implemented.

4.4 Risk Reduction

There should be systems for identifying, evaluating, and reducing the risk posed by transport. These systems should consider the hazards of chemicals, the methods of containing, handling, transporting and storing them, the likelihood of accidents or releases occurring in transit, and human and environmental exposure along the selected routes of movement.

Similar systems should be applied to new projects at an early stage.

4.5 Specification of Packages, Tanks, and Other Equipment

There should be systems for ensuring that packages, transport tanks, and other equipment are appropriate for the chemical being shipped, comply with all testing, certification, and other legal requirements, and are free of leaks and visible defects. Systems should also specify criteria for the cleaning and return of transport tanks and returnable containers, and for the disposal of cleaning residues.

4.6 Incident Reporting and Investigation

There should be systems for the reporting and investigation of all transport incidents, including accidents, releases, and potentially hazardous situations, and for identifying and implementing measures for prevention of future incidents.

4.7 Codes and Regulations

There should be systems for the interpretation, monitoring for changes, and application of all relevant National and International regulations, industry codes, and company standards covering transport.

4.8 Control of Operations

There should be systems to ensure that day–to–day operations such as the filling and emptying of transport tanks, the loading and unloading of vehicles, and the use of storage facilities are carried out in compliance with relevant regulations, codes and standards, to protect personnel and to minimize emissions to the environment.

4.9 Training

There should be systems to ensure that training needs are identified and satisfied in an appropriate and adequate manner so that transport is carried out safely and with proper regard for environmental protection. Such systems should include the provision of appropriate induction training of new employees.

Systems should ensure that training is regularly reviewed so that all employees are competent at all times to carry out the duties for which they are responsible. In particular it should be reviewed whenever there are material changes in the work carried out by employees or in the regulations, codes, or standards relevant to that work.

4.10 Selection and Auditing of Contractors

There should be systems for the selection and auditing of third party contractors involved in transport and associated activities. Such systems should emphasize safety performance, environmental protection performance, regulatory compliance, and equipment maintenance and training and should ensure regular review of safety and environmental protection performance and the monitoring of necessary improvements.

4.11 Data and Information

There should be systems to ensure the collection of appropriate chemical data and their assessment as a basis for hazard classification, labelling, documentation, packaging, transport, and storage. Such systems should ensure the provision of labels, placards, documentation, transport emergency information, and technical and safety information to contractors, customers, and others, as necessary and in the appropriate format. Guidance on the safe disposal of unused products, residues, and empty packages should also be provided.

4.12 Emergency Response

There should be systems for responding promptly to accidents, releases, and emergencies occurring during transport. These systems should take into account the risks involved and should include responsibilities, communications and preplanning, training of those involved, provision of necessary information, and co-operation with Emergency Services and company and industry emergency response schemes.

5 CONCLUSIONS

The necessity for predictably good management performance has been examined.

Legislation, however comprehensive, will not be effective unless understood and enforced by line managers who need to be clearly aware of the relative roles of themselves and their advisers in managing safety.

Line managers carry the main responsibility for managing safety and for setting an example of concern and commitment at every level. There is an advisory and support role for professionals (including safety professionals) which can profitably include auditing, measuring, and reporting safety performance.

A formal structure of management practices has been suggested consisting of safety policies, standards, and operational procedures supported as necessary by guidance.

The use of such a structure to clarify objectives, organizational responsibilities, and safety standards supported by monitoring of performance and auditing to ensure compliance should provide a robust basis for high standards of safety management.

6 ACKNOWLEDGEMENTS

Although the views expressed in this chapter are the author's own, he would like to acknowledge the help and assistance received during many long discussions with colleagues within his own company, in many other chemical and transport companies, in Trade Association Committees, on the ACDS Transport Sub-Committee, and also from many publications and papers on various aspects of topics covered.

7 REFERENCES

1 HSE Accident Prevention Advisory Unit, 'The Management of Health and Safety', The Industrial Society, 1988.

2 Advisory Committee on Major Hazards, Third Report, HMSO, 1984.

3 I.C. Canadine and G. Purdy, 'The Transport of Chlorine by Road and Rail in Britain – A Consideration of the Risks', Proceedings of 6th International Symposium 'Loss Prevention and Safety Promotion in the Process Industries', Oslo, Norway, June 1989.

4 I.C. Canadine, *Manufacturing Chemist*, 1989, **60**, No. 1 Jan., 33, No. 2 Feb., 54.

5 UK Press Reports, March 1987.

6 The Dangerous Substances (Conveyance by Road in Road Tankers and Tank Containers) Regulations, (SI 1981/1059 HMSO), 1981.

7 The Classification, Packaging and Labelling of Dangerous Substances Regulations, (SI 1984/1244 HMSO), 1984.

8 The Road Traffic (Carriage of Dangerous Substances in Packages *etc.*) Regulations, (SI 1986/1951 HMSO), 1986.

9 European Agreement Concerning the International Carriage of Dangerous Goods by Road (ADR), HMSO, 1990.

10 International Maritime Dangerous Goods Code, IMO, London, 1990.

11 Technical Instructions for the Safe Transport of Dangerous Goods by Air, International Civil Aviation Organization, Montreal, Canada 1991.

12 'Human Factors in Industrial Safety', 9HS(G) 48, HMSO, 1989.

13 'Guidance on Safety, Occupational Health and Environmental Protection Auditing', Chemical Industries Association, London, 1991.

14 V.C. Marshall, 'Major Chemical Hazards', Ellis Horwood, Chichester, UK, 1987.

15 'The Peterborough Explosion – A report of the investigation by the Health and Safety Executive into the explosion of a vehicle carrying explosives at Fengate Industrial Estate, Peterborough on 22 March 1989', HMSO.

16 'Transportation Code of Practice – Standards Guide', Canadian Chemical Producers Association, Ottawa, Canada, 1988.

17 'Responsible Care – Guidance for Chief Executives', Chemical Industries Association, London, 1990.

18 'Responsible Care – Distribution Code of Management Practices', Chemical Manufacturers Association, Washington DC., 1990.

19 Health and Safety at Work *etc.* Act, HMSO, 1974.

13

Regulations and Health Standards in the Management of Risks in the Chinese Chemical Manufacturing Industry

S. Li

SCIENTIFIC INFORMATION CENTRE, CHINESE ACADEMY OF PREVENTIVE MEDICINE, 10 TIAN TAN XI LI, BEIJING 100050, CHINA

1 INTRODUCTION

The safe use of chemicals can only be guaranteed if management techniques relating to chemical risks are strengthened. For this reason, it is necessary to formulate comprehensive measures for both control of hazards and their management. It is equally vital to have co-operation between various government departments, including the Ministry of Chemical Industry, Ministry of Agriculture, Ministry of Public Health, and the Environmental Protection Agency. Such co-operation will ensure the establishment of environmental (air, water, and soil) quality monitoring systems, workplace environmental and health monitoring systems, and systems to alert attention to the hazards associated with toxic chemicals. Additionally, scientific research related to environmental protection has to be given high priority. There are both biological and medical problems encountered in health departments, whereas environmental protection involves both environmental and health problems. Thus, the management of chemical risks cannot be the responsibility of a single department; there has to be co-ordination and concerted actions between a number of departments. This leads to the paramount requirement of reliable and effective information exchange. Environmental pollution is not constrained by boundaries; information exchange must relate to the international requirements prescribed for toxic chemical monitoring, toxicological research, environmental transport of chemicals, legislation and standards, and the necessary and associated management. These techniques and requirements all have a significant role in the establishment of control functions, management techniques, and the prescription of legislation and health standards.

In this chapter several issues concerning legislation and standards for environmental hygiene and workplace health aspects in China are described. In order to achieve the safe use of chemicals without environmental pollution and without damage to the human body, it is not sufficient to depend on health departments or the Environmental Protection Agency. It has to be the joint target for all related departments within control and management functions to prevent environmental pollution, to safeguard human health within every department's own unique specialities and potential management responsibilities and to shoulder jointly the absolute requirements to protect mankind. Only in this way can the safe use of chemicals be realized. Scientists of every country are now facing new challenges on environmental problems. The effects of management will be verified through practice.

2 REGULATION AND HEALTH STANDARDS

The constantly increasing quantity and variety of chemicals produced globally and an increasing awareness that chemicals have the potential to affect adversely human health and the environment require the development of proper and adequate measures to control and test chemicals. Therefore, the problems of safe use of chemicals against deleterious effects of potentially toxic chemicals have been dealt with by all countries. The enactment and promulgation of national regulations and standards for chemicals are one of the most important requirements for chemical safety.

The People's Republic of China's government pays great attention to the programme for environmental protection, including the safe use of chemicals. In accordance with Article II of the Constitution of the People's Republic of China, the State protects the environment and natural resources and prevents and eliminates pollution and other hazards to the public.

The Environmental Protection Law adopted in 1979 stipulates active prevention and control measures to obviate noxious substances from factories, mines, businesses, and urban life, including waste gases, waste water, waste residues, dust, garbage, radioactive material, *etc.* in addition to noise, vibration, and offending odours from pollution and damage to the environment (Article 16); it is stressed that highly effective low toxicity and low residue agricultural pesticides must be developed (Article 21); it points out that the amount of permissible harmful gases and dust in the working environment must conform with the standards for industrial hygiene specified by Chinese Law (Article 22) and Registration and control of toxic chemicals must be strictly adhered with. Highly toxic substances must be securely contained to prevent leakage during storage and transportation (Article 24).[1]

2.1 Maximum Allowable Concentration

In order to assure human health, to prevent contamination from chemicals, in 1956, the Chinese government promulgated the 'Health Standard for Industry and Enterprise Design', which has been revised in 1961 and 1979. The Maximum Allowable Concentration (MAC) of 111 harmful chemicals in the atmospheres of factories has been specified in document (TJ 36–79) (Table 1), and some additional associated material giving the MACs of toxic chemicals in the workplace atmosphere introduced (Tables 2 and 3)[2]; for ambient air pollutants the 35 MACs of chemicals given in Table 4 and

53 MACs of substances harmful to surface water given in Table 5 have been prescribed in the above document.[1]

In recent years, MACs of chemicals in soil have been established in China, such as the Hygienic Standard for arsenic in soil (GB*8915–88) and Hygienic Standard for copper in soil (GB 11728–89).[4,5]

Table 1 *Maximum allowable concentration of toxic chemicals in the workplace atmosphere (TJ 36–79)*

Toxic chemicals	*Maximum allowable concentration* (mg m^{-3})
Acetone	400
Acetonitrile	3
Acrolein	0.3
Acrylonitrile (skin)	2
Ammonia	30
Amyl acetate	100
Amyl alcohol	100
Aniline, methylaniline, and dimethylaniline (skin)	5
Arsenic trioxide and pentoxide	0.3
Arsine	0.3
Benzene (skin)	40
Beryllium and its compounds	0.001
Bromomethane (skin)	1
Butene	100
Butyl acetate	300
Butyl alcohol	200
Butyl aldehyde	10
Cadmium oxide	0.1
Caprolactam	10
Carbon disulphide (skin)	10
Carbon monoxide	30
Carbon tetrachloride (skin)	25
Caustic alkali (as NaOH)	0.5
Chlorine	1
Chlorobenzene	50
Chloroethene	30
Chloronaphthalene and chlorodiphenyl (skin)	1
Chloropicrin	1
Chloroprene (skin)	2
Chromium trioxide, chromate, and dichromate (as CrO_3)	0.05
Cyclohexane	100

*GB denotes the initials in the Chinese phonetic alphabet meaning national standard.

Cyclohexanol	50
Cyclohexanone	50
Decalin (decahydronaphthalene) and tetralin (tetrahydronaphthalene)	100
Dichlorodiphenyltrichloroethane (DDT)	0.3
Dichloroethane	25
1,3-Dichloropropanol (skin)	5
Diethyl ether	500
Dimethoate (Roger) (skin)	1
Dimethylamine	10
Dimethyldichlorosilane	2
Dimethylformamide (DMF) (skin)	10
Dinitro- and trinitro-benzene and its homologues (dinitrobenzene, trinitrobenzene, *etc.*) (skin)	1
Diphenyl ether	7
Dipterex (skin)	1
Divinyl (diethene)	100
Epichlorhydrin (skin)	1
Ethyl acetate	300
Ethylene oxide	5
Formaldehyde	3
Furfural (Furfuraldehyde)	10
Hexachlorocyclohexane	0.1
γ-Hexachlorocyclohexane	0.05
Hydrogen chloride and hydrochloric acid	15
Hydrogen cyanide and cyanates (as HCN) (skin)	0.3
Hydrogen fluoride and fluorides (as F^-)	1
Hydrogen sulphide	10
Iodomethane (skin)	1
Lead (dust)	0.05
Lead (fume)	0.03
Lead sulphide	0.5
Malathion (skin)	2
Manganese and its compounds (as MnO_2)	0.2
Mercuric chloride	0.1
Mercury, metal	0.01
Mercury, organic compounds (skin)	0.005
Metasystox (skin) (RN:8022-00-2)	0.2
Methyl acetate	100
Methyl alcohol	50
Methylamine	5
Methylparathion (skin)	0.1
Monomethylamine	5
Mononitrobenzene and its homologues (nitrobenzene, nitrotoluene, *etc.*) (skin)	5
Molybdenum (insoluble compounds)	6
Molybdenum (soluble compounds)	4
Nickel carbonyl	0.001

Nitrochloro- and dinitrochloro- compounds of benzene (nitrochlorobenzene, dinitrochlorobenzene, *etc.*) (skin)	1
Nitrogen oxide (as NO_2)	5
Ozone	0.3
Parathion (skin)	0.05
Pentachlorophenol and its sodium salts	0.3
Phenol (skin)	5
Phosgene	0.5
Phosphine	0.3
Phosphorus pentoxide	1
Propenol (skin)	2
Propyl acetate	300
Propyl alcohol	200
Pyridine	4
Selenium dioxide	0.1
Solvent gasolines	350
Styrene	40
Sulphur dioxide	15
Sulphuric acid and sulphur trioxide	2
Systox (demeton) (skin)	0.02
Tetraethyl-lead (skin)	0.005
Thimet (skin)	0.01
Toluene	100
Toluene-2,4-di-isocyanate (TDI)	0.2
Trichloroethene	30
Trichlorosilane	3
Triethyltin chloride (skin)	0.01
Tungsten and tungsten carbide	6
Turpentine	300
Vanadium-ferroalloy	1
Vanadium pentoxide (dust)	0.5
Vanadium pentoxide (fume)	0.1
Xylene	100
Yellow phosphorus	0.03
Zinc oxide	5
Zirconium and its compounds	5

Table 2 *Additional revised MACs after 'TJ 36-79' (promulgated from 1983 through 1989)*

Toxic chemical	*MAC* (mg m^{-3})
Acrylamide (skin)	0.3
Aluminium	
Metal	4
Oxides	6
Alloys	4
Allyl chloride	2

Antimony and its compounds	1
Chloroathalonil	0.4
Cobalt and its oxides	0.1
Copper (dust)	1
Copper oxide (fume)	0.2
1,2-Dichloroethane	15
m-Dihydroxybenzene (resorcinol)	10
Dimethylacetamide (skin)	10
Ethylene chlorohydrin (skin)	2
Ethylene oxide	2
Ethylenediamine	4
Glycidyl methacrylate	5
Liquified petroleum gas	1000
Lithium hydride	0.05
Methyl methacrylate (skin)	30
Methyl acrylate (skin)	20
Phosphamidon (skin)	0.02
Phosphorus trichloride	0.5
Raffinate (50–220 °C)	300
Solvent gasolines*	300
Sulphur hexafluoride	6000
Trichlorfon (dipterex) (skin)*	0.5
Tricresyl phosphate (skin)	0.3
Vanadium and its compounds*	
Metal, vanadium-ferroalloy and carbides	1
Vanadium pentoxide (dust)	0.1
Vanadium pentoxide (fume)	0.02

*Denotes the revised MAC.

Table 3 *Additional revised MACs intended for promulgation after 1989*

Toxic chemical	*MAC* (mg m^{-3})
Carbon dioxide	18000
Carbon disulphide*	5
Chloromethane	40
Chlorprene*	4
Dibutyltin dilaurate (skin)	0.2
Deltamethrin	0.03
Ethylbenzene	50
Ethylene glycol	20
Fenvalerate	0.05
Hydrazoic acid	0.2
Kitazin (skin) (RN:13286-32-3) (Phosphorothioic acid, *OO*-diethyl *S*-(phenylmethyl) ester)	1
Magnesium oxide (fume)	10

Naphthalene	50
Omethoate (skin)	0.3
Sodium azide	0.3
Sulphuryl fluoride	20
Thallium (skin)	0.01
Trifluromethyl hypofluoride	0.2

*Denotes the revised MAC.

Table 4 *Maximum allowable concentration of toxic chemicals in the atmosphere of population centres*

Toxic chemical	*MAC* (mg m^{-3})	
	Maximum	*Average over 24 hours*
Acetaldehyde	0.01	
Acetone	0.80	
Acrolein	0.10	
Acrylonitrile		0.05
Ammonia	0.20	
Aniline	0.10	
Arsenic (calculated as As)	0.003	
Benzene	2.40	0.80
Carbon disulphide	0.04	
Carbon monoxide	3.00	1.00
Chlorine	0.10	0.03
Chloroprene	0.10	
Chromium	0.0015	
Dipterex	0.10	
Dust fallout	3–5 tonne km^{-2} day^{-1}†	
Epichlorhydrin	0.20	
Fluoride (calculated as F^-)	0.02	0.007
Formaldehyde	0.05	
Hydrogen chloride	0.05	0.015
Hydrogen sulphide	0.01	
Lead and its inorganic compounds (calculated as Pb)		0.0015[3]
Manganese and its compounds (calculated as MnO_2)	0.03	0.01
Mercury		0.0003
Methyl alcohol	3.00	1.00
Methylparathion	0.01	
Nitrobenzene	0.01	
Nitrogen oxides (calculated as NO_2)	0.15	
Phenol	0.02	
Phosphorus pentoxide	0.15	0.05
Pyridine	0.08	

Styrene	0.01	
Sulphur dioxide	0.50	0.15
Sulphuric acid	0.30	0.10
Suspended particulates	0.50	0.15
Xylene	0.30	

†Exclusive of local background level.

Table 5 *Recommended maximum allowable concentrations for toxic chemicals in surface waters*

Toxic chemical	*MAC* (mg l^{-1})
Acetaldehyde	0.05
Acetonitrile	5.0
Acrolein	0.1
Acrylonitrile	2.0
Aniline	0.1
Antimony	0.05
Arsenic	0.04
Benzene	2.5
Beryllium	0.0002
BHC	0.02
Butyl xanthogenates	0.005
Cadmium	0.01
Carbon disulphide	2.0
Chlorobenzene	0.02
Chromium, trivalent	0.5
hexavalent	0.05
Cobalt	1.0
Copper	0.1
Cyanides	0.05
DDT	0.2
Demeton (E–059)	0.03
Dichlorobenzenes	0.02
Dimethoate	0.08
Dinitrobenzenes	0.5
Dinitrochlorobenzenes	0.5
Fluoride	1.0
Formaldehyde	0.5
Free chlorine	None (calculated from chlorine demand of surface water)
Hexachlorobenzene	0.05
Hexanolactam (RN:105–60–2) (2*H*–azepin–2–one, hexahydro–)	Calculated from BOD of surface water
Hydrazine hydrate	0.01
Isopropylbenzene	0.25
Lead	0.1

Malathion (4049)	0.25
Mercury	0.001
Methylparathion (methyl E-605)	0.02
Molybdenum	0.5
Nickel	0.5
Nitrochlorobenzenes	0.05
Parathion (E-605)	0.003
Petroleum (including kerosene and gasoline)	0.3
Phenols (volatile	0.01
Picric acid	0.5
Pyridine	0.2
Selenium	0.01
Styrene	0.3
Sulphides	None (calculated from dissolved oxygen of surface water)
Tetrachlorobenzene	0.02
Tetraethyl-lead	None
TNT	0.5
Trichlorobenzene	0.02
Turpentine	0.2
Vanadium	0.1
Zinc	1.0

For health purposes, the Ministry of Public Health enacted the 'Health Standards for Drinking Water' in 1956. In 1985, the standards were revised and reviewed; the new edition is being prepared to take into account achievements in environmental medicine and practical experience. It contains a list of exposure limits for the 15 toxic chemicals issued in 1985 (Table 6). This standard plays a unique role in helping to provide safe water for people all over the country as well as preventing the spread of water-borne diseases. Promoted by the International Drinking Water Campaign (International Drinking Water Supply and Sanitation Decade 1981-1991, WHO), rural water supplies in China were improved on the basis of this standard.

Table 6 *Maximum allowable concentration of toxic chemicals in drinking water in China*[6]

Toxic chemical	*MAC* (mg l^{-1})
Arsenic	0.05
Benz[*a*]pyrene*	0.01 μg l^{-1}
Benzene hexachloride (BHC)*	0.005
Beryllium	0.0002[7]
Cadmium	0.01
Carbon tetrachloride*	0.003
Chloroform*	0.06

Chromium (hexavalent)	0.05
Cyanides	0.05
DDT*	0.001
Fluoride	1.0
Lead	0.05
Mercury	0.001
Nitrate (calculated as N)	20
Selenium	0.01
Silver	0.05

*Tentative standard.

In 1982, the Provisional Law of the People's Republic of China on Food Hygiene was adopted. The major aims were to assure good food hygiene, to prevent the contamination of foods and presence of factors that are harmful to human health, to guarantee the physical health of the people, and to strengthen the physique of the whole population. This law provided for the prohibition of the production and handling of foods containing toxic or harmful substances, or that have been contaminated by toxic or harmful substances, which could be deleterious to human health. The production, handling, and use of food additives must meet the hygienic standards and regulations governing the use of food additives.

In recent years, the production and use of cosmetics has been increased. The Ministry of Public Health and the Ministry of Light Industry has promulgated national regulations for the control and evaluation of the safe use for human health for cosmetics, such as Hygienic Standard for Cosmetics (GB-7916-87). A list of maximum allowable concentrations of chemical substances for use in cosmetics is shown in Table 7.

Table 7 *Maximum allowable concentrations for chemicals for use in cosmetics*[8]

Substance	*MAC (%)*
Alkali sulphides	2
Alkaline earth sulphides	6
Aluminium fluoride	0.15 (when mixed with other fluorides, its total MAC ≯0.15)
Ammonia	6
Ammonium fluoride	0.15 (when mixed with other fluorides, its total MAC ≯0.15)
Ammonium fluorosilicate	0.15 (when mixed with other fluorides, its total MAC ≯0.15)
Ammonium monofluorophosphate	0.15 (when mixed with other fluorides, its total MAC ≯0.15)
Boric acid	0.5
Butylated hydroxyanisole	0.15

Butylated hydroxytoluene	0.15
Calcium fluoride	0.15 (when mixed with other fluorides, its total MAC ≯0.15)
Calcium monofluorophosphate	0.15 (when mixed with other fluorides, its total MAC ≯0.15)
Cantharicides tincture	1
Chlorates of alkali metals	3
Diaminophenols	10 (calculated as free radicals)
Dichloromethane	35 (when mixed with trichloromethane, its total MAC ≯35%)
Dichlorophen (INN)	0.5
Dihydroxyacetone	5
Hexadecylammonium	0.15 (when mixed with other fluorides, its total MAC ≯0.15)
3-(*N*-Hexadecyl-*N*-2-hydroxyethylammonio)-propylbis(2-hydroxyethyl)-ammonium	0.15 (when mixed with other fluorides, its total MAC ≯0.15)
Hydrogen peroxide	12
Hydroquninone	2
1,3-Bis(hydroxymethyl)-imidazolidene-2-thione	2
Magnesium fluorosilicate	0.15 (when mixed with other fluorides, its total MAC ≯0.15)
Methanol	0.2
6-Methylcoumarin	0.003
Methyl phenylenediamines, their *N*-substituted derivatives, and their salts with the exception of *p*-methyl-*m*-phenylenediamine and its salts	10 (calculated as free radicals)
α-Naphthol	0.5
Nitromethane	0.3
Octadecenylammonium fluoride	0.15 (when mixed with other fluorides, its total MAC ≯0.15)
Oxalic acid, its esters, and alkaline salts	5
Phenol and its alkali salts	1
m- and *p*-Phenylenediamines, their *N*-substituted derivatives and their salts; *N*-substituted derivatives of *o*-phenylenediamines	6 (calculated as free radicals)
N,*N'*,*N'*-Tris(polyoxyethylene)-*N*-hexadecylpropylenediamine dihydrofluoride	0.15 (when mixed with other fluorides, its total MAC ≯0.15
Potassium fluoride	0.15 (when mixed with other fluorides, its total MAC ≯0.15)

Potassium fluorosilicate	0.15 (when mixed with other fluorides, its total MAC ≯0.15)
Potassium monofluorophosphate	0.15 (when mixed with other fluorides, its total MAC ≯0.15)
Potassium or sodium hydroxide	2
Propyl gallate	0.01
Pyrogallol	5
Quinine and its salts	0.2
Quinolin-8-ol and bis-(8-hydroxyquinolinium) sulphate	0.3
Resorcinol	5
Selenium disulphide difluoride	0.5
Silver nitrate	4
Sodium fluoride	0.15 (when mixed with other fluorides, its total MAC ≯0.15)
Sodium fluorosilicate	0.15 (when mixed with other fluorides, its total MAC ≯0.15)
Sodium monofluorophosphate	0.15 (when mixed with other fluorides, its total MAC ≯0.15)
Sodium nitrite	0.2
Stannous fluoride	0.15 (when mixed with other fluorides, its total MAC ≯0.15)
Thioglycollic acid, its salts and esters	2 (calculated as thioglycollic acid, its salts and esters)
Tosylchloramide sodium (sodium derivative of *N*-chloro-*p*-toluenesulphonamide trihydrate)	0.2
Tribromsalan (3,4,5-tribromosalicylanilide)	1
1,1,1-Trichloroethane (methylchloroform)	35
Water-soluble zinc salts with the exception of zinc *p*-hydroxybenzene-sulphonate and zinc pyrithione	1
Zinc 4-hydroxybenzene sulphonate	6

In the Hygiene Standard for Cosmetics Document there is an introduction of a list of substances banned for use in cosmetics, allowable application rates, limited use conditions, required, explanations for labels, and allowable concentrations for preservatives in cosmetics.[8]

2.2 Classification of Health Hazard Levels from Occupational Exposure to Toxic Substances[9]

This standard is used for the classification of health hazards resulting from occupational exposure to toxic substances. It is a basic standard in the scientific management for labour protection.

2.2.1 Definition. Toxic substances to which the labour forces may be occupationally exposed include substances present in raw materials, final products, semi-final products, intermediate products, by-products, and impurities which could enter the human body during preparation via the respiratory tract, skin, and oral route and cause hazards to health.

2.2.2 Principles. The classification of health hazards resulting from occupational exposure is carried out according to the following six criteria: acute toxicity, adverse effects resulting from acute intoxication, adverse effects during chronic intoxication, consequences of chronic intoxication, carcinogenicity, and MAC.

The principle of the classification is to carry out comprehensive comparisons and overall assessments; the classification is subject to the results of the assessment by the majority of the six criteria. However, for certain toxic substances, classification could be made according to the major hazardous criteria, such as acute toxicity, chronic toxicity, or carcinogenicity.

2.2.3 Basis. Acute toxicity. The LC_{50} lowest value was obtained after inhalation or LD_{50} after skin and oral exposure in animals was used to assess the acute toxicity.

Observations from acute intoxication. This is a qualitative criterion which includes the frequency and outcome of acute intoxication and is further divided into four levels, *i.e.* often, sometimes, rare, and never.

Evidence during chronic intoxication. In principle, this is assessed by the incidence rate for chronic intoxication for highly exposed workers. If the incidence rate for chronic intoxication is not available, the incidence rate of intoxication symptoms or signs can be used.

Outcome of chronic intoxication. After the cessation of exposure, the outcome is divided into four levels, *i.e.* progressive, non–curable, curable, and spontaneous recovery. Using the results from animal experiments the outcome was also assessed according to the characteristics of lesions (progressive, non–reversible, or reversible) and the pathological–physiological characteristics of the target organs (repair, regeneration, or functional conservation).

Carcinogenicity. According to the IARC (International Agency for Research on Cancer) publications or other well–accepted carcinogenicity data, this is divided into human carcinogens, suspected human carcinogens, animal carcinogens, and non–carcinogens.

Table 8 *Classification of health hazards resulting from occupational exposure to toxic substances*[9]

Criteria	Classification I Extremely hazardous	II Highly hazardous	III Moderately hazardous	IV Slightly hazardous
Acute toxicity				
Inhalation LC_{50} (mg m^{-3})	<200	200–	2000–	>20000
Via skin LD_{50} (mg kg^{-1})	<100	100–	500–	> 2500
Via oral LD_{50} (mg kg^{-1})	< 25	25–	500–	> 500
Acute poisoning				
Occurrence	Often	Sometimes	Occasionally	No report
Prognosis	Bad	Good		
Chronic poisoning				
Incidence	High (≥5%)	Medium (<5%)	Occasionally	No report medium incidence of symptoms (≥10%)
Outcome of chronic poisoning after cessation of exposure	Progressive non-curable	Curable	Recoverable	Spontaneous recovery, no serious outcome
Carcinogenicity	Human carcinogen	Suspect human carcinogen	Animal carcinogen	Non-carcinogenic
MAC (mg m^{-3})	<0.1	0.1–	10–	>10

Maximum Allowable Concentration. This is assessed according to the Hygienic Standards, *i.e.* MAC for Toxic Substances in the atmosphere of the workplace.

The degree of health hazard resulting from occupational exposure to toxic substances is classified as extreme hazard, high hazard, medium hazard, and slight hazard, according to the classification shown in Table 8.

2.2.4 The classification for different professions. According to this classification standard, the health hazards resulting from occupational exposure to 56 common toxic substances in China were classified and these are listed in Table 8.

The health effect of the same toxic substances used in professions other than that listed in Table 9 was assessed according to their atmosphere concentrations in the workshop, incidence of poisoning, and duration of exposure. If the atmosphere concentration in the workshop often exceeds the MAC of the Hygienic Standard, but the incidence of poisoning or symptoms is less than the corresponding value for this classification standard, the classification could be lowered by one order of magnitude.

When workers are exposed to a number of toxic substances, it was classified according to the health effects resulting from the most hazardous toxic substance.

Table 9 *Classification of health hazards resulting from occupational exposure to toxic substances and examples from different manufacturing processes*[9]

Toxic substance	*Examples of different manufacturing processes*
Class I (extremely hazardous)	
Mercury and its compounds	Mercury smelting, chlorine and caustic soda production
Benzene	Adhesive containing benzene, use and production
Arsenic and its inorganic compounds*	Metal mineral (tin) containing arsenic, mining and smelting
Vinyl chloride	Polyvinyl chloride polymer production
Chromate, dichromate	Chromate and dichromate production
Phosphorus	Phosphorus production
Beryllium and its compounds	Beryllium smelting, beryllium compounds production
Parathion	Parathion production, storage and transport
Nickel carbonyl	Nickel carbonyl production

Octafluoroisobutylene	Difluorochloromethane pyrolysis and treatment of its waste
Chloromethyl ether	Bischloromethyl ether, chloromethyl ether, and ion exchange resin production
Manganese and its inorganic compounds	Manganese mining and smelting, manganese iron and manganese steel production, permanganic welding rods production
Cyanide	Sodium cyanide and polyacrylates production

*Excluding non-carcinogenic, inorganic arsenide.

Class II (highly hazardous

Trinitrobenzene	TNT production
Lead and its compounds	Lead smelting, battery production
Carbon disulphide	Carbon disulphide and viscose silk production
Chlorine	Liquid chlorine and caustic soda production and electrolysis of sodium chloride
Acrylonitrile	Acrylonitrile and polyacrylonitrile production
Carbon tetrachloride	Carbon tetrachloride production
Hydrogen sulphide	Sulphide colours production
Formaldehyde	Phenol resin and urea-formaldehyde resin production
Hydrogen fluoride	Electrolytic aluminium and hydrofluoric acid production
Pentachlorophenol and its sodium salt	Pentachlorophenol and sodium pentachloro-phenol production
Cadmium and its compounds	Cadmium smelting and cadmium compounds production
Trichlorfon	Trichlorfon production and storage
Chloropropene	Epoxy chloropropene and sodium propylene sulphonate production
Vanadium and its compounds	Vanadium-iron mineral mining and smelting
Bromomethane	Bromomethane production
Dimethyl sulphate	Dimethyl sulphate production, storage, and transport
Metallic nickel	Nickel mineral mining and smelting
Toluene di-isocyanate	Polyurethane production
Epichlorohydrin	Epichlorohydrin production
Hydrogen arsenide	Non-ferrous metals containing arsenic production
Dichlovos	Dichlovos production and storage
Phosgene	Phosgene production
Chlorobutadiene	Chlorobutadiene production and polymerization
Carbon monoxide	Gas production, coal coking
Nitrobenzene	Nitrobenzene production

Class III (moderately hazardous)

Styrene	Styrene production, glass fibre resin and forced plastic production
Methanol	Methanol production
Nitric acid	Nitric acid production and storage
Sulphuric acid	Sulphuric acid production and storage
Hydrochloric acid	Hydrochloric acid production and storage
Toluene	Toluene production
Xylene	Spray–paint
Trichloroethylene	Trichloroethylene production, metal degreasing
Dimethylamide	Dimethylamide production, butadiene rubber synthesis
Hexafluoropropylene	Hexafluoropropylene production
Phenol	Phenol and phenol resins production
Nitrogen oxides	Nitric acid production

Class IV (slightly hazardous)

Petroleum solvent	Rubber (*e.g.* tyres and shoes) production
Acetone	Acetone production
Sodium hydroxide	Caustic soda and paper manufacture
Tetrafluoroethylene	Polytetrafluoroethylene production
Ammonia	Ammonia and nitrogen fertilizer production

2.3 Health Protection Zones

2.3.1 Guidelines. A standard is used to determine the health protection zone between industrial premises and residential areas, in order to protect human health and the environmental quality of the atmosphere in residential areas.

This standard is used when building new industrial premises and for reconstruction projects in plains and slightly hilly areas. The existing relevant industrial factories should conform to this standard. In complex topographical areas the establishment of health protection zones would be made in accordance with a comprehensive environmental quality evaluation assessment. The authority responsible for industrial premises should discuss and determine the requirements for adequate public health and environmental assessment.

The industrialist should adopt effectively the progressive economic and rational production technologies and equipment requirements for controlling environmental aspects, and management and maintenance must be strengthened in order to accomplish these requirements. In addition, the irregular emissions of wastes must be decreased to a minimum. The concentrations of environmental pollutants from chimneys must comply with the State standards.

Table 10 *Some Health Protection Zone Standards*[10]

Standard No	*Industrial operations*	*Wind speed* ($m\ s^{-1}$)	*Distance* (m)
GB 11654–88	Paper mill using vitriol as a raw material	<2	1000
		2~4	800
		>4	600
GB 11655–89	Neoprene plants	<2	2000
		2~4	1600
		>4	1200
GB 11656–89	Phosphorus plants	<2	1000
		2~4	800
		>4	600
GB 11657–89	Copper smelter (closed blast furnace)	<2	1000
		2~4	800
		>4	600
GB 11660–89	Iron–making plants	<2	600
		2~4	500
		>4	500
GB 11661–89	Coke–oven plants	<2	1400
		2~4	1200
		>4	1000
GB 11662–89	Sintering plants	<2	1400
		2~4	1000
		>4	800
GB 11663–89	Sulphuric acid plants	<2	600
		2~4	600
		>4	400
GB 11664–89	Calcium magnesium phosphate fertilizer plants	<2	1000
		2~4	800
		>4	600
GB 11665–89	Calcium superphosphate plants	<2	800
		2~4	600
		>4	600

In order to determine a health protection zone the following factors should be considered: wind direction, wind speed, and topographic position, *etc.* in order to reduce the level of atmospheric pollution in residential areas to the lowest practical value.

2.3.2 Terminology. Health Protection Zone means the minimal distance between the boundary of the factory, producing harmful emissions and the boundary of the residential area.

2.3.3 Standard content. The Health Protection Zone Standards shown in Table 10 were determined with average wind speeds in the past 5 years in a local area.

For determining certain of the standards the levels of production of the relevant industrial plants were also considered (see Table 11).

Table 11 *Health Protection Zone Standards including production levels*[10]

Standard No	*Production industrial operations*	*Production levels*	*Wind speed* (m s^{-1})	*Distance* (m)
GB 11658-89	Polyvinyl chloride resin plants	t yr^{-1}		
		<10000	<2	1000
			2~4	800
			>4	600
		≥10000	<2	1200
			2~4	1000
			>4	800
GB 11659-89	Lead storage battery plants	kva		
		<10000	<2	600
			2~4	400
			>4	300
		≥10000	<2	800
			2~4	500
			>4	400
GB 11666-89	Small nitrogenous fertilizer plants	synthetic ammonia (10000 tonnes yr^{-1})		
		<2.5	<2	1200
			2~4	800
			>4	600
		≥2.5	<2	1600
			2~4	1000
			>4	800

In addition to the above, to establish a Health Protection Zone Standard for oil refineries, the scale of production of the industrial plants and the concentration of sulphur in the raw oil were also considered (see Table 12).

Table 12 *Health Protection Zone Standards for oil refineries*[11]

Standard No.	*Production scale* (10000 t yr^{-1})	*Sulphur in raw oil* (%)	*Wind speed* (m s^{-1})	*Distance* (m)
GB 8195-87	≥250	≥0.5	<2	1500
			2~4	1300
			>4	1000
		<0.5	<2	1300
			2~4	1000
			>4	800
	<250	≥0.5	<2	1300
			2~4	1000
			>4	800
		<0.5	<2	1000
			2~4	800
			>4	800

3 MANAGEMENT AND CONTROL

In order to implement the Chinese national health policy of 'Put prevention first', and to raise the efficiency in public health practice, including environmental protection, the Chinese State Council has approved the establishment, within the Ministry of Public Health, of the Chinese Academy of Preventive Medicine (CAPM) in 1983. The five basic tasks of CAPM are summarized as follows:

To carry out fundamental and applied research on preventive medicine, and co-ordinate preventive medicine research programmes throughout the country;

To provide technical assistance to the provincial health institutions and train public health professionals for the provinces;

To engage in surveillance, monitoring, and supervision of health and epidemic prevention, and to produce quarantine programmes;

To develop the scientific basis for establishing regulations, criteria, and appropriate public health priorities and policies;

To collect, retrieve, analyse, review, and exchange information on preventive medicine.

At present, within CAPM, there are eight Institutes:

Institute of Parasitic Diseases;
Institute of Virology;
Institute of Epidemiology and Microbiology;
Institute of Occupational Medicine;
Institute of Nutrition and Food Hygiene;
Institute of Environmental Hygiene and Engineering;
Institute of Environmental Health Monitoring;
Institute of Food Safety Control and Inspection.

Except for the Institute of Parasitic Diseases, which is located in Shanghai, all the Institutes are in Beijing.

CAPM is the research centre, technical assistance centre, and training centre in the field of preventive medicine in China; its activities play a significant role in environmental medicine and occupational medicine.

In order to improve the authority and scientific level of the standard of health, the National Committee for Health Standards (NCHS) under the Ministry of Public Health, was established in 1981; it is a scientific and technical committee. The basic tasks of NCHS are to evaluate the health criteria for harmful substances and to make proposals and suggestions for the guidelines policies, research plans, and programmes for national health standards and regulations. NCHS consists of experts, including hygienists, toxicologists, chemists, physicians, engineers, *etc.* NCHS includes the following subcommittees:

Occupational Health Standard Subcommittee;
Environmental Health Standard Subcommittee;
Food Hygiene Standard Subcommittee;
School Health Standard Subcomittee;
Diagnostic Criteria of Occupational Diseases Subcommittee;
Radiation Hygiene and Protection Subcommittee;
Diagnostic Criteria of Radioactive Diseases Subcommittee;
Diagnostic Criteria of Infectious Diseases and Disinfection Subcommittee;
Diagnostic Criteria of Endemic Diseases Subcommittee.

The NCHS's major requirement is the provision of adequate health standards. The second is to take into consideration the most feasible and cost effective methods for improving health standards. Full use must be made of technical and advanced scientific achievements from abroad, the employment of past experience from experimental work and field observations, and the study of existing standards issued by the same developed countries.[1,2]

Under the Ministry of Public Health there are sanitary–antiepidemic stations including a reasonably complete system of public health in China, which was established in the 1950s at provincial and municipal levels. According to the 1990 Chinese medical statistical data, the total number of sanitary–antiepidemic stations is 3618 at various levels.[13] In addition to the above, there are institutes of labour hygiene and occupational disease in many provinces and cities. Investigation and monitoring of the adverse effects of toxic chemicals on humans to supervise and control the implementation of relevant health standards, and to prevent chemical poisoning due to potentially toxic chemicals, are the major tasks of sanitary–antiepidemic stations and institutes of labour hygiene and occupational disease.

Furthermore, at present, in China there is the National Environmental Protection Agency (EPA). It was established at the end of the 1970s, and has rapidly developed in the past five years. In recent years, certain environmental quality standards have been promulgated by the Chinese EPA, including 'Ambient Air Quality Standards', 'Surface Water Quality Standards', Emission Standards for Pollutants from Industrial Premises', *etc.*[14–16] These standards play an important role in environmental protection programmes in China. There is also an EPA system at provincial and municipal levels. It consists of regional EPAs, institutes of environmental protection, and the environmental monitoring stations. These organizations also have a significant function in the control of the safe use of toxic chemicals and reduction in potentially toxic chemicals in the environment.

A general introduction on pesticides, environmental pollution, and human health in China, and relevant information concerned with risk assessment of chemicals in the environment in China, were included in the RSC publications 'Chemistry, Agriculture and the Environment' and 'Risk Assessment of Chemicals in the Environment'.[17,18]

4 CONCLUSIONS

The establishment of health legislation and standards is one of the key measures to protect human health from hazards relating to chemicals. Obviously, chemical safety is a programme like 'systems engineering' involving chemical engineering, hygiene, environmental protection, agriculture, light industry, *etc.* Every specific department is required to undertake measures in the safe use of chemicals. These measures are complex: containment technology, legislation, standards, management, and appropriate means of monitoring and surveillance.

Only with a comprehensive understanding of the health hazards of the chemicals can safety in use of chemicals be possible. Improvement in the environmental concepts among people and knowledge of the safe use of chemicals will also reduce risks due to chemicals and increase safety measures. In short – 'safe use of chemicals' is a multidisciplinary problem. In health departments, risk assessments of chemicals, health criteria of chemicals in the environment, and potential risks of chemicals to human health (especially carcinogenicity, teratogenicity, and mutagenicity research) must be emphasized. As a result of further research on the toxicological assessments of chemicals, attention must be paid to strengthen research on investigation of environmental epidemiology. Populations at high risk, such as the elderly, infants, young children, and pregnant women should be first selected as research subjects.

Once comprehensive measures for environmental protection are taken, legislation is strengthened, and the environmental concepts in people's minds have increased, the safe use of chemicals is possible and they will not become health hazards and potential risk factors liable to damage the human environment. Hence, it is very beneficial to increase the exchange of information between countries. In addition, it is suggested that the international organizations, those who have responsibilities to ensure the safe use of chemicals, such as WHO, UNEP, FAO, and related organizations (IRPTC, IPCS), should strengthen support and the advice to the 'developing countries' in the area of 'Environment and Health' and should give attention to improvement of personnel training, so as to illustrate optimal means to utilize UN documents on health protection to avoid chemical hazards.

It is believed that only with combined efforts can chemicals be used safely.

5 REFERENCES

1 G. Sun, S. Wang, and S. Li, 'Environmental Medicine', Science and Technology Publishing House, Tianjing, 1987, pp. 408–429.

2 'Maximum Allowable Concentrations for Chemical Substances and Permissible Exposure Limits for Physical Agents at the Workplace, (1979–1989)', ed. Occupational Health Standard Subcommittee, China National Technical Committee of Health Standards, 1990, pp. 4–14.

3 Sanitary Standard for Lead and its Inorganic compounds in the Atmosphere, China–GB 7355–87.

4 Hygiene Standard for Arsenic in Soil, China–GB 8915–88.

5 Hygiene Standard for Copper in Soil, China–GB 11728–89.

6 Hygiene Standards for Drinking Water, China–GB 5749–85.

7 Hygiene Standard for Beryllium in Drinking Water Sources, China–GB 8161–87.

8 Hygiene Standards for Cosmetics, China–GB 7916–87.

9 Classification of Health Hazard Levels from Occupational Exposure to Toxic Substances, China–GB 5044–85.

10 Health Protection Zone Standards for Industrial Premises, Part 1, China–GB 11654–11666–89.

11 Health Protection Zone Standard for Oil Refineries, China–GB 8195–87.

12 Y. Pang, 'Introduction to Health Standards in China', 1988, p. 5 (unpublished).

13 1990 China Public Health Statistical Bulletin, 'Health News', Beijing, 1991, No. 3386.

14 Ambient Air Quality Standard, China–GB 3095–82.

15 Surface Water Quality Standard, China–GB 3838–1988.

16 Comprehensive Emission Standards for Pollutants from Industrial Premises, China–GB 8978–88.

17 S. Li, 'An Epidemiological Approach for the Risk Assessment of Chemicals Causing Human Cancer and other Disorders', in 'Risk Assessment of Chemicals in the Environment', ed. M.L. Richardson, The Royal Society of Chemistry, London, 1988, pp. 207–221.

18 S. Li, 'Pesticides, Environmental Pollution and Human Health in China', in 'Chemistry, Agriculture and the Environment', ed. M.L. Richardson, The Royal Society of Chemistry, Cambridge, 1991, pp. 389–409.

14

Ranking Industrial Risks: Decision Support Tools for Risk Management

Robert W. Johnson

BATTELLE, PROCESS RISK MANAGEMENT SERVICES, 505 KING AVENUE, COLUMBUS, OHIO 43201, USA

1 INTRODUCTION

Risk management is a multi-faceted activity that often results in decisions on how corporate resources should be spent to reduce risk. This particularly applies when considering the handling or manufacturing of hazardous and toxic chemicals. Safety problems that appear minor on the surface may have catastrophic potential. We often treat only the immediate symptoms of a problem instead of addressing its potential. By not considering the full potential of the problem, we leave ourselves vulnerable to possibly catastrophic impacts.

Risk ranking methods assist in risk management decisions by providing a framework to assess where the most important risks reside within a group of industrial process plants. Although based on the techniques of quantitative risk analysis, the methods do not require detailed assessments of plant risk to provide useful results for risk management. The methods rely on descriptions of plant equipment and operational characteristics, chemical toxicity and volatility data, meteorological data, and distance to the plant's neighbours to determine which plants have the highest potential for exposing individuals to harmful effects. The results of the methods assist the decision-maker in assigning risk management resources and developing risk management strategies.

2 RISK RANKING APPROACHES

Battelle has developed and successfully applied two approaches for focusing on acute risks from accident events within industrial process plants. These approaches are compared in Table 1.

Table 1 *Comparison of two risk ranking approaches*

Parameter	*Relative risk ranking*	*Absolute risk ranking*
Type of risks included in ranking	Single category only, such as toxic vapour inhalation risks	Several categories combined in same ranking
Measure of risk employed	Purely relative to set of processes being ranked	Absolute measure, such as fatalities per year
Level of effort required	Low	Medium to high

3 RELATIVE RANKING

The relative ranking approach provides a ranking of individual processes or systems based on the relative risk of one type of risk. The example of toxic vapour exposure risk will be used throughout this chapter on relative risk ranking. While absolute risk levels are not specified, relative ranking can nevertheless identify the major contributors to potential toxic vapour exposures.

Risk, in this context, is calculated in the classical way as the product of the frequency of occurrence of toxic vapour releases and the severity of the release consequences, as shown in equation form.

$$\text{RISK} = \text{FREQUENCY} \times \text{SEVERITY}$$
$$(\text{exposures year}^{-1}) = (\text{releases year}^{-1}) \times (\text{exposures release}^{-1})$$

This risk function relies on relative measures of frequency and severity for all processes or systems included in the comparison. In addition, each contributor to frequency and severity is weighted according to its importance to potential off-site toxic vapour exposure. Thus, the risk equation becomes a set of factors, with exponents for each factor to provide the weighting:

$$\text{RISK} = A^a \times B^b \times C^c \times D^d \times E^e \ldots$$

where *A, B, C, D, E* ... are contributing factors to the frequency and the severity of a given kind of process accident.

3.1 Information Requirements

Historical data describing process safety incidents (including near misses) and data such as tanker unloading frequencies can indicate the relative frequency of releases for different processes. The severity of consequences can be taken as a combination of several additional factors which affect the severity: annual usage rate; average on-site inventory; spill containment and secondary containment; process/storage temperature; chemical volatility; inhalation toxicity of chemical; post-release mitigation systems; distance to, and vulnerability of, surrounding populations; and, population density.

This information can be obtained by completion of process information survey forms. They can be filled out by a member of the manufacturing staff with no previous risk analysis experience. Follow-up telephone interviews by the analyst compiling the risk ranking can assure that the process information database is complete.

3.2 Example Result

An example result of a relative risk ranking is shown in Figure 1, where it can be clearly seen that much more is to be gained by attempting to reduce risks in one of the three highest-ranked plant/chemical combinations than in another further down in the ranking.

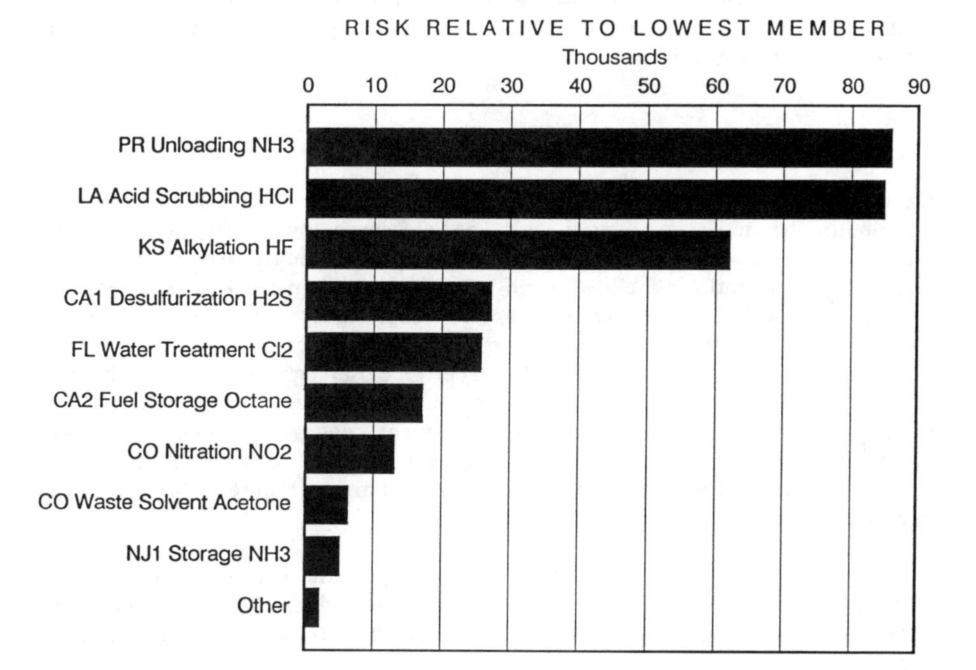

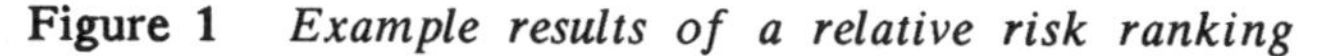
Figure 1 *Example results of a relative risk ranking*

3.3 Example Application

Battelle has applied relative ranking to existing manufacturers' facilities. In a recent application, the client's objectives were to (i) identify which chemicals and processes presented the greatest potential for toxic vapour exposure, and (ii) establish priorities for focusing limited resources where they would do the most good in terms of risk control.

The scope of this relative ranking application encompassed all of the client's plants, including the handling of eighteen toxic chemicals in 114 plants at fifty distinct sites worldwide. The basic approach consisted of compiling and combining data from each plant that handled more than one tonne year^{-1} of the chemicals. The data chosen for the study were those factors that have a direct effect on either the release frequencies or the consequence severities. Prior quantitative risk assessments, as well as interviews with plant operating staff, formed the basis for judging which factors to include in the ranking. The completed data base for this application contains more than 1100 items.

Results from the ranking processes showed that over 99% of the potential for off-site toxic vapour exposures was attributable to only four chemicals; namely, oleum, chlorine, sulphur trioxide, and ammonia. More detailed examination of the individual processes revealed that this risk was due to twelve individual processes at only ten sites. These results directly contributed to the client's development of risk management strategies for future business operations.

4 ABSOLUTE RANKING

Battelle's absolute risk ranking approach identifies process units that contribute the most to overall risk, as measured by a liability/loss rate measured in monetary units year^{-1}. The process units are ranked with respect to potential liability from process accidents such as reactive chemical fires/explosions, vessel rupture explosions, and toxic releases. The process units indicated by the ranking method as having the greatest potential liability are candidates to receive closer scrutiny, using hazard review methods such as hazard and operability studies. In addition, the ranking method identifies elements of plant design or operation that contribute prominently to risk. Thus, the method can point to likely candidate actions for risk control recommendations. Figure 2 presents the overall strategy for absolute risk ranking.

Risk ranking using an absolute measure of risk can combine liability loss rates for casualties, property damage, and business interruption to obtain an overall loss rate. Steps employed in implementing this strategy are as follows:

i) Select 'process units' (specific unit operations within each plant area) to be included in the risk ranking;

ii) Complete an initial survey form for each of the process units, describing the process equipment and operational characteristics;

iii) Based on the information in the initial survey and a site visit and/or interviews, calculate a predicted exposure radius for toxic release, reactive chemical fire/explosion, and pressure explosion hazards;

iv) Complete a second survey form to obtain the replacement value of equipment and average number of people within each area of exposure;

v) Develop order–of–magnitude estimates of the likelihood of the significant loss–potential incidents;

vi) Combine the results of the surveys with credit factors for safety systems and devices included in the process design, such as isolation valves, fire protection, and emergency relief devices;

vii) Calculate a predicted loss rate (*e.g.* dollars year^{-1}) for each process unit with respect to casualties, property damage, and business interruption losses;

viii) Combine the results for all of the process units to obtain an overall risk ranking; and

ix) Analyse the risk–contributing factors for the process units showing the highest overall risk, and propose recommendations for reducing risk and/or for conducting more in–depth hazard evaluations.

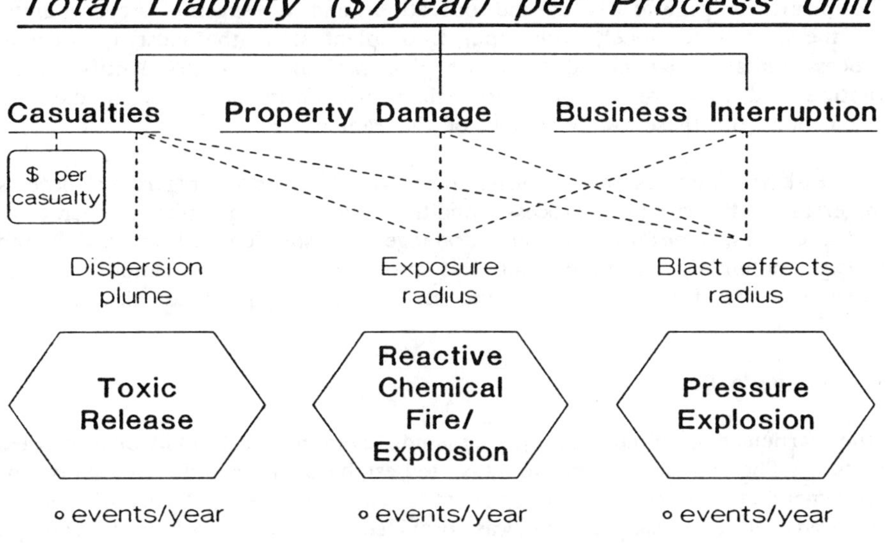

Figure 2 *Absolute risk ranking – overall strategy*

Risk summaries can also be prepared which present individual rankings for hazard types (toxic releases, *etc.*), with the overall risk ranking combining the results of these three summaries.

The fire/explosion radius and the associated property damage and business interruption losses are calculated by Battelle using 'Dow's Fire & Explosion Index Hazard Classification Guide'.[1] Battelle has developed its own computer code to assist in the calculation of the Dow Index, making the analysis more rapid and consistent.

The vessel rupture radius can be calculated based on the blast energy released by sudden expansion of a compressed gas, as would occur in a bursting vessel explosion. The radius is taken as the maximum distance to a side-on overpressure value, typically around 2 psig (14 kPa).

The toxic release radius is calculated based on the maximum distance to a short-term toxic inhalation limit such as an Emergency Response Planning Guideline (ERPG).[2] The vapour release rate used to calculate the toxic release radius can be either an all-vapour release, a flashing liquid release, or evaporation from a boiling or quiescent pool.

4.1 Information Requirements

Absolute risk ranking requires a significant amount of information. This information is normally obtained either (a) on-site by means of a site visit, or (b) by completion of detailed survey forms.

The purpose of the site visit is three-fold: to become familiar with the plant, its general layout and operation, and its specific process units; for the analyst to select, with input from plant staff, the most appropriate process units to be included in the risk ranking; and to identify vessel rupture and toxic release scenarios which will form the basis of the risk calculations for these two elements of the ranking.

Risk ranking surveys request detailed information regarding process materials and quantities, process conditions such as operating temperatures and pressures, types of reactions, drainage and spill control, corrosion and leakage potential, operating and test procedures, and mitigation systems. They can be filled out by plant staff with no previous risk analysis experience.

4.2 Example Result

One particular absolute risk ranking study examined 120 individual process units. The client's objective was to establish a quantitative basis for implementing risk control strategies at a corporate level. All information required for the risk ranking was provided by the client's staff through completion of risk ranking surveys forms. Figure 3 shows part of the results obtained from this application, in the form of a graphical risk profile. The results indicate not only which process units contribute most to the overall risk, but also what incident type is most important (in this case, reactive chemical fire/explosion events).

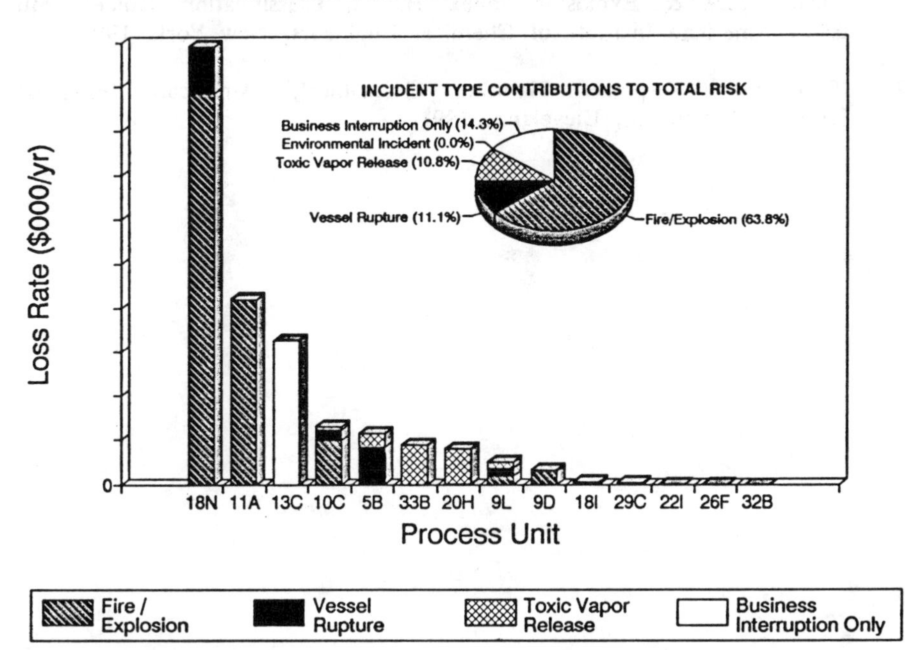

Figure 3 *Example results of an absolute risk ranking*

5 CONCLUSIONS

Accidents at industrial process plants can pose serious risks to industry and the public, and all types of industrial plants are receiving increased public scrutiny. This scrutiny is motivating the management of chemical manufacturing companies toward developing more comprehensive, systematic, and responsible strategies to control risk from potential accidents, through the use of formal risk management programs.

An important initial element in risk management is identification of the accident potential within a corporation's business operations. Once this potential is recognized, positive steps can be taken to control risk.

Risk ranking methods are a logical first step to focus corporate risk management efforts when dealing with multiple plants and diverse business operations. Based on proven risk analysis techniques, the methods use distinct plant and process characteristics to describe loss potential. The plant-specific rankings allow management to set distinct priorities for controlling acute industrial risks in order to better protect their business investments and their neighbours in the surrounding community.

6 REFERENCES

1 'Dow's Fire & Explosion Index Hazard Classification Guide', 6th Edn., American Institute of Chemical Engineers, New York, 1987.

2 'Emergency Response Planning Guidelines', American Industrial Hygiene Association, Cleveland, 1991.

Risk Management from Waste

15

Managing the Risk from Atmospheric Emissions: A Case Study for the Netherlands

K. F. de Boer and A. H. Bakema

NATIONAL INSTITUTE OF PUBLIC HEALTH AND ENVIRONMENT PROTECTION (RIVM), ANTHONIE VAN LEEUWENHOEKLAAN 9, PO BOX 1, 3720 BA BILTHOVEN, THE NETHERLANDS

1 INTRODUCTION

In 1985 the Dutch Priority Programme on Acidification was initiated to provide an answer to the increasing interest of policy-makers on the effects of air pollution, in particular on natural ecosystems.[1] As part of this research programme, an integrated regional model was developed that would allow for the evaluation of the effectiveness of alternative abatement strategies. The impact of 'acid rain' on ecosystems was also modelled. This study used scientific results obtained in other studies of the Programme, and all these were integrated into one model.

The model was developed during 1985–1991. A structure based on modules, that could be developed and used independently, was chosen. The modules were developed at the National Institute of Public Health and Environmental Protection (RIVM) and several other research institutes (Staring Center, the Dutch Research Institute for Nature Management, the Research Institute for Forestry and Urban Ecology, the Center for Agrobiological Research, and Resource Analysis). A group of five people (the Integrated Modelling Department of the Environmental Forecasting Bureau of RIVM), amongst whom are the authors, acted as the project team. They had the responsibility to connect all the modules into a coherent systems model. A few modules were developed within the project team itself. Some scenario analysis results will be shown in this chapter.

2 MODEL OVERVIEW

The basic conceptual outline of the Dutch Acidification System (DAS) model is given in Figure 1. The boxes represent the modules of the DAS model. The regional division of the Netherlands is a compromise between the desired level of detail, the availability of regional data, and limitations on the accuracy with which the air transport model could produce local time-averaged data. This resulted in 20 regions of about 60 x 60 km^2 in size. For emissions only a division of the remainder of Europe was made, based on the contribution of each area to the depositions and concentrations in the Netherlands. From this study 19 European regions were chosen. The regionalizations are given in Figures 2 and 3, respectively.

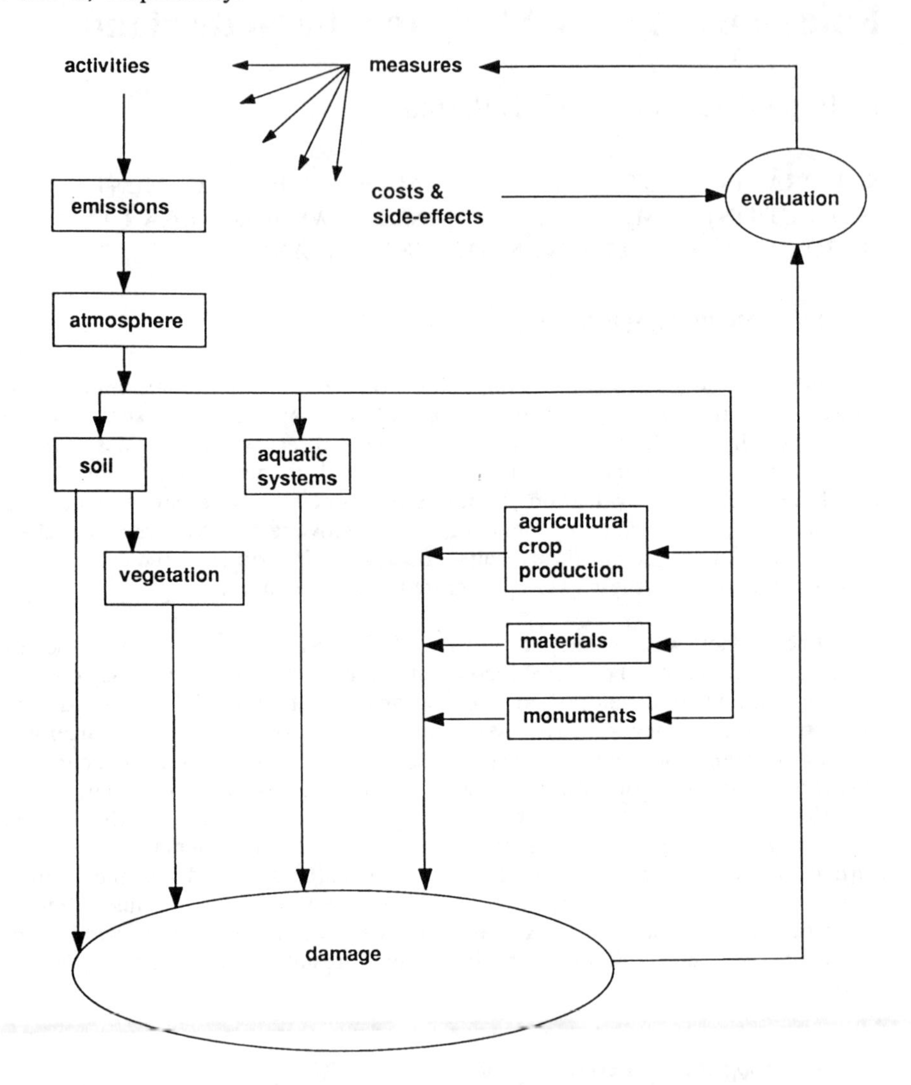

Figure 1 *Conceptual outline of the Dutch Acidification Systems model*

The time span of the model is about one hundred years, from its inception in 1950. Historical data were included to allow initialization of some receptor systems. The time interval used within the model is set at one year, although within some of the ecological models a smaller time interval is used.

Figure 2 *Regionalization of the Netherlands. Within the DAS model these areas function both as source and receptor for acidifying substances*

Figure 3 *Regionalization of Europe. Within the DAS model these areas function only as source for acidifying substances*

Air pollutants include sulphur dioxide (SO_2), nitrogen oxides (NO_x), and ammonia (NH_3). Some of the modules use additional compounds, such as ozone (O_3) (forests, agricultural damage) and the cations sodium, potassium, magnesium, and calcium (soils, forests) and the anion chloride. For ozone a future trend was estimated, while the depositions of actions and anions were kept constant at their measured values.

The model consists of several modules, each covering a part of the causality chain. There is a module to generate emission data, one calculating air transport flows, concentrations, and depositions, and several modules which describe the reactions of receptor systems to the input of acidifying substances. Included in the receptor systems are: soil, forests (represented by Douglas fir), heathland, agricultural crops, building materials, and monuments. A model for shallow isolated lakes (heathland pools) is under development. The following sections deal with these modules in more detail.

2.1 The Emissions Module

Because acidification is a transboundary, long-distance issue, emissions are needed from the whole of Europe. In the scenario studies emissions of SO_2, NO_x, and NH_3 are needed for all European countries for the period 1950–2050.

The air transport module (see Section 2.2) requires a distinction in the stack height of emissions. This information too has to be provided by the emissions module. Some receptor systems require deposition and/or concentration data for each year, so emissions need to be generated for every year. Since for the period 1950–1980 the yearly emissions were not known, linear interpolation was used to fill in gaps. Finally, only anthropogenic emissions are used in the model. These are an order of a magnitude higher than natural emissions (like NO_x from forests, SO_x from sea spray), so the error was considered negligible.[2]

2.1.1 Dutch emission data. To contract meaningful scenarios, emissions are entered per economic sector. The sectors used in the DAS model are given in Table 1. They make a pragmatic allocation of stack heights to emissions possible: all emissions from a single sector are from one stack height only. Three classes of stack heights are used: low (<50 m), high (>100 m), and medium high (between 50 and 100 m).

Recent emission data are available mostly on a national basis only, but for use within DAS this is not adequate. The air transport module requires emissions in the 20 Dutch regions (see Figure 2). To overcome this problem, results of the Netherlands Organization for Applied Scientific Research (TNO) emissions inventory[3] were used to build a distribution matrix of emissions per sector over the 20 areas. The model has been designed to accommodate changes in the distribution matrix, but lack of data has prevented the use of this feature to date.

Table 1 *Economic sectors used within DAS*

Economic sector	*Elevation of emissions*
Refineries	high
Powerplants	high
Chemical industry	medium
Primary metal industry	medium
Fertilizer industry	medium
Agriculture	low
Private cars	low
Truck, buses, *etc.*	low
Ships	low
Domestic	low
Other industry	medium
Other low sources	low

2.1.2 Emission data for other European countries. An inventory of emissions from European countries has been made, based on sources such as Economic Commission for Europe (ECE),[4] Organization for Economic Co-operation and Development (OECD),[5] Co-operative Programme for Monitoring and Evaluation of the Long-Range Transmission of Air Pollutants in Europe (EMEP),[6] Photochemical Oxidants and Acid Deposition Model Application (PHOXA),[7] and International Institute for Applied Systems Analysis (IIASA).[8] Detailed information on stack heights and regional distribution of emissions was derived from a database of European emissions available at the RIVM Air Research Laboratory (the PHOXA-database[7]).

2.2 The Air Transport Module

Modelling air transport is a complex matter. Many models have been developed for this purpose in the past, and new ones are being developed today. The RIVM Air Research Laboratory has developed the TREND model.[9,10] This is a statistical atmosphere transport model which can describe transport, conversion, and deposition of air pollutants from local sources as well as from more distant sources. Since this model consumes significant computing time and power to run, it is not ideal for use within DAS. Therefore the Air Research Laboratory developed a set of transport matrices from this model, which can be used with certain restrictions.[11]

The matrices include a linear relationship between emissions and concentration or deposition of a compound, and interactions between compounds in the atmosphere are neglected. Research of the relevant atmospheric processes has shown that such linear relations do not hold for acidifying substances. Apart from the interactions in the atmospheric chemistry, interactions play a role in deposition processes. For example, the oxidation rate of SO_2 and NO_x is also dependent on the concentration of OH radicals, ozone, and H_2O_2. The concentrations of these oxidants are again dependent on the concentration of NO_x and VOC (Volatile Organic Compounds). Ammonia and sulphur dioxide are possibly both related to the dry deposition rate of the latter. The acidity in cloud and rain water determines the solubility of SO_2, and therefore also its wet deposition.

Based upon sensitivity studies performed with the TREND model it was, however, concluded that the deviations due to non-linearities remain within the accuracy limits of the model, if the emission regime deviates no more than 30% from the regime for which the matrices were constructed. When the deviation becomes longer, the error becomes greater, but will stay below 30% (extremes excluded).

The matrices have been constructed using an 8-year averaged meteorology. The TREND model, on which the matrices are based, has been calibrated for the year 1980.

2.3 The Soil Module

The REgionalized Soil Acidification Model (RESAM)[12,13] is used to evaluate the long-term effect of acid deposition on forest soils. RESAM is a process-oriented soil-chemistry model, including geochemical weathering of the soil, cation exchange reactions, nitrogen transformations, and nutrient cycling by the vegetation. The model calculates the concentrations of the components H, Al, Ca, Mg, K, Na, NH_4, NO_3, SO_4, Cl, HCO_3, and RCOO in the soil solution, and the element levels in the solid phase. The driving variables of the model are the quantity and composition of yearly total acid deposition. Results from the model include the concentrations of these components in the various soil layers distinguished within the model.

Within the DAS framework, the model is applied for all of the Dutch receptor areas (Figure 2). Seven tree species and five major sandy soil types are considered. The resulting forest-soil combinations cover roughly 65% of all forested areas in the Netherlands. For each receptor area, the surface areas of soil-vegetation combinations are known. Only soil-vegetation combinations with a surface area larger than 25 hectares are taken into account, to reduce the number of computations. Results are mostly presented as the percentage of the forested area where a concentration or a ratio of concentrations exceeds some critical level. Thus, an overall insight is obtained in the soil situation in all acidification areas.

2.4 The Forest Module

The forest module (SOILVEG)[12,14] calculates the effect of acid deposition on stands of Douglas fir. The model contains both biological and soil processes. The emphasis is on the reaction of tree growth to acid deposition and soil acidification. A simplified version of the RESAM model is used to quantify the soil processes. Effects in the trees occur through two mechanisms: direct effects of the concentration and deposition of acidifying chemicals in the leaves or needles, and indirect effects through the soil solution and uptake of chemicals by the roots. In SOILVEG both effects are taken into account.

Output variables of the model are among others, growth, and total mass of the Douglas fir stand, and nutrient concentrations in the various tree compartments. Four sandy soil types are distinguished. Calculations are carried out for each of these soil types in all receptor areas. Results are presented for each soil type separately. Uncertainty analysis of SOILVEG has indicated that uncertainties in the decomposition kinetics of organic matter are the major source of uncertainties in the model output.[15]

2.5 The Heathland Modules

Two modules to calculate effects of acid deposition on heathland are used in the DAS model, one for dry (CALLUNA)[16] and one for wet heathland (ERICA).[12,17] Heathland in the Netherlands suffers from acidic deposition in a number of ways. First, acidification itself causes some rare heathland species to disappear. Secondly, the soil enriching effect of nitrogen deposition causes grasses to prosper at the cost of the original heathland vegetation which is adapted to an environment poor in nutrients. Thirdly, the chance of occurrence of pests like the heather beetle is increased at a higher nitrogen content of the heather vegetation.

Of these three mechanisms, only the second (soil enrichment) is taken into account in the module ERICA, which describes the effects on *Erica tetralix* (cross-leaved heath) and *Molinia caerulea* (purple moor-grass), whilst the module CALLUNA describes the effect on *Calluna vulgaris* (heather) and *Deschampsia flexuosa* (wavy-hair grass); the occurrence of pests is also modelled. Pests play a minor role in wet heathland. In both modules, the effect of increased nitrogen supply is described through its influence on the competitive strength of the grass and heather species, leading to a shift in abundance. Management of heathland is assumed to take place by sod-cutting once every 20 to 25 years. Nearly all existing vegetation is then removed.

In CALLUNA, the probability that a heather beetle plague occurs is positively related to the nitrogen content of the vegetation. When such a plague occurs, the amount of heather present is reduced to 1% of its original value. The stochasticity of heather beetle occurrence is undertaken by performing 100 simulations for each situation, and subsequent averaging of the resulting abundance of the heather and grass species.

2.6 The Damage Modules

Within DAS, simple dose-effect relationships are used to evaluate the damage to agricultural crops, to construction materials, and to stone monuments.

Damage to agricultural crops is calculated in the module AGRIPROD[12] and is expressed as a yield reduction due to ambient ozone and SO_2 concentrations, using specific damage functions.[12,18]

Economic damage to construction materials (calculated in module MATERIALS[12]) is caused by increased corrosion, resulting in higher maintenance costs and earlier replacement. Materials taken into account are steel, zinc, and their combinations. Damage is related to the ambient concentration of SO_2 with a specific damage function.[12,19]

Stone monuments suffer from increased weathering rates due to acid deposition. In the module MONUMENTS[12] a linear relationship with the dry deposition of SO_x is used.[12,19]

Although these relatively simple relationships can be used to compare the relative impact of different scenarios, they suffer from limited data and large uncertainties.[18,19]

3 MODEL USE

The DAS model has recently been used in two scenario studies. Four different scenarios have been analysed in the past few months. Three were developed within the Dutch Priority Programme on Acidification (the DPPA scenarios), and one was developed for the second National Environmental Survey by RIVM[20] (the NES scenario). Some results from these analyses are presented here. It should be noted that for the historical period small differences exist between the two studies. This is due to changing insight in historical emissions, mainly those of ammonia. The last year for which data are available is the year 1989. Emissions for 1990 and later have therefore been estimated. Historical data used in these studies are not described here, but can be found in reports from the Dutch Priority Programme on Acidification.[12]

3.1 The Scenarios

The anthropogenic emissions of acidifying compounds have shown a major increase in the period 1960–1980. Subsequently, the emission of SO_2 has decreased, due to the fact that several European countries are achieving significant abatement of these emissions. The emission of NO_x has stabilized since 1980, but the emission of ammonia has increased. Within the framework of the United Nations Economic Commission of Europe (UN/ECE), national reduction plans for SO_2 and NO_x have been registered.[21] These plans have been used as the basis for the NES scenario.

In the DPPA scenarios, however, the European emissions are based on estimates by the Dutch Environment Ministry[22,23] of the maximal reduction to be expected from each of the countries. These scenarios are therefore relatively optimistic. Table 2 summarizes the scenarios.

Table 2 *Reduction percentages for the year 2000 (relative to 1980) as used in the Dutch Priority Programme on Acidification (DPPA) and in the National Environment Survey (NES). Only the reductions for countries near to the Netherlands are given*

	SO_2		NO_x		NH_3	
	DPPA	*NES*	*DPPA*	*NES*	*DPPA*	*NES*
Netherlands	84	80	56	35	67	55
Belgium	80	70	50	53	25	8
Germany	80	73	50	33	25	0
France	80	62	50	55	25	0
UK	45	50	50	19	25	0

The emission reductions in the Netherlands for the NES scenario are based on expert judgement on the effectiveness of the abatement measures, announced in the National Environmental Policy Plan.[22] The realization of the abatement strategies in other countries has not been evaluated. It is assumed that the reduction plans as registered by the ECE are realistic and will be met.

In the DPPA study three deposition pathways have been used for the period 2000–2050. These are not based on emissions but rather on total acid deposition. The total acid deposition was assumed to remain constant (scenario 1), to fall to a level of 1400 mole ha^{-1} yr^{-1} in 2010 (scenario 2), or to fall to 700 mole ha^{-1} yr^{-1} in 2050 (scenario 3). In all three scenarios the ratios between the compounds which contribute to the total acid deposition (SO_2, NO_x, and NH_3) were retained as those calculated for the year 2000 (based on emission data). The DPPA scenarios were used to characterize the behaviour of the receptor systems. Figure 4 shows the resulting total acid deposition, averaged over the Netherlands.

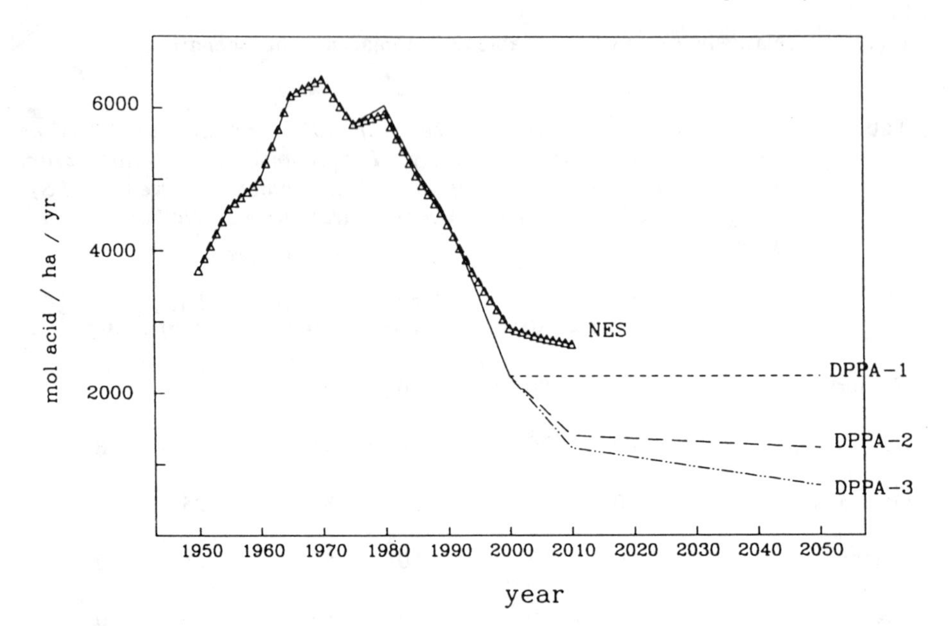

Figure 4 *Total potential acid deposition in the Netherlands, as calculated with the DPPA and the NES scenarios* (mole H^+ ha^{-1} yr^{-1})

3.2 Results

Precise predictions of the future state of one particular ecosystem cannot be made with the DAS model. The simulation results provide an overall view of the plausible and likely effects of the scenarios on the whole of the Netherlands. Various environmental, ecological, and economical risks associated with the scenarios can thus be evaluated. As an example, some of the results of the DPPA and NES scenario studies are presented below. The main purpose of these scenario studies has been the comparison between different policy scenarios and the comparison between the situation at different years in one scenario.

3.2.1 Effects on the soil. In general, when a reduction of acid deposition occurs, the model calculates a relatively quick improvement in the soil conditions: the pH increases, aluminium and nitrate concentrations decrease, and ratios between aluminium and calcium concentration and between ammonia and potassium concentration also decrease. However, nitrate concentration and the ammonia:potassium ratio show a delayed reduction, due to mobilization of nitrogen from the litter layer.

As an example of the model results, exceedance of the critical aluminium:calcium ratio of 1.0 is given in Figure 5, for two years in the NES scenario. Acidifying components of deposition are exchanged with Al^{3+} while Ca^{2+} and other cations are leached out.

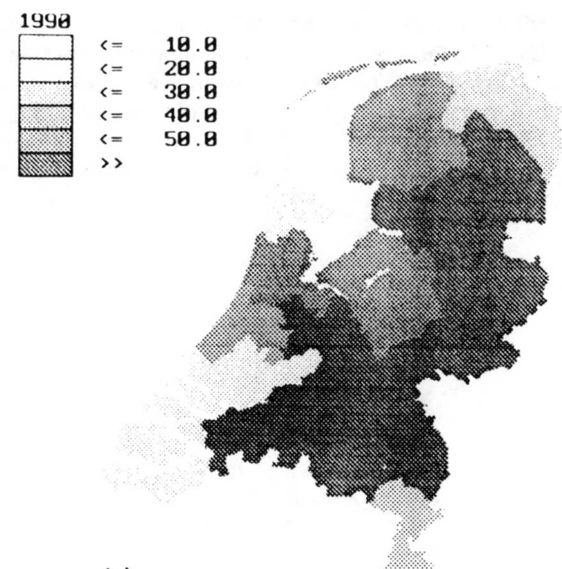

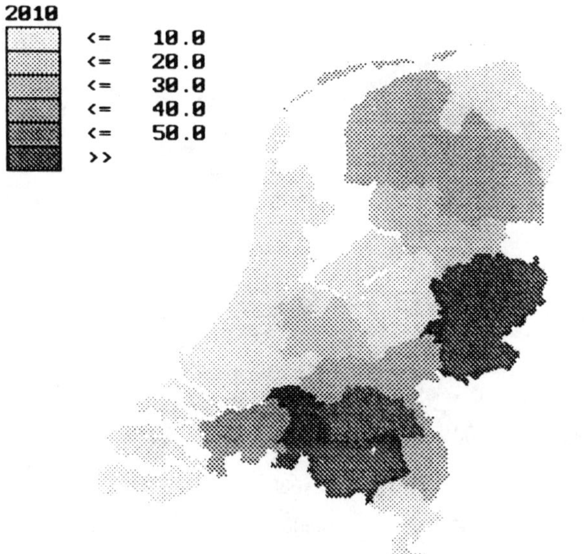

Figure 5 *Percentage area where the Al/Ca ratio exceeds the critical value of 1.0, for the 20 distinguished receptor areas in the Netherlands, in (a) 1990 and (b) 2010, as calculated with the NES scenario*

When the concentration of Al^{3+} becomes larger than the concentration of Ca^{2+}, uptake of Ca by the vegetation is hampered, leading to lower vitality. Al/Ca ratios are vastly reduced in nearly all areas. In Figure 6 the nationwide exceedance percentages of Al/Ca ratios are presented for the period 1990–2050 for the three DPPA scenarios. The differences between scenarios are comparatively small, but the reduction in exceedance percentage is very significant.

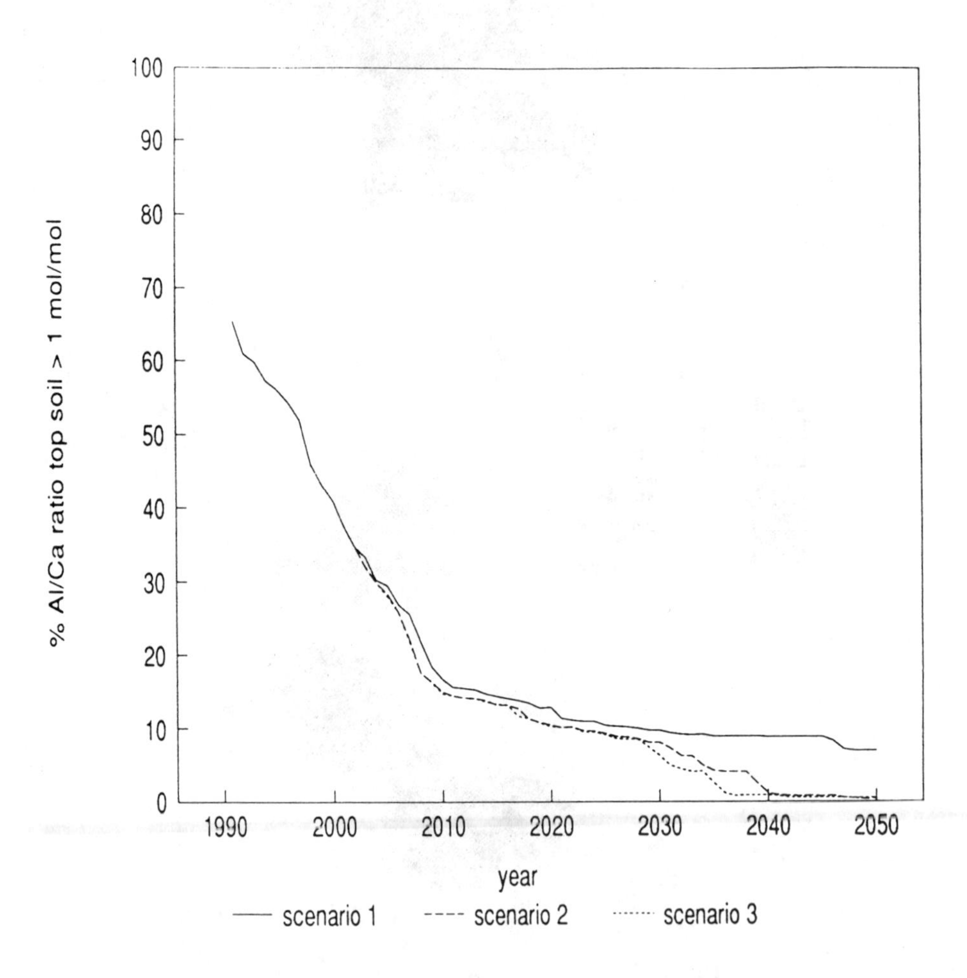

Figure 6 *Percentage area where the Al/Ca ratio exceeds the critical value of 1.0, for the Netherlands as a whole, as calculated with the three DPPA scenarios*

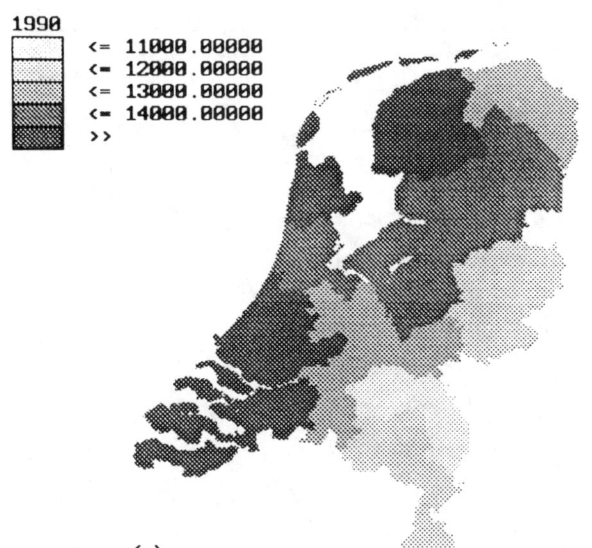

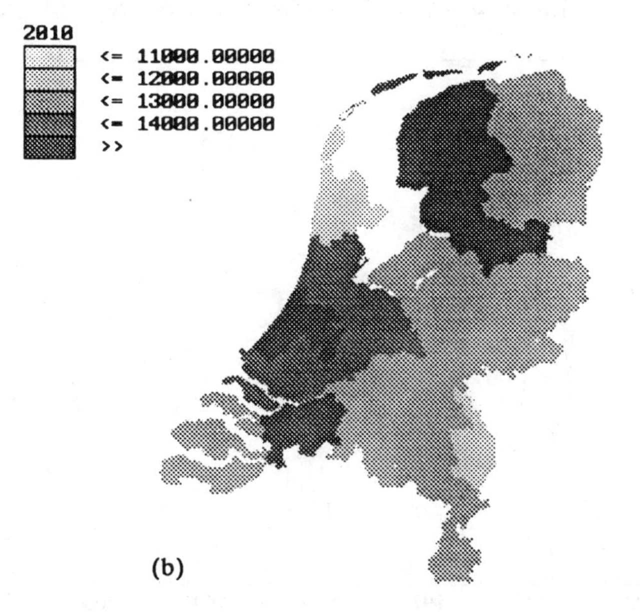

Figure 7 *Needle biomass of Douglas fir on sandy soils, for the 20 distinguished receptor areas in the Netherlands, in (a) 1990 and (b) 2010, as calculated with the NES scenario*

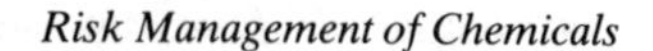

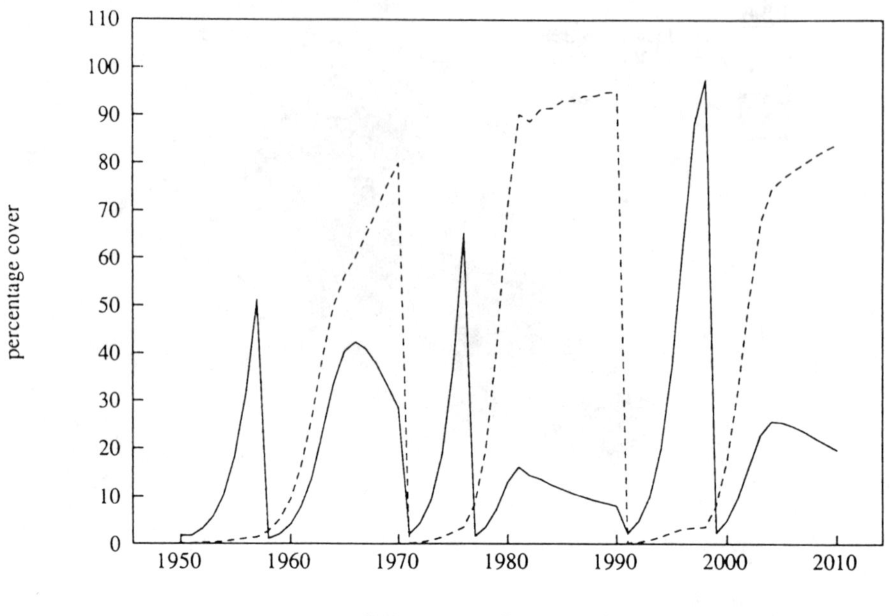

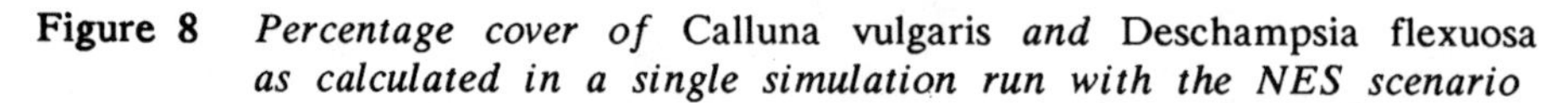

Figure 8 *Percentage cover of* Calluna vulgaris *and* Deschampsia flexuosa *as calculated in a single simulation run with the NES scenario*

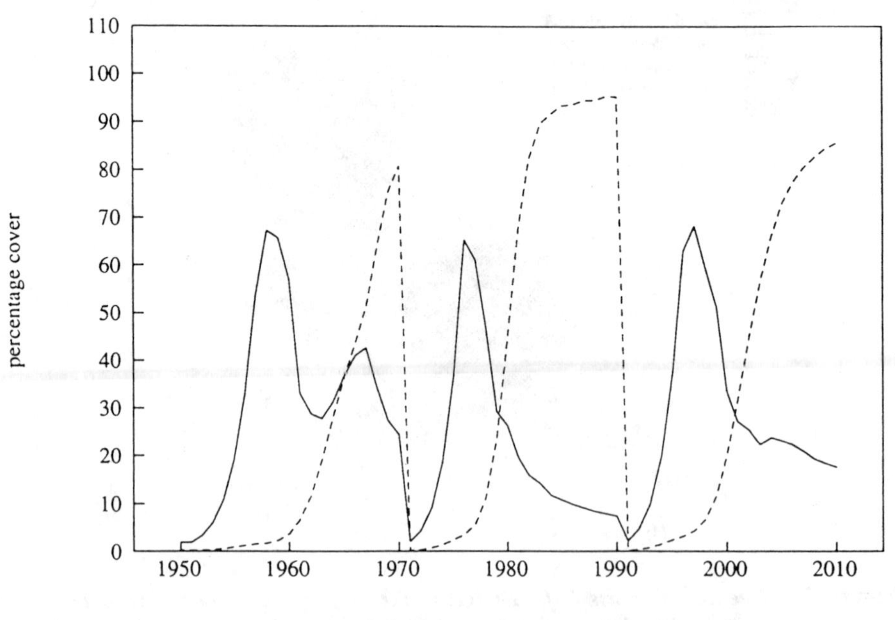

Figure 9 *Percentage cover of* Calluna vulgaris *and* Deschampsia flexuosa, *average values of 100 simulation runs with the NES scenario*

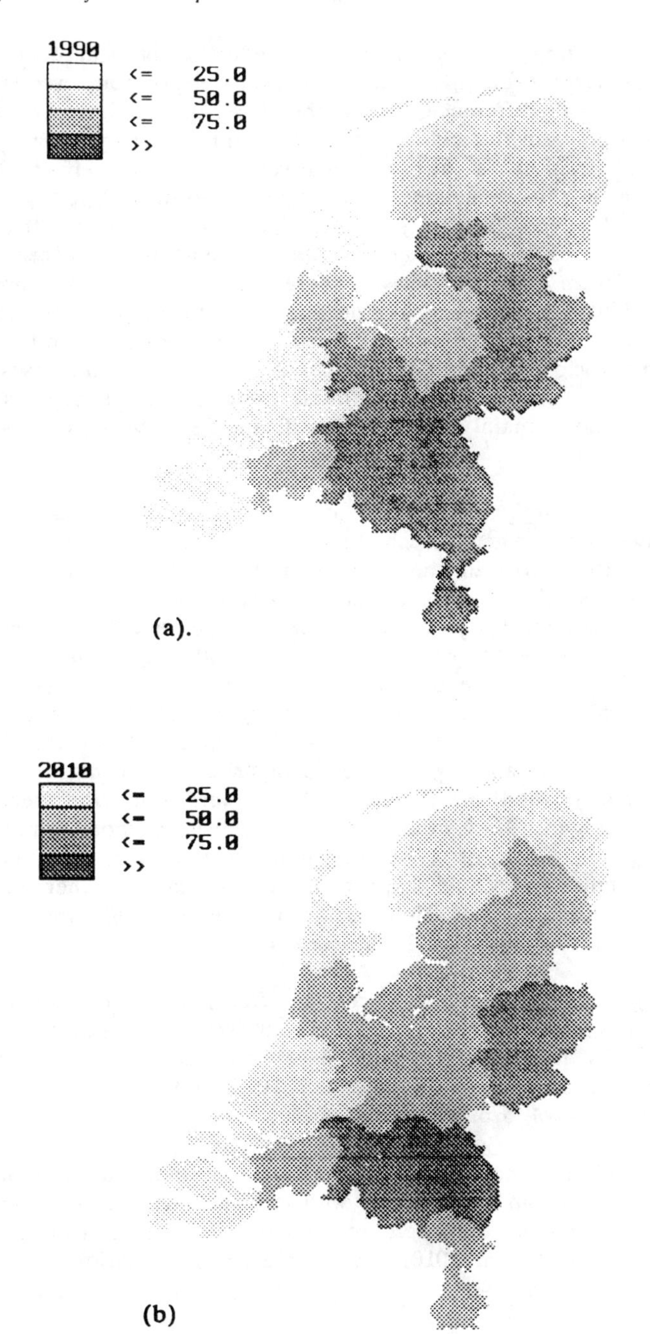

Figure 10 *Percentage cover of* Deschampsia flexuosa, *average during the first 20 years after sod-cutting, in (a) 1990 and (b) 2010, as calculated with the NES scenario*

3.2.2 Effects on forests. The effect of a reduction of acid deposition on forests depends largely on the present level of deposition, and therefore shows large regional differences. In the most polluted areas reduction leads to increased growth and optimal leaf nitrogen concentrations. In areas where deposition is already relatively low, the effect is more complicated, since the reduction of nitrogen deposition leads to reduced nutrient availability for the trees. In Figure 7 the needle biomass is shown in 1990 and 2010 as calculated with the most realistic scenario. In the relatively clean coastal areas, where forest growth is limited by nitrogen availability, needle biomass declines due to reduced availability of nitrogen. In the South-Eastern areas (where most forests in the Netherlands are located), needle biomass remains equal or increases slightly during this period, since forest growth is limited in these regions by the availability of cations mainly, and suffers from the adverse effects of soil acidification.

3.2.3 Effects on heathland. Results of the CALLUNA module are shown for the Veluwe area, where much heathland occurs. Results are shown for the period 1950-2010 in the NES scenario, with sod-cutting every 20 years. Figure 8 shows the calculated cover percentages of *Calluna vulgaris* and *Deschampsia flexuosa* of one single simulation run. The drastic reduction of both species in 1970, 1990, and 2010 is due to sod-cutting. The occurrence of heather beetle plagues can clearly be seen in the years 1957, 1976, and 1998, where nearly all heather plants die, but the grass remains unaffected, and in fact profits from the increased availability of nutrients and light. For comparison of scenario results the average of 100 simulation runs is used. This average for the Veluwe area is presented in Figure 9. Although these results are more representative for the average condition of a heather field in the area, they no longer describe the events that could happen in any specific heather field. In Figure 10 the results of such simulations are presented for the year 1990 and 2010 as calculated with the NES scenario.

3.2.4 Damage. Predicted differences in effects with the scenarios used are very small, due to the fact that all scenarios assume a major reduction in acid deposition between 1990 and 2010. For agricultural crops the yield reduction in 1990 and 2010 is estimated at 2-3%, due mainly to the effect of ozone concentration.

For damage to construction materials and monuments, damage is already low in 1990 and compared with the damage between 1960 and 1970, due to a major reduction of sulphur dioxide concentration and deposition. The damage in 2010 is negligible in all scenarios.

3.3 Obstacles

When using the DAS model for scenario analysis several obstacles were encountered. One of the most serious is the lack of a solid database. In the emissions module, where data from many different sources have to be assembled when new scenarios are developed, data-checking proved to be very difficult.

The DAS model has also grown to be a rather unbalanced model. Some modules have become very large, computing-time-consuming simulation models (RESAM, SOILVEG), while others are relatively simple and fast-running modules (*e.g.* air transport). The large modules incorporate many detailed processes. The uncertainty in the input does not always warrant this level of detail.

Uncertainty analysis has been carried out for most of the modules individually, but so far no uncertainty analysis has been carried out over the total DAS model. This makes it impossible to estimate the uncertainty in effects that result when calculating the whole cause-effect chain, from emissions to receptor systems.

Another limitation is that, apart from the occurrence of heather beetle plagues in the CALLUNA module, multiple stresses are not considered in the DAS model. Therefore the model can be used to assess the effects of acidification only. Susceptibility of forests to, for instance, drought and frost, is, however, known to be influenced by high nitrogen contents also. This type of multiple stress cannot be evaluated with the DAS model in its current form.

4 CONCLUSIONS

The Dutch Acidification Systems model has proven to be a valuable tool in evaluating abatement strategies and scenarios for the acidification problem, most explicitly by its ability to calculate the possible effect of emission reductions on the various receptor systems. Furthermore, it has proved to integrate knowledge from different institutes into one model. Yet, some aspects of the model still need to be improved, as outlined above.

The Integrated Modelling Department is therefore now starting to construct a new modelling environment, that should give the user the possibilities to estimate the consequences of various policies by using information from a database. This model, called EXPECT (EXPloring Environmental Consequences for Tomorrow)[24] will incorporate parts of the DAS model, but the large simulation models will be simplified. Furthermore, multiple stresses and other influences like eutrophication and desiccation will be incorporated in the different modules where appropriate. The output of EXPECT will be verified and compared with the output of more detailed models such as RESAM and SOILVEG.

The EXPECT model will be a policy assessment tool that allows fast and easy analysis of large numbers of different policy options, whereas DAS and other models will remain in use to evaluate scenarios in more detail.

5 REFERENCES

1 'Final Report Second Phase Dutch Priority Programme on Acidification', ed. G.J. Heij and T. Schneider, Dutch Priority Programme on Acidification, Report No. 200–09, 1991.

2 R. Thomas, W.G. van Arkel, H.P. Baars, E.C. van Ierland, K.F. de Boer, E. Buysman, T.J.H.M. Hutten, and R.J. Swart, 'Emissions of SO_2, NO_x, VOC, and NH_3 in the Netherlands and Europe in the period 1950–2030; the Emission Module in the Dutch Acidification Systems Model', Report No. 758472002, National Institute of Public Health and Environmental Protection, 1988.

3 TNO, 'Gegevens Emissieregistratie le en 2e Ronde' (Data from the Emission Inventory System, first and second inventory), TNO, Delft, 1985.

4 Economic Commission for Europe. 'National Strategies and Policies for Air Pollution Abatement', United Nations, Geneva/New York, 1987.

5 OECD, 'Emission Inventory for OECD–Europe', AMPG/MAP, ENV/ AIR/87.8, draft, May 10, 1988.

6 H. Dovland and J. Saltbones, 'Emissions of Sulphur Dioxide in Europe in 1980 and 1983', EMEP/ECE, revised August 1987.

7 TNO, 'Photochemical Oxidants and Acid Deposition Model Application within the Framework of Control Strategy Development Emission Database, PHOXA Report No. 1', PHOXA, TNO, Apeldoorn, 1987.

8 IIASA, 'Rains, Enem version 3.b', Laxenburg, Austria, 1987.

9 J.A. van Jaarsveld and D. Onderdelinden, 'Trend; an Analytical Long-term Deposition Model for Multi–scale Purposes', Report No. 228603009, National Institute of Public Health and Environmental Protection, 1990.

10 W.A.H. Asman and J.A. van Jaarsveld, 'A Variable–resolution Statistical Transport Model for Ammonia and Ammonium', Report No. 228471007, National Institute of Public Health and Environmental Protection, 1990.

11 T.N. Olsthoorn and F.A.A.M. de Leeuw, 'Berekening van de Zure Depositie op Nederland op Basis van Overdrachtsmatrices, (Calculation of the acid deposition on the Netherlands on the basis of transfer matrices; in Dutch)', Report No. 758805005, National Institute of Public Health and Environmental Protection, 1988.

12 A.H. Bakema, K.F. de Boer, G.W. Bultman, J.J.M. van Grinsven, C. van Heerden, R.M. Kok, J. Kros, J.G. van Minnen, G.M.J. Mohren, T.N. Olsthoorn, W. de Vries, and F.G. Wortelboer, 'Dutch Acidification Systems Model – Specifications', Dutch Priority Programme on Acidification, Report No. 114.1–01, 1990.

13 W. de Vries and J. Kros, 'Lange Termijn Effecten van Verschillende Depositiescenario's op Representatieve Bosbodems in Nederland, (Long term effects of different deposition scenarios on representative forest soils in the Netherlands; in Dutch)', Wageningen, Staring Centre, Report No. 30, 1989.

14 J.J.M. Berdowski, C. van Heerden, J.G. van Minnen, J.J.M. van Grinsven, and W. de Vries, 'SOILVEG: A Model to Evaluate Effects of Acid Atmospheric Deposition on Soil and Forest', Dutch Priority Programme on Acidification, Report No. 114.1–02, 1991.

15 J.J.M. van Grinsven, P.J.M. Janssen, J.G. van Minnen, C. van Heerden, J.J.M. Berdowski, and R. Sanders, 'SOILVEG: A Model to Evaluate Effects of Acid Atmospheric Deposition on Soil and Forest. Volume 2. Uncertainty Analysis', Dutch Priority Programme on Acidification, Report No. 114.1–03, 1991.

16 R. Bobbink, G.W. Heil, and M.B.A.G. Raessen, 'Atmospheric Deposition and Canopy Exchange in Heathland Ecosystems', Dept. of Plant Ecology and Evolutionary Biology, University of Utrecht, 1990, pp. 80.

17 F. Berendse, 'De Nutriëntenbalans van Droge Zandgrondvegetaties in Verband met de Eutrofiëring via de Lucht. Deel 1. Een Simulatiemodel als Hulpmiddel bij het Beheer van Vochtige Heidevelden (The nutrient balance of dry sandy soil vegetations in relation with eutrophication via air. Part 1. A simulation model as management tool of wet heathland; in Dutch)', Center for Agrobiological Research, Report No. m.743, 1988.

18 L.J. van der Eerden, A.E.G. Tonneijck, and J.H.M. Wijnands, 'Crop Loss Due to Air Pollution in the Netherlands', *Environ. Pollut*, 1988, **53**, 365–376.

19 H.J. Gosseling, A.A. Olsthoorn, and J.F. Feenstra, 'Schade aan Materialen door Verzuring (Damage to materials due to acidification; in Dutch)', IvM Report No. W90–002, 1990.

20 National Institute of Public Health and Environmental Protection. 'Tweede Nationale Milieuverkenning 1990–2010 (Second National Environmental Survey 1990–2010; in Dutch)', National Institute of Public Health and Environmental Protection, 1991.

21 United Nations, 'European Sulphur and Nitrogen Budgets for 1988 and 1989 (provisional)', Geneva: Economic Commission for Europe, EB. AIR/GE, 1/16/Add.1, 1991.

22 Ministry of Housing, Physical Planning, and Environment, 'To Choose or to Lose; the National Environmental Policy Plan', Second Chamber, session 1988–1989, 21137, Nos. 1–2.

23 Ministry of Housing, Physical Planning, and Environment, 'Bestrijdingsplan Verzuring (Abatement Strategy Acidification; in Dutch)', Second Chamber, session 1988–1989, 18225, No. 31.

24 L.C. Braat, A.H. Bakema, K.F. de Boer, R.M. Kok, R. Meijers, and J.G. van Minnen, 'EXPECT: An Integrated Model System for Scenario Analysis and Environmental Impact Assessment', National Institute of Public Health and Environmental Protection, 1991, in press.

16

Managing Risk from Chemicals in Liquid Effluents

Cliff S. Johnston

INSTITUTE OF OFFSHORE ENGINEERING, HERIOT–WATT UNIVERSITY, RICCARTON, EDINBURGH EH14 4AS, UK

1 BACKGROUND

Over recent years there has been a dramatic increase in public and political awareness of environmental issues. Such attention has developed at all levels, through international conventions, North Sea Conferences, and EC and individual member Government initiatives. Resultant pressures on industry will continue to increase during the 1990s, with environmental management playing a key role within evolving Company strategies.

Several of these issues stand out, such as the growing concern of the pollution of the North Sea,[1] with three North Sea Conferences (in 1985, 1987, and 1990) focusing attention on the need for stricter controls of discharges including the target of a 50% reduction in inputs of 'dangerous substances' over the period 1985–1995.

Within the UK there have been significant changes to the legislation relating to environmental protection, and also the reorganization of the water industry in England and Wales. The Water Act (1989) resulted in the formation of the National Rivers Authority (NRA), to act as an independent body responsible for discharges to controlled waters (lakes, reservoirs, rivers, and coastal waters) in England and Wales. The River Purification Boards have similar powers in Scotland. The Water act has largely developed (given teeth to) the control procedures of the Control of Pollution Act (1974) as they relate to water pollution.

Thus, in the 1990s, industry faces much tighter regulatory control of wastewater treatment and effluent discharge. Obviously this will involve considerable additional cost to industry through application of the 'polluter pays principle', thence to all of us as consumers as effluent treatment costs become incorporated in the price of the product.

Although some industries may claim more innocence than others, attitudes to wastewater treatment have been extremely superficial. Few have had an understanding of the composition of their wastewaters, yet this is the essential basis to practical wastewater treatment design and operation. Hence effluent treatment and disposal have tended to be dealt with as necessary add-ons to the production process. Similarly, enforcement of pollution control legislation by regulatory bodies has been very slow and superficial (*e.g.* Control of Pollution Act 1974), and since the creation of the Water Authorities (England and Wales) and reorganization of the River Purification Boards (Scotland) in 1974 emphasis has focused on income generation from trade effluent charges, rather than controlling the actual environment impact.

Secrecy, often under the guise of protecting proprietary information *etc.*, has kept the public largely uninformed of the risks from effluent discharges; and penalties for non-compliance have been (and still are) trivial if they are to enforce legal compliance.

Traditionally, wastewater treatment has focused on the control of 'gross' effluent parameters such as suspended solids, BOD, and COD; seldom on specific pollutants with the exception of some metals.

With the relatively low level of priority given to wastewater treatment, the development and application of new processing technologies has been slow. Similarly, traditional monitoring and control systems have tended to be slow and usually involved laborious laboratory-based methods.[2] Consequently response to changes (even dramatic) in wastewater streams was very slow, with frequent consequent treatment plant overloading and failures.

2 IDENTIFYING THE RISK

2.1 The Environment at Risk

The ultimate target in controlling the quality of liquid effluents is the protection of the receiving environment and its other users. Central, therefore, to any effluent discharge authorization must be the understanding of the receiving environment, recognition of sensitive components resulting in the defining of acceptable Environmental Quality Objectives (EQOs), and setting of relevant Environmental Quality Standards (EQSs). Integral to the EQSs set in the UK are the maximum emission limits for specific 'dangerous' pollutants, as established under various EC Directives (EC Directive 76/464/EEC).

However, central to the UK concept of the EQO is the ability of the receiving environment to cope with the pollutant concerned, *e.g.* dilution and natural biodegradation potential, and the subsequent downstream use of the receiving water, as set by the controlling authority; *e.g.* river water may be used for a coarse fishery or a source of drinking water, the latter obviously demanding very high effluent standards.

In controlling standards of effluent quality most emphasis is placed on monitoring chemical components, particularly the families of known dangerous chemicals (below) and gross measures such as Biological Oxygen

Demand (BOD). Subsequent environmental (effects) monitoring can be very varied, often reflecting interests/experience available, *e.g.* benthic infaunal population studies.[3]

Increasingly, efforts have been given to the development of procedures for the assessment of effluent toxicity, mainly involving short-term test procedures, *e.g.* with algae and shellfish larvae.[4] These tests would be employed in addition to specific chemical procedures, particularly in assessing risk from complex effluents. Currently, several laboratories are involved in ecotoxicological test protocol development and appraisal (*e.g.* a major Paris Commission exercise on oilfield chemicals used in North Sea offshore oil operations).

2.2 Dangerous Substances

Over recent years the Paris Commission (PARCOM), which is concerned with marine pollution from land based sources, has established lists (List I and II) of 'dangerous' substances, based on:

Toxicity
Persistence
Bioaccumulation potential.

List I substances include:

Organohalogen compounds
Organophosphorus compounds
Organotin compounds
Substances which are carcinogenic
Mercury and its compounds
Cadmium and its compounds
Persistent mineral oils and hydrocarbons of petroleum origin
Persistent synthetic substances.

Following the 1987 North Sea Conference it was agreed that a reduction of 50% of 'dangerous substances' entering the North Sea should occur between 1985 and 1995 (eventually 100%) but no final agreement on a common list of such substances was achieved until 1990 when a list of 36 substances was established. In the meantime (1988) the UK had established its own list of 26 substances, the UK Red List,[5] and industrial processes discharging **significant** amounts of Red List substances would be expected by Her Majesty's Inspectorate of Pollution (HMIP), to operate the best available technology not entailing excessive cost (BATNEEC), in order to minimize the amounts of these substances discharged.

Such application of the BATNEEC approach is being linked to the setting of environmental quality standards (EQS) for effluent. This will involve assessment of effluent toxicity and risk of bioaccumulation.

The Paris Commission has initiated the formulation of international standards based on both List I and II substances and requirements for the monitoring of the marine environment. It is also promoting the development of procedures for the assessment of chemical (and effluent) toxicity and biodegradability.

3 SOURCE OF THE PROBLEM(S)

3.1 Processing Problems

The final solution to problems created by the discharge of chemicals in liquid effluents must be the major reduction of 'dangerous substances', without transferring any significant problem(s) to other pollutant inputs to the environment, *e.g.* atmospheric emissions (the concept of Integrated Pollution Control, integral to the Environmental Protection Act 1990 in the UK). Theoretically at least, the ultimate target would be zero emission, but economic constraints on final polishing of effluents are usually prohibitive.

In considering the effluent processing problem three broad, inter-dependent, stages should be addressed:

i) The generation of wastewater stream(s) from the primary manufacturing processes;

ii) The processing of individual or bulk wastewater streams;

iii) The disposal/use of the products of wastewater processing,

(a) recycling/other use
(b) effluent/waste disposal.

If the final level of effluent contamination is to be minimized three options must be considered (or more likely a combination of all three):

i) Stoppage at source, in the primary manufacturing process(es);

ii) Generation of a wastewater in as treatable a form as possible; followed by

iii) Use of the latest technology, *i.e.* BATNEEC must be continually 'primed' with new technology to reduce the economic constraints.

Typically, in chemical plants over 95% of the raw materials end up in the product or marketable by-products transported to the customer, although considerably lower figures apply for several natural product processing systems. Many of the losses (typically the 5%) enter wastewater streams, as losses via drains and piped process lines. With present processing technologies some of these losses are inevitable, but in many cases they (and certainly higher losses) result from:

i) Use of old technologies and processes, frequently also demanding excesses of both water and energy;

ii) Use of inadequate control, particularly incorporating feedback from wastewater monitoring;

ii) Attitude and skill weaknesses at management and operative levels, often with lack of communication at all levels.

Similarly, these three factors are frequently responsible for poor wastewater treatment capabilities.

3.2 The Need for a Full Process Audit

Before any practical wastewater treatment/effluent disposal option can be considered it is essential to undertake a full process audit to identify nature/levels of potential pollutant inputs to wastewaters and final effluent streams. Whether dealing with an existing system or a proposed new plant this, ideally, should include a full mass balance of the chemicals/ processes involved, incorporating assessments of different plant options. Section 5 gives a brief outline of the problems of attempting to assess the fate of the wide spectrum of chemicals used in the offshore processing of oil and gas.[6]

If involved in the design of a new manufacturing plant, data from the previous use of similar plant and experimental/theoretical studies on chemical separation and partition efficiencies should enable estimates of possible wastewater stream composition to be established, at least best to worse case scenarios. These, in turn, provide feedstock data with which to design/assess suitable wastewater options. Such audit data are also vital in considering appropriate monitoring and control systems.

In dealing with an existing manufacturing plant, but considering new wastewater treatment systems, a full practical audit is possible, incorporating identification of actual wastewater inputs and detailed characterization of all wastewater streams. It is important to obtain some measure of the variation in such wastewater characteristics, particularly the nature and frequency of major variations likely to upset subsequent processing.

Although it, perhaps, seems obvious that a thorough knowledge of feedstock characteristics should be essential for both the design and operation of wastewater treatment plant, this has been lacking to date in most industries. Some of the more detailed studies of complex wastewaters have been in the oil refining industry, which has recognized for many generations the need to treat separate process wastewater streams prior to general treatment, *e.g.* to remove sulphides and phenols.[7]

4 THE FUTURE - NEED FOR AN INTEGRATED SOLUTION

4.1 Total System Approach

If the final level of effluent contamination is to be greatly reduced, the problems of potential waste generation and treatment must be central, in fact critical to the design and operation of the complete manufacturing process (Figure 1).

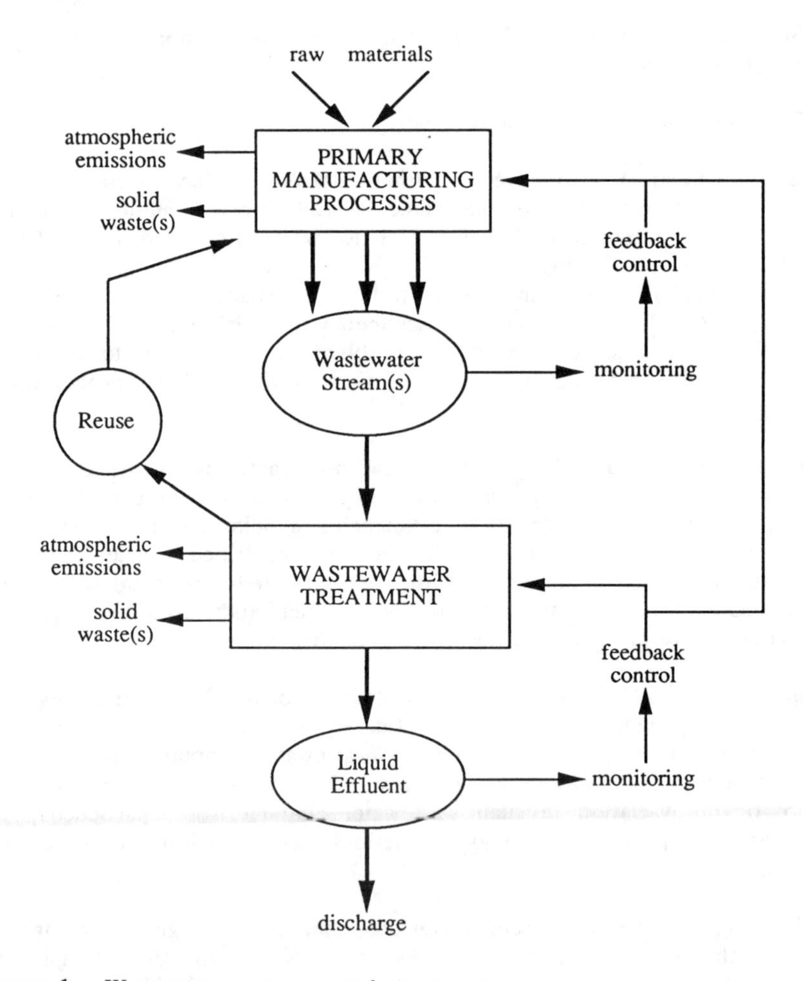

Figure 1 *Wastewater sources and treatment*

Wastewater treatment plant should be considered like any other part of the processing plant. Demands on effluent quality represent the design specifications for the 'product'. If this target specification is to be achieved it is then necessary:

i) To specify feedstock characteristics, *i.e.* **wastewater characteristics**;

ii) To assess available processing options.

To ensure reliable performance of installed wastewater processing plant it is essential to incorporate rapid, on-line monitoring of both wastewater influent(s) and final effluent, with feedback control to primary processing and wastewater processing, respectively.

Monitoring of flowrates, nature, and loading of individual wastewater streams will not only identify and help to control specific in-plant sources of pollutants but will assist in allocation of treatment costs internally in the company. This may also provide some incentive for 'departments' to reduce 'losses to drains' *etc*.

4.2 Trends in Wastewater Treatment and Monitoring Technology

Obviously this section could be a book in its own right;[8] therefore only some general points/examples will be made.

4.2.1 Processing Plant. Central to all effluent treatment is the separation of contaminating solids and dissolved chemicals from water, usually linked to the oxidation of organic matter, as in the traditional sewerage treatment system[8] (Figure 2).

In chemical processing plants wastestreams may be very complex, requiring more extensive/advanced treatment if contaminants, particularly dangerous chemicals, are to be removed/reduced (*e.g.* oils, heavy metals, and the spectrum of complex toxic, often persistent organic compounds such as organochlorine compounds). With larger chemical companies a wide range of advanced removal systems has been employed, involving such separation methods as chemical precipitation, adsorption, ion exchange, electrodialysis, and membrane filtration.[9]

As examples of trends in such processing technology, two key areas of plant development justify specific mention:

i) Use of hydrocyclones for reliable, high speed removal of oils from effluents;

ii) Application of membrane filtration systems for removal of a wide range of dangerous chemicals, with concomitant production of 'clean' water available for reuse.

Traditional systems for oil:water separation are extensively described in the literature, *e.g.* from refinery effluents[7] and offshore operations.[10] Primary and secondary treatment systems usually involve gravity separation ranging from simple tank storage/separation, through plate separators (API), to gas/air flotation cells, the latter often involving use of coagulants

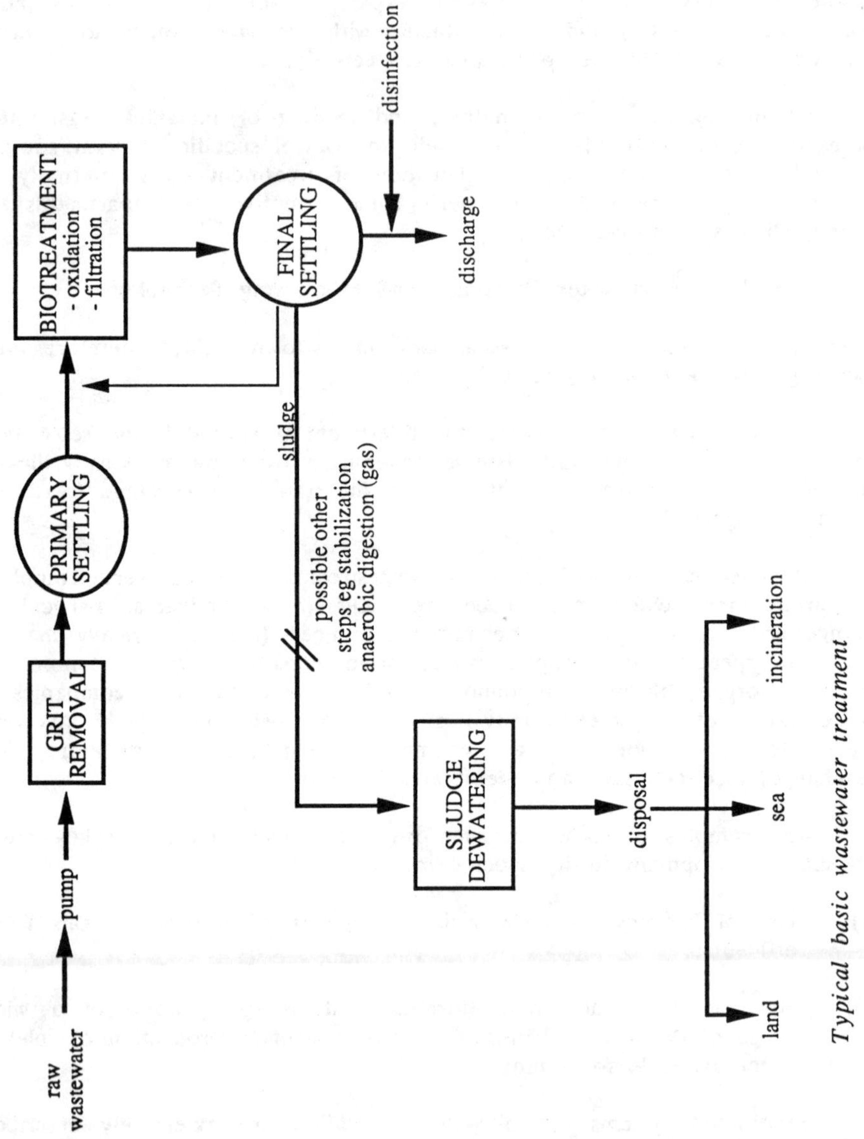

Figure 2 *Typical basic wastewater treatment*

and/or flocculants. Final 'polishing' is then by biological oxidation. The recent introduction of hydrocyclones has greatly speeded up initial separation processes giving reliable performance in plant considerably smaller and lighter than the traditional systems.[11]

However, perhaps the greatest potential lies in the use of membrane filtration techniques for the treatment of a wide range of contaminated wastewaters, which in addition make it possible to extract water for recycle and concentrate problem chemicals suitable for reuse or disposal.[12]

The method is not only applicable to dangerous chemical removal but has massive potential for treating large wastewater volumes such as laundry wastewater, liquid manure effluents, *etc.* For example, the use of membrane systems to treat laundry water can result in the recycle of more than 60% of the total water used and provide a 50% energy saving.[13]

4.2.2 Monitoring Systems. Until recently, monitoring of wastewaters/ effluents concentrated on general parameters of water quality such as suspended solids, BOD, and COD, primarily to ensure compliance with discharge controls.

There is now a demand for much more detailed and reliable monitoring systems:

i) To meet stricter compliance requirements for discharge to the sewer or environment;

ii) To provide plant performance data for control of operations.

Although the analytical bases to both requirements are similar, the latter requires rapid data generation in a form that can be linked automatically to feedback control systems.

The development of many new sensors and instrumentation systems offers considerable opportunities for detailed on-line pollutant characterization, backed up as necessary by detailed but slower laboratory analyses (although even many laboratory systems can now be automated and undertaken rapidly).[2] These techniques offer opportunities for considerable improvement in the control of treatment plant operation, including rapid response to changing influent wastewater chemistry.

As indicated earlier in the paper, compliance monitoring of liquid effluents is becoming increasingly demanding, certainly covering specified dangerous substances (*e.g.* the UK Red List).[5] Such chemical monitoring of effluent discharges is increasingly integrated within wider environmental effects monitoring, with particular opportunities for the development of ecotoxicological assessments, including direct measures of effluent toxicity.

5 A CASE STUDY RÉSUMÉ

Audit of potential chemical contamination of oily water effluents from offshore oil and gas operations

Rather than use a traditional, perhaps more widely understood, land-based industry to demonstrate the need for process audit, the offshore oil and gas industry has been selected.[6] It demonstrates a very complex audit requirement, directly relevant to current Paris Commission concern regarding pollutant inputs to the North Sea.

The offshore production of oil and gas in the North Sea involves a secondary recovery strategy employing water injection (sometimes gas) into the reservoir to maintain reservoir pressure to expel the hydrocarbons from the bearing rock strata. The water used is seawater taken adjacent to the production platform. This injection water requires treatment prior to injection downhole, including filtration to avoid reservoir plugging and chemical treatment to reduce problems from:

Corrosion
Bacterial activity
Scale formation (insoluble deposition in formation, pipework, and plant).

Subsequent production of oil and gas requires the separation (in pressure stages high, medium, and low) of three phases:

Gas
Oil
Water.

Particularly at the high-pressure separation stage there is risk of foaming and emulsion formation requiring chemical dosing. Separated gas requires treatment notably drying and 'sweetening' (mainly removal of hydrogen sulphide) before export. Similarly, oil prior to shipment by pipeline requires treatment primarily to protect the pipeline from corrosion but also to enhance flow (if a heavy or particularly a waxy crude oil is involved).

In the separation of exportable gas and oil from the produced fluids, remaining water will be contaminated with oil and possibly upstream chemicals. Traditionally this oily water is 'cleaned' to permitted levels with such systems as plate separators and/or induced gas flotation (IGF) systems,[10] but increasingly hydrocyclones are being employed.[11]

When water 'cuts' (*i.e.* the percentage water in the produced fluids) increase as a field ages, it is assumed that many of the chemicals used in both the injection water and produced fluids processing will enter the oily water treatment system.

Current attention of the industry is focusing on attempts to audit the full system (Figure 3) to derive estimates of such chemicals entering the oily water treatment system feedstock and their likely carry–over to discharged effluent.

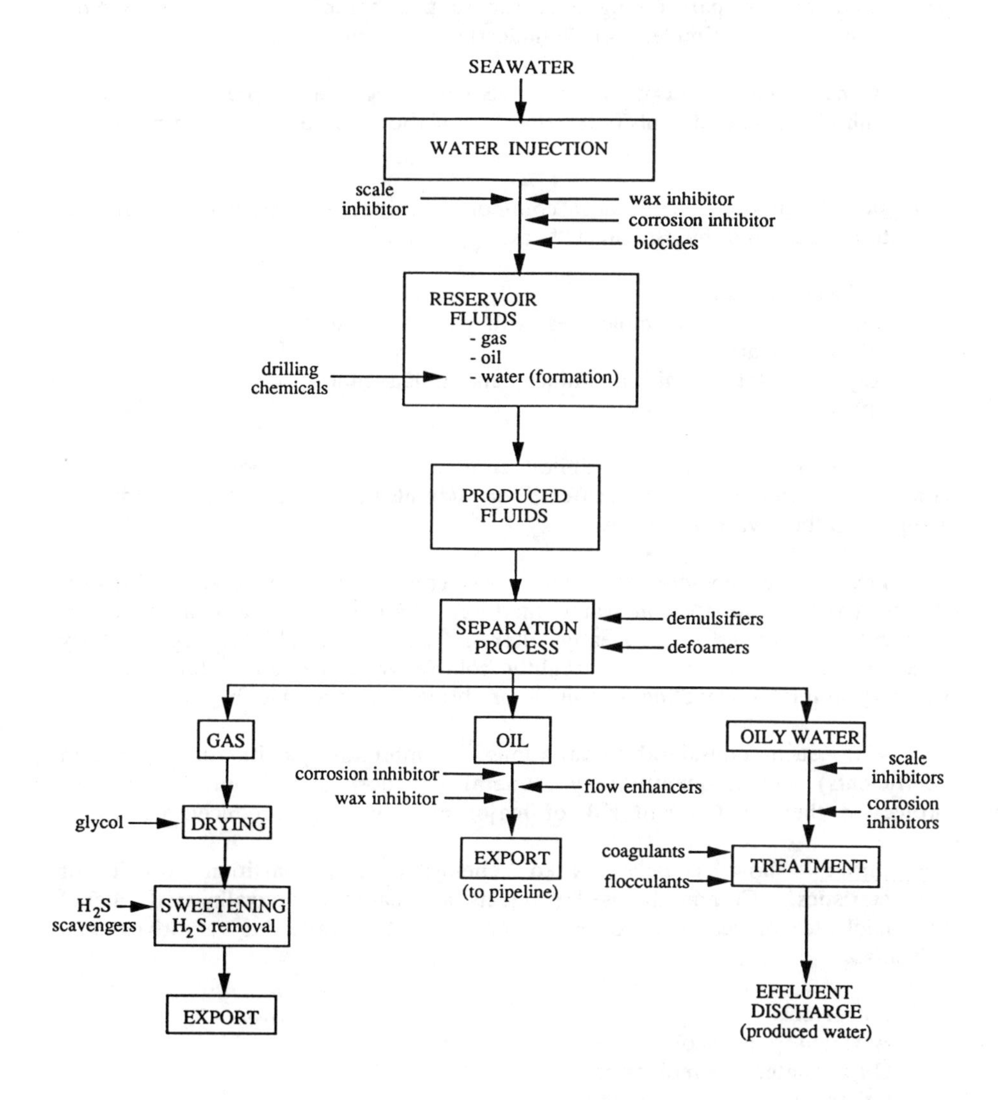

Figure 3 *Chemical usage in offshore oil:gas production*

With such a three-phase fluid system (particularly the oil–water) this presents major problems.

Central to the problem are:

i) The complex formulations (and frequent lack of detailed chemical description) of the proprietary products employed;

ii) The lack of partitioning data for such a complex multiphase system, even for the simpler better understood chemicals.

Considering just two of the product categories employed, corrosion inhibitors and demulsifiers, the formulation complexity is immediately obvious.

Corrosion Inhibitors. The most commonly employed types are nitrogenous in nature, classified by Bregman[15] as:

Amides/imidazolines
Salts of nitrogenous molecules with carboxylic acids
Nitrogen quaternaries
Polyoxyalkylated amines, amides, and imidazolines
Nitrogen heterocyclics.

These are the active inhibitor components in a product, often also containing antifoamers, emulsifiers, or demulsifiers and other modifying agents together with a solvent.

The active components can vary considerably in their solubility, thence partition in the oil–water mixture. For example a change in an imidazoline structure can greatly alter its relative solubility; *e.g.* a fatty acid imidazoline is only very slightly soluble in brine but soluble in oil, whereas simple imidazoline is soluble in brine but insoluble in oil.[15]

With such considerable variations in inhibitor partitioning (partition coefficients) and frequently very 'generalized' dosing procedures to ensure protection there is frequent risk of major carry–over to produced waters.

Demulsifiers show equally varied chemistry and partition coefficient characteristics. Commonly used products are based on oxyalkylated phenol formaldehyde resins blended with some of the following additives and solvents:

Polyglycols
Acylated polyglycols
Oxyalkylated alkanol amines
Other oxyalkylated phenols
Aryl sulphonates
Fatty acids
Fatty alcohols

Aromatic hydrocarbons, *e.g.* xylene or light petroleum 'cut'
Isopropanol/propanol
Methanol.

Both oxyalkylation and the solvents assist in the water solubilization of the active component(s).

In the UK sector of the North Sea current control of these oilfield chemicals is based on a voluntary notification scheme set up by the Department of Energy.[17] Classification is based on their known chemistry, toxicity, and potential persistence/biodegradation. However, the key test requirement (at present) is an assessment of lethal toxicity by a standardized 96 hour LC_{50} test, primarily using the brown shrimp, *Crangon crangon*.

Environmental risk from the use of such oilfield chemicals has been actively considered by the Paris Commission, and currently there are detailed toxicity test protocols being assessed by several North Sea states for possible implementation in new control requirements.

For industry, therefore, the proactive approach is not only to anticipate stricter controls with new test protocols but to undertake in-house audits (and produced water characterization) to attempt to assess the extent of any potential problem.

6 CONCLUSIONS

The considerable increase in stringency of effluent discharge controls will certainly provide stimulus for the development of:

i) New technologies for effluent processing and monitoring;

ii) Improved methods for measuring environmental quality, particularly at the bridge between effluent compliance monitoring and longer term ecological effects monitoring;

iii) Corporate environmental management strategies, with industries taking more proactive roles in environmental management;

iv) More direct public awareness and involvement, with the consequent requirement for access to data.

However, it will cost a lot! In some cases improved processing efficiency (including recycle of materials and energy economies) may recoup some cost but overall most increased costs will be passed through as increased product costs to the consumer.

To most companies economic considerations will play a key role in decisions as to whether they pay for discharge of wastewater to the sewer systems or develop in-house effluent treatment capabilities.[8] With many larger companies wider corporate environmental considerations may lead them to in-house treatment. Those with complex and difficult pollutants may have no choice. However, it is likely that smaller companies will continue to discharge to the sewer system, faced not only with increasing charges but tighter demands on their wastewater quality. if the pace of change is rapid many companies will fail to survive unless given direct assistance. A good understanding and non-confrontational relationship between companies and controlling authorities will be essential, not to avoid the stricter controls set on paper but to find ways of achieving them in practice.

7 REFERENCES

1 Institute of Offshore Engineering, 'Input of Contaminants to the North Sea from the United Kingdom', Report for the UK Department of the Environment, Heriot-Watt University, Edinburgh, 1985.

2 'Standard Methods for the Examination of Water and Wastewater', 16th Edn., American Public Health Association, American Waterworks Association and Water Petroleum Control Federation, 1985.

3 C.G. Moore, 'The Use of Community Structure in Pollution Monitoring', in 'Environment Toxicology', Interim Document 13, Health Aspects of Chemical Safety, World Health Organization, Copenhagen, 1983, pp. 283-305.

4 R. Butler and N.J. Grandy, 'Discharge Control and Monitoring by Biological Techniques', Water Research centre Report No. PRS 2436-M, 1990.

5 C.D. Byrne 'Selection of Substances Requiring Priority Action', in 'Risk Assessment of Chemicals in the Environment, ed. M.L. Richardson, The Royal Society of Chemistry, London, 1988, pp. 398-434.

6 C.S. Johnston, J.C. Side, and S.R.H. Davies, 'The Use of Environmental Audit in Offshore Operations', in 'Microbial Problems in the Offshore Oil Industry', ed. E.C. Hill, J.L. Sherman, and R.J. Watkinson, Institute of Petroleum, London, 1987, Wiley, pp. 147-164.

7 M.R. Beychok, 'Aqueous Wastes from Petroleum Plants', Wiley, London, 1967, 370pp.

8 M.J. Hammer, 'Water and Wastewater Technology', 2nd Edn., Wiley, New York, 1986, 536pp.

9 C.A. Wentz, 'Hazardous Waste Management', McGraw Hill, Intl. Edn., New York, 1989.

10 'Oily Water Discharges – Regulatory, Technical & Scientific Considerations', Applied Science, London 1980, 225pp.

11 T. Cornitius, 'Advances in Water Treating Solving Production Problems', *Offshore*, March 1988, p. 27.

12 P.H. Ferguson, 'Membrane Separation Techniques for Aqueous Effluent and Product Recovery', in 'Effluent Treatment and Disposal', EFCE Publ. Series No. 53, The Institution of Chemical Engineers, London.

13 M. Winge, 'Waste Water Minimization by Membrane Filtration', in 'Industrial Waste Water Treatment in the 1990s', IBC Conference, London, 1990.

14 Paying for Water Pollution Control, UK Department of Environment, Report, February 1989.

15 J.I. Bregman, 'Corrosion Inhibitors', Macmillan, New York, 1963.

16 J.A. Haselgrave, 'Partitioning of Corrosion Inhibitors in Water/hydrocarbon Systems', presented at the 2nd International Symposium on 'Chemicals in the Oil Industry', University of Manchester, 1985.

17 R.A.A. Blackman, 'The UK Notification Scheme for the Selection of Chemicals for Use Offshore', presented at the 2nd International Symposium on 'Chemicals in the Oil Industry', University of Manchester, 1985.

17
Managing the Risk from Solid Waste Disposal

G. A. Garland

WHO REGIONAL OFFICE FOR SOUTH EAST ASIA, WORLD HEALTH HOUSE, NEW DELHI, 1.10.002, INDIA

1 INTRODUCTION

Solid waste is receiving considerable attention in the United States,[1,2] the European Community,[3] and developing countries.[4,5] Solid waste may be described as any material thrown away which is not a wastewater discharge or an atmospheric emission. Using this description, unwanted pesticide thrown in the trash-can is a solid waste, while pesticide poured down the drain is not. Cyanide gas thrown away in cylinders is a solid waste while the same gas escaping to the atmosphere is not. Very complicated official definitions of solid waste emerge when regulatory officials attempt to include intent in the definition. Drums of unused chemicals resting in the back acres of an industrial plant may pose some risk yet be considered 'materials in storage' by the plant owner. 'Used' oil being spread on a dirt road to contain dust may pose a risk yet be considered a resource by the applicator. Solid waste, then, may be characterized more by how it originated than by its physical state. The definition of solid waste is important to public officials as they are generally held responsible for its management. This would include provision of solid waste disposal facilities. In the absence of laws or regulations to the contrary, this suggests that the community is responsible for anything which arises at the local dump. While this may conjure up a nightmare as anything of any degree of hazard may be thrown away and be transferred to a community land disposal site, it is an optimistic view. As Flinthoff[4] suggests, in developing countries substantial proportions of wastes accumulate in courtyards or vacant lots, are thrown into rivers or ditches, or are blown away by the wind. Fly tipping, dumping by the side of the road, in a farmer's field, or in a wetland are not uncommon in developed countries as well. While less respectable, these practices also amount to solid waste disposal for which community officials may be called to account. This chapter adopts the view that responsibility for managing the risk of solid waste disposal rests ultimately with public officials - national, local, or in-between. Moreover, the discussion must go beyond the technology of controlling risk from land disposal of solid waste - for example, the probability of how many liners of a certain thickness and composition releasing some amount of a material over some period of time - a problem to which artificial intelligence has been applied successfully. This chapter will discuss approaches which public officials may use for dealing with various issues in managing the risk from solid waste disposal, including safe land disposal; hazardous waste; hospital waste; integrated solid waste management; public education and participation; and pressing issues in developing countries.

2 SAFE LAND DISPOSAL

If any material whatsoever can be brought to a community land disposal site, potential for fires, explosion, or exposure to toxic or infectious materials is difficult to reduce. Moreover, any liner at the site would have to be universally compatible. To some extent, control of materials entering the site is impossible as inspecting large truckloads of waste is impractical. Later sections will deal with hazardous waste and hospital waste. In addition, knowledge of other sources of waste in the community and public education may help to deter the dumping of wastes which are incompatible with the design and operational plan of the site.[6]

The United States Environmental Protection Agency (US EPA) has proposed rules for safe disposal of solid waste which include the following items:[7,8]

i) Restrictions on locating landfills at, on, or near airports, floodplains, wetlands, fault areas, seismic impact zones, and unstable areas;

ii) Operating requirements, including procedures for excluding hazardous waste, daily cover, disease vector control, explosive gases, air criteria, access control, run-on and run-off control, surface water requirements, liquids management, and record keeping;

iii) Design criteria which establish a risk-based performance standard based on lifetime cancer risk and including liners, leachate collection systems, and final cover as necessary to meet the performance standard at the waste management unit boundary;

iv) Ground Water Monitoring, which is approved by the State, is installed at the unit boundary, yields representative samples of the uppermost aquifer, has well casings, and performs throughout the life of the monitoring programme;

v) Corrective action plans which include assessment of the need for corrective measures, remedy selection, and corrective action programme implementation;

vi) Closure that minimizes post-closure release of leachate and explosive gases, minimizes the need for further maintenance, and ensures protection of human health and environment;

vii) Post–closure care for a minimum of 30 years which maintains the final cover and containment system, leachate collection, ground water monitoring, and gas monitoring; and

viii) Financial assurance for closure, post–closure care, and corrective action for known releases.

New regulations for landfill in New York State require double composite liners to provide a minimum of six layers of protection between the trash and the underlying ground water with dual leachate collection systems and leak detection systems. Landfills in New York State will be required to conform to rigid siting restrictions to prevent them from being built where they may have an impact on sensitive environments such as principal and primary aquifers or regulated wetlands.[9] This may represent the ultimate in containment sites. An alternative concept, attenuate and disperse sites, allows slow release of leachate from the landfill and relies on various attenuation mechanisms to reduce the polluting characteristics of the leachate. This latter concept is hard to sell to the public,[10] although it recognizes that containment sites are fighting a battle with gravity which they may lose in the long run.

3 HAZARDOUS WASTE MANAGEMENT

To prevent hazardous waste from being dumped indiscriminately or disposed of with household wastes, an effective national regulatory programme must be implemented with the following elements: definitions of hazardous wastes; a registration or notification programme for hazardous waste generators, transporters, and facility operators; requiring hazardous waste generators to identify hazardous wastes, assure proper packaging and labelling for transport, and enable tracking wastes to their final destination; requirements for safe waste transport and a manifest or trip ticket system; permits or licences for hazardous waste facilities; and control of waste import and export.[11] While household hazardous waste is usually excluded from such national control programmes, various measures can be taken to lessen risk from that source.[12]

4 HOSPITAL WASTE

Although clinical wastes from hospitals are among the most hazardous and potentially dangerous in the community, clinical waste is not classified as a hazardous or special waste throughout Europe and North America.[13] Nevertheless, public officials may well want to consider measures for keeping it from community solid waste disposal facilities. Categories of clinical waste include pathological waste, infectious waste, sharps, pharmaceutical waste, chemical waste, aerosols and pressurized containers, and radioactive wastes. Sharps, including needles or syringes, are receiving great public attention because of possible AIDS infection.

Several references are available to assist community decision-makers in managing hospital waste.[14–17]

5 INTEGRATED SOLID WASTE MANAGEMENT

Integrated solid waste management takes advantage of the fact that the solid waste stream is made up of distinct components that can be managed and disposed of separately using a combination of techniques and programmes.[2] It has already been suggested that relatively high risk components – hazardous and hospital waste – might get special treatment. If one agrees that solid waste disposal facilities must be designed for the worst that can happen, then attention to managing even relatively low risk, inert materials becomes important in order to conserve space in relatively expensive, safe land disposal facilities. Moreover, if components of the waste stream which pose less risk to human health and environment can be diverted, management attention can be focused on safe disposal of those components requiring it. Finally, to the extent that components of the waste stream can supplant demand for raw materials or energy, then the sustainability of life on the planet is promoted.

Remaining components of the solid waste stream can be classified by source as domestic, commercial (restaurants, stores, offices), industrial non-process wastes, non-hazardous industrial process wastes, public services (parks, wastewater sludge, street cleaning), and demolition/construction. All of these could end up at the waste disposal site. Restaurant wastes could be fed to animals, relatively homogenous wastes from stores, offices, and industrial non-process wastes could be recycled, public service wastes might be composted, and demolition/construction waste might be used for cover materials. In North America and Europe, great attention is being paid to options for domestic solid waste.

In order to take full advantage of integrated solid waste management, authorities need comprehensive knowledge of the solid waste stream, possible management technologies, and possibilities for waste minimization/resource recovery. 'Urban Solid Waste Management' Annex, pp. 259–261, provides a useful check list of information and data collection requirements.

The US EPA suggests a hierarchy of integrated waste management with source reduction at the top. Source reduction measures include product reuse (reusable shopping bag, returnable beverage containers), reduced material volume (lighter containers, buying in bulk, using concentrates), reduced toxicity (substitution for lead and cadmium in inks and paints), increased product lifetime (tyres that last twice as long mean half as many thrown away), and decreased consumption (altered buying practices to reflect environmental consciousness).

Recycling of materials such as paper, glass, aluminium, ferrous metals, plastics, batteries, and even used oil and tyres is the next level in the hierarchy. Successful recycling must include separation and collection, reprocessing or manufacturing, and reuse – separation and collection are only the first steps. Participation in recycling programmes depends on convenience, so methods of collection must be carefully considered. For example, source collection of recyclables is likely to elicit greater participation than drop–off programmes which require householders to bring materials to a specified collection centre.

Composting of yard (lawn/garden waste such as leaves, grass clippings, weeds), and possibly food waste is another recycling possibility either in the backyard or at community compost operations.[18] Not only is yard waste bulky, it also forms organic acids on decay which enhance the mobility of other landfill constituents. Note that this suggestion for composting is for yard waste (and possibly food waste) only. Thus problems with lack of saleability because of pieces of glass or other contaminants are minimized as is the need to grind mixed solid waste. In developing countries where yard waste and food waste may comprise half of the solid waste stream while the other half is largely rock and dirt, the total mixture has proven difficult or impossible to compost. Successful recycling is far more likely if separation of the components of the waste stream can be maintained from the source. Helpful classification of recycling systems is given in 'Urban Solid Waste Management' (pp. 59 and 245).

Waste combustion and land filling share the lowest rung of the hierarchy. Modern combustion facilities not only reduce the quantity of waste for land disposal, they also recover energy. They should be designed in conjunction with source reduction, recycling, and composting programmes as these will affect the quantity and heating value of the combustor feed stream. Waste combustors require considerable capital investment and long–term planning is essential. Proper pollution control for emissions and ash management are necessary to protect human health and the environment. Several discussions of the factors included in municipal waste combustion are available.[2,3] Knowledge of materials which will raise difficulties at landfills may help set the agenda for waste reduction or recycling programmes. All levels of the hierarchy will have some role to play in a properly integrated approach to solid waste management.

6 PUBLIC EDUCATION/PUBLIC PARTICIPATION

Although public officials are the focus of attention in managing risk from solid waste disposal, especially when something goes wrong, everyone is responsible and has a role to play. In keeping with the view adopted in this chapter, public officials must take the initiative to inform citizens in their roles as consumers of goods and generators of waste, as well as commercial establishments and industry. Citizens might be made aware of environmentally friendly products, the need to avoid litter and indiscriminately throwing away their solid waste, and the role they need to play in separation and storage of solid wastes to enhance recycling and promote a healthy environment. Commercial establishments such as fast food outlets can reduce excess packaging and consider reusable items. Restaurants might connect with animal feeders. Offices, with relatively homogeneous waste, may make arrangements for recycling paper with minor efforts to maintain separation of various grades of paper. Industry can be encouraged to use secondary materials, make products that last longer and are less toxic, and take measures to keep their solid waste streams separate so they can be more readily recycled.

Many techniques of public education have been suggested.[19,20] Perhaps the most effective method for enrolling various parties in playing their part in managing risk from solid waste disposal is to involve them meaningfully in making decisions that affect them. Again, many suggestions have been made for achieving public participation, especially regarding facility siting.[19–23]

7 PRESSING ISSUES IN DEVELOPING COUNTRIES

As water–borne diseases are a leading cause of death in developing countries and human excrement plays a key role in facilitating these deaths, one can argue that human excrement is the most hazardous substance in developing countries. Lack of proper sanitation leads in some cases to human excrement being part of the solid waste stream. Animal wastes from streets may be a related issue. Removing human excrement from the solid waste stream should be a top priority in developing countries without proper sanitation in urban areas.

In some cities in developing countries, solid wastes are simply thrown in the street. They may be swept up daily (perhaps along with human and animal excrement) and put in communal bins for 'temporary' storage. After some time, the waste may be shovelled out and hauled to a dump. This system maximizes exposure to the health hazards of solid waste (see, for example, 'Urban Solid Waste Management', pp. 20–33). Incidence of parasitic infection, respiratory and skin diseases, and pulmonary TB among solid waste workers as well as any ragpickers or scavengers is evidence of the high–risk nature of this system.[23] Multiple handling of the solid waste is also maximized. Replacement of such systems should also receive urgent attention, along with health monitoring for workers and scavengers.

Residents of slums or squatter settlements typically receive little or no service. Organizing self–help through community participation has proven successful with a modest amount of encouragement and funding, possibly in co–operation with non–governmental organizations. Cointreau makes several suggestions for implementing improved service to the urban pool.[24]

Institutional arrangements in developing countries suffer from fragmentation and poor financing.[25,26] Many different actors may be responsible for collection, vehicle maintenance, treatment and disposal facilities, and so forth with the result that vehicle down–time is high, co–ordination is poor, and ineffectiveness is widespread. Charges for solid waste services are a fraction of the cost of operations and only a fraction of that is actually collected. As a result, subsidies from national or regional governments must be sought and timely decisions are largely impossible. Privatization may be a partial solution, especially where public sector salaries are very low and work habits are irregular. A strong public sector is, however, needed in any case in order to negotiate and monitor proper contracts with the private sector.

It may be quite useless to attempt to address these issues piecemeal in the absence of political commitment. Tangible evidence of political commitment includes sound policies and clear plans which have the endorsement of decision makers and the support of the public. Clear policies for municipal solid waste management should address:

Scope of Service. In particular, should hospital waste, hazardous waste, and human excrement be handled separately and not normally by the municipal solid waste management system?

Waste Minimization and Resource Recovery. What is the role of manufacturers, distributors, and the generators of solid waste in waste minimization and resource recovery?

Level of Service. How often will solid waste be picked up, from where, and from whom?

Generator Participation. What is expected of generators of solid waste regarding storage practices, separation requirements, and so forth?

User Charges. Will users be expected to pay for services, especially commercial establishments? Should some users subsidize others?

Institutional Arrangements. What institutional arrangements can be used to recognize that solid waste management is a public utility which requires sound management and adequate financing?

Enforcement. What provisions can be made to ensure appropriate co–operation from all participants?

Stakeholders. How will all stakeholders be given an opportunity to participate in decisions such as scope and level of service, user changes, institutional arrangements, and enforcement?

Co–ordination Mechanism. What co–operation mechanisms are needed with housing, water supply, drainage, road, land use, transport, and health services?

Sound plans for municipal solid waste management should:

Anticipate the amount and type of solid waste to be managed in light of population growth, urbanization and expanding municipal limits;

Project the funds, both for capital requirements and for operations, necessary for adequate solid waste management and identify sources of funds;

Identify necessary processing and disposal facilities;

Identify necessary storage/collection equipment and maintenance facilities and programs;

Describe the role of generators;

Identify measures to promote waste minimization and resource recovery;

Specify the requisite legal and institutional base;

Recognize the contribution of and accommodation for waste pickers;

Describe how services will be delivered for poor areas;

Identify programs for staff training, labour relations, health monitoring, and user grievances;

Identify demonstration projects to establish the feasibility of system improvements;

Recognize the need for staged development as resources permit.

8 CONCLUSIONS

Although the 'city dump' or 'town tip' has been the focus of past criticism and has been the cause of extensive and expensive environmental damage, managing the risk from solid waste disposal must look well beyond safe land disposal. Of course safety of land disposal is important and several measures have been suggested toward that end. As the public sector is ultimately held responsible for all forms of 'solid waste disposal', a broader view must be taken. The most risky part of solid waste must be identified and special measures for safe management of this hazardous waste must be taken. Measures for the more problematic hospital wastes must also be recommended. Local institution(s) must be properly equipped for managing the risk of solid waste disposal, including adequate financing, statutory powers, and role definition. Proper sanitation is an indispensible adjunct. Integrated solid waste management should be implemented with due consideration to privatization, the role of NGOs, service to the poor, public education, worker health and safety, and the informal sector.

9 REFERENCES

1 'The Solid Waste Dilemma: An Agenda for Action', EPA/530-SW-89-019, US EPA, Washington, DC, February 1989.

2 'Decision-Makers Guide to Solid Waste Management', EPA/530-SW-89-072, US EPA, Washington, DC, November 1989.

3 'Urban Solid Waste Management', World Health Organization, Regional Office for Europe, Copenhagen, and Instituto per i Rapporti Internazionali di Sanita, Firenze, 1st Edn., 1991-1993.

4 F. Flinthoff, 'Management of Solid Wastes in Developing Countries', WHO Regional Publications South-East Asia Series No. 1, World Health Organization Regional Office for South-East Asia, 2nd Edn., 1984.

5 S.J. Cointreau, 'Environmental Management of Urban Solid Waste in Developing Countries', Urban Development Technical Paper Number 5, The World Bank, Washington, DC, June 1982.

6 'The Safe Disposal of Hazardous Wastes, the Special Needs and Problems of Developing Countries', ed. R. Batstone, J.E. Smith, Jr., and D. Wilson, World Bank Technical Paper 93, The World Bank, Washington, DC, USA, 1989, **1**, 1.

7 Proposed Rule, *Federal Register*, Tuesday, August 30 1988, p. 33314.

8 'Decision-Makers Guide to Solid Waste Management', EPA/530-SW-89-019, US EPA, Washington, DC, February 1989, pp. 107-116.

9 N.H. Nosenchuck, 'Are Landfills and Incinerators Part of the Answer? Three Viewpoints', *EPA Journal*, March/April 1989, **15** (2), 26.

10 'Urban Solid Waste Management', World Health Organization, Regional Office for Europe, Copenhagen, and Instituto per i Rapporti Internazionali di Sanita, Firenze, 1st Edn., 1991-1993, p. 131.

11 W.S. Forester and J.H. Skinner, 'International Perspectives on Hazardous Waste Management', Academic Press, London, 1987, p. 3.

12 'Household Hazardous Waste: Bibliography of Useful References and List of State Experts', EPA/530-SW-88-014, US EPA, Washington, DC, March 1988.

13 'Urban Solid Waste Management', World Health Organization, Regional Office for Europe, Copenhagen, and Instituto per i Rapporti Internazionali di Sanita, Firenze, 1st Edn., 1991-1993, p. 180.

14 'Urban Solid Waste Management', World Health Organization, Regional Office for Europe, Copenhagen, and Instituto per i Rapporti Internazionali di Sanita, Firenze, 1st Edn., 1991–1993, pp. 180–210.

15 'Guidelines for Infectious Waste Management', EPA/530–SW–86–014, US EPA, Washington, DC, May 1986.

16 'Management of Waste from Hospitals: Report on a WHO Meeting', Bergen, 28 June–1 July 1983, WHO, Copenhagen.

17 M.G. Lee, 'The Environmental Risks Associated with the Use and Disposal of Pharmaceuticals in Hospitals', in 'Risk Assessment of Chemicals in the Environment', ed. M.L. Richardson, The Royal Society of Chemistry, London, 1988, pp. 491–504.

18 D.R. Brunner and D.J. Keller, 'Sanitary Landfill Design and Operation', US Government Printing Office, Washington, DC, 1972.

19 'Urban Solid Waste Management', World Health Organization, Regional Office for Europe, Copenhagen, and Instituto per i Rapporti Internazionali di Sanita, Firenze, 1st Edn., 1991–1993, pp. 239–241, 251.

20 'Decision–Makers Guide to Solid Waste Management', EPA/530–SW–89–019, US EPA, Washington, DC, February 1989, pp. 125–130.

21 'Sites for our Solid Waste: A Guidebook for Effective Public Involvement', EPA/530–SW–90–019, US EPA, Washington, DC, April 1990.

22 C. Furedy, 'Resource Paper: Emerging Concepts of Citizen Participation, Co–operation, and Education for Responsive Solid Waste Management in Asian Cities', presented at International Expert Group Seminar on Policy Responses Towards Improving Solid Waste Management in Asian Metropolises, Bandung, Indonesia, 4–8 February 1991.

23 A.D. Bhide, 'Regional Overview of Solid Waste Management', WHO Regional Office for South–East Asia, January 1990 (Draft, p. vi).

24 S.J. Cointreau, 'Financial Arrangements for Viable Solid Waste Management Systems in Developing Countries', presented at Third International Expert Group Seminar on Strategies for Developing Responsive Solid Waste Management in Asian Metropolises, Bandung, Indonesia, 4–8 February 1991.

25 A.D. Bhide, 'Regional Overview of Solid Waste Management', WHO Regional Office for South–East Asia, January 1990, pp. 2–12.

26 S.J. Cointreau, 'Financial Arrangements for Viable Solid Waste Management Systems in Developing Countries', presented at Third International Expert Group Seminar on Strategies for Developing Responsive Solid Waste Management in Asian Metropolises, Bandung, Indonesia, 4–8 February 1991, pp. 2–5.

18
Risk Management by Waste Minimization

B. Samimi-Eidenbenz

INTERNATIONAL CONSULTING CENTER FOR ENVIRONMENTAL TECHNOLOGY AND NUTRITION INDUSTRY (ICCT), PO BOX 6909, 8023 ZÜRICH, SWITZERLAND

1 INTRODUCTION

Recently, industrial societies have become aware of the serious consequences of their generation of hazardous wastes. Great efforts are currently in hand to create strategies and technologies for removing such hazardous wastes as otherwise they can effect irreversible changes on the quality of the environment and may pose a threat to human and animal health. Although administrative, legal, and technical procedures have been established for the disposal of hazardous wastes, other areas such as waste avoidance or minimization and reuse of wastes as raw materials or as energy sources need to be addressed. Public concern over the potential effects of wastes also needs to be allayed.

Principles behind strategies for waste management need to be consistent with the scientific, technical, and legal constraints affecting waste handling. Detailed tactics, including development of new processes, will be undertaken normally by the manufacturers creating the wastes.

The new ideology of developing clean techniques and cleaning up existing technologies has to be conducted within a framework set by decision makers. Thus, strategies developed by the United Nations (UN) and OECD/EEC/EFTA joint commissions with industrial support should create a situation where there is an internationally consistent approach to the management of industrial wastes.

The real aim in order to reduce environmental pollution should be to change industrial production technologies to minimize waste production. Waste minimization requirements involve consumers and their attitudes to wastes. Thus, the cycle 'production – consumption – waste' must be examined as an entirety when non- or low-waste concepts are being created, or low-waste technology(ies) are being developed.

This chapter aims to describe methods for waste minimization and reuse in the management of environmental and human health risks, including methods for decreasing risk by conversion of toxic waste into less toxic forms.

2 GENERAL CONSIDERATION OF THE NEED AND ECONOMICS OF WASTE MINIMIZATION

Various UN/OECD/EEC studies have proposed that introducing environmental protection measures results in a slight increase in the rate of growth of GNP.[1] The OECD study describes a direct economic relationship between minimization of waste and risk management. The document indicates:

i) Investment on environmental protection is recovered ultimately as alternatively a high cost must be paid to alleviate the damage so caused;

ii) The new markets for environmental protection equipment and advice affect significantly levels of employment. They stimulate a range of industrial (engineering) companies to provide facilities, equipment, and technical services related to environmental and industrial safety;

iii) New 'clean' technologies create new markets. Old highly polluting factories will be forced to close as they cannot generate the capital required for pollution control systems.

Paradoxically, therefore, investment in environmental and industrial safety techniques is consistent with economic growth. Policies for the management of hazardous wastes are an essential part of this process of protecting the environment. They need to mesh with the policies concerned with the continued use of hazardous substances, the conservation of natural resources, and the protection of both health and the environment. Such policies have to be developed at a national level. Agreed criteria for identifying and classifying hazardous waste are a prerequisite for an effective waste management programme, and these should be universally applicable.

Because wastes are heterogenically complex and are in many cases an unknown chemical mixture, their classification is not simple. Complex waste mixtures often have dangerous or hazardous properties, but are rarely acutely toxic. Individual component needs to be considered when preparing a realistic waste plan. Such plans are best developed by producers prior to disposal.

The primary information on the waste needs to indicate physical and chemical properties of individual components and adequate analytical chemical data to characterize the waste. Sufficient information on possible chemical reactions in the waste and likely environmental behaviour is also essential. Additionally, the chemical reactivity, mobility, solubility, and persistence of the potentially hazardous components must be considered in relation to the properties of the sites into which they are to be disposed.

2.1 Definitions of Waste

Waste can be defined as something which the producer no longer wants. (See also chapter by J.–M. Devos.) It appears that waste is matter, present at the wrong place and the wrong time.[1,2] In reality, waste is an economic factor. It has value, either as a source of raw material or as an energy source. The systematic utilization of waste by means of separation and reuse, *i.e.* recycling, may be worth US$ 110–135 tonne^{-1}.

Pure chemical wastes have a low calorific value (*ca.* 950–1100 kW tonne^{-1}. The residues after incineration often have physical, chemical, or biological characteristics which then require special handling and disposal procedures in order to ensure environmental safety.

Risks may arise directly from hazardous waste (as pollution sources) directly or indirectly by transfer to the environmental media and surrounding areas. Additional risks may arise when a wide variety of wastes are deposited in the environment and are in contact with each other or with the environment.

The International Environmental Bureau's (IEB) International Symposium on Waste Management and Risk Assessment Affairs proposed a general definition for hazardous wastes, *viz.* 'special wastes', toxic or hazardous wastes, which may damage the initial character of the natural environment or may affect the natural function or natural behaviour of environmental media, or may provoke risk to the health of the environment.[2]

The Symposium proposed that the term 'special waste' and its definition are universally accepted. Special wastes are hazardous wastes if they have the following properties:

i) The composition of various fractions includes compounds and substances which have toxic effects;

ii) The concentration of each substance or compound and their physico–chemical reactivity are such as to give rise to toxic materials;

iii) The potentially hazardous material interacts with surrounding environmental media to yield toxic materials.

2.2 The Cycling of Material Resources and Industrial Products

The inter–relationship between natural resources, production, consumption, waste production, and waste recycling with reference to waste management and waste minimization cycles is illustrated in Figure 1.[3]

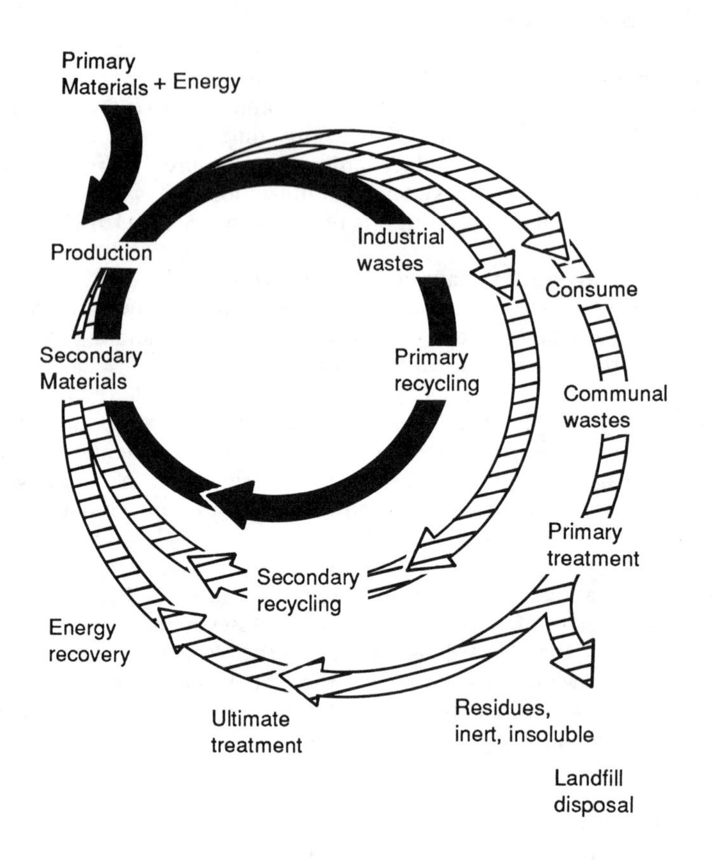

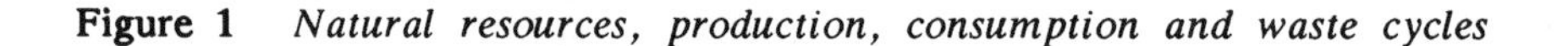
Figure 1 *Natural resources, production, consumption and waste cycles*

The trend in the world economy is to maximize the use of materials, at the right time, at the right place, and to use the correct starting material to avoid stress on natural resources. If the waste is treated as an economic factor it may become a good alternative to raw materials and energy sources.

The philosophy that recycling is an act that seems to promise a magical restoration of squandered resources is ecologically attractive. However, the principle of entropy implies that the degradation of sophisticated modern products into new materials will always need input of energy.[4] Because the recovery of some waste fractions requires more energy and plant investment costs than can be generated from its use, the costs associated with treating ill health arising from inappropriate handling of waste or environmental clean up will need to be taken account of in the justification for waste minimization by way of recycling and recovery.

It is therefore both the view of the Green Movement and in everyone's interest that the recycling, recovery, and minimization of wastes is and will be economically and ecologically attractive, and that investment in development of relevant techniques is appropriate.

The risk potential for environmental change or damage at a waste disposal site is clearly related to the concentration of the substances which may cause reactions with surrounding media. Materials that may be produced include intermediary products, such as volatile gaseous hydrocarbons at landfill disposal sites.

3 INTEGRATED HAZARDOUS WASTE MANAGEMENT POLICY

An integrated hazardous waste management policy considers waste from its point of generation through a variety of handling, including separation, reduction, collection, transport, and further treatment.

Priority should be given to minimization and recovery options:

i) Waste minimization policies can be developed after critical examination of waste generating processes, at least to minimize the quantity of hazardous waste, or to reduce the level of hazard of the waste for ultimate disposal;

ii) The potential for reuse as raw material for recycling or for recovery of secondary products from the waste;

iii) The availability of environmentally acceptable disposal of non-recyclable or non-available hazardous wastes.

Risk reduction strategies mean avoiding waste through reuse, recycling, and keeping different classes of waste separate. The waste may be used internally or externally, but ultimately some will have to be eliminated or disposed of safely.

Waste management must be considered as an integral part of production. It is possible to consider it in a new manufacturing process. When the introduction of new plant or a new technique is considered, those responsible for its management should, at the design stage, look critically at every wastestream which could arise. Therefore, an effective waste management concept must be based on the principles of avoidance, minimization, and utilization. Many modern suppliers consider this an integral part of their development programme and advertise it as a marketing policy.

The 'no-waste' system does not exist. However, processes such as recycling of solvents, acids, or oils leave minimum ultimate residues. These procedures are known as optimization of economically and environmentally safe actions. The ultimate result is that by application of clean technologies, the production of hazardous waste is reduced to a minimum quantity, but not 'zero', because except for slightly contaminated solvents all other waste components remain as residues in slags and sludges. In the case of liquid waste recycling a small loss of initial quantity is obtainable. Such recycling processes could be installed on site, or using external recycling and recovery plants.

3.1 Specific Aspects in Waste Minimization Management

In certain cases the use of internal recycling or recovery systems offers greater advantages than external recycling or ultimate disposal.

One of the most important problems in waste management policy is the avoidance of any emotional or political difficulties which may hinder efforts to achieve clean industrial production.

Ideally, the products must not only be produced with low or non-waste technologies, but must be clean themselves, *i.e.* produce little non-hazardous waste. For example, in the agro-industrial sector, biological pest control methods, still under development, could be particularly beneficial. Comparative studies illustrate that in major fields of agriculture, the choice of chemical or biological pest control agents depends on the availability, selectivity, and specificity of the agent and the question of quantity.[5,6]

By eliminating chemical wastes, the 'cleaner-production/clean-technology' concept is being adapted to a practical situation. One of the problems of waste minimization management is that of secondary waste from the use and pre- or post-treatment of primary wastes. Classic examples are the generation of dioxins during incineration under specific conditions, and the presence of high concentrations of chromium(III) in sludges and liquid wastes from the tanning industry. However, during recent years, several processes have been developed to receive the solid wastes from the manufacture of chemicals, pharmaceuticals, and cosmetic products.

Corning *et al.*[7] stated that there are two principal specific problem areas - sulphides and chromium, which arise in the treatment of liquid wastes in the tanning industry. Alkaline sulphides, which are themselves less problematic, can, under acid conditions, evolve hydrogen sulphide, which can corrode sewers and damage sewage treatment works.

The incineration of total waste is unacceptable, but the incineration of residues arising from pre- or post-treatment of utilizable or recyclable waste is a competitively effective method of disposal that yields minimal final residues (maximum 3.5-5.0% ash) and is attractive on account of its potential for heat and steam generation.

To reduce the costs, biological degradation processes and bio-gas recovery systems can be used, but the mixed tannery effluents may contain residues of synthetic tannery agents, disinfectants, lubricants, and dyestuffs, which may inhibit optimum biological activity. It has been reported that under aeration, using integral oxygen digestion systems and further neutralization, a primary oxidative conversion to innocuous products has been developed (Table 1).

Table 1 *Performance of tannery oxidation ditch*

Biochemical parameter	*Concentration in feed* mg l^{-1}	*Concentration in effluent* mg l^{-1}	*Removal* %
BOD_5	595	9	98.5
COD	1164	106	90.9
Total nitrogen	125	5	96.0
Chromium	24	0.7	97.1

4 USE OF WASTE FOR ENERGY PRODUCTION

During 1973–1979, with the sudden and dramatic price increase of energy, all petroleum products had an equally dramatic impact on raw material and end-product prices. In some cases, as raw material prices rose, it was possible to develop new strategies to obtain the maximum possible utilization of raw materials, involving the use of waste to generate energy.

The classic case was the drastic change in the economic resource cycles for organic chemicals. The price of organic solvents rose as much as tenfold, owing to the increase in oil prices. This led to recycling technologies related to the recovery of solvents and the recovery of energy or engine oil to become more attractive. The recycling of plastics and burning of spent solvents and oils as fuel supplements were good examples of energy saving.

This new situation resulted in a rapid decrease in the quantity of high calorific liquid, semi-solid, and solid wastes for disposal.

4.1 The Energy Consumption Unit as an Ecological-Economic Factor

In some industrial countries, the oil price increase also had serious repercussions in the waste disposal industry causing the redesign of some conventional incinerators as the need arose to develop new plant for integrating incineration of combined solid/liquid waste with recycling or reuse. It required a completely new approach to the future planning of environmental technologies and for the preparation of actual waste management programmes with special respect to energy consumption.

In general, waste containing hazardous substances should be identified, classified, and subjected to the same safety and handling precautions as other hazardous substances, unless evidence is sufficient to demonstrate that the risk involved is substantially less than the estimated risk potential.

It is understood that relevant management strategies require a special skill which can only be acquired partially by academic studies. As well as theoretical knowledge, practical experience in handling is required, in relevant topics such as sampling, analytical chemical techniques, and interpretation.

4.2 The Estimation of Risk Potential during Production

Several factors, such as the role of production capacity and the quantity of waste generated and the volume of residues, correlate with an increasing rate of the expected risk potential. The production capacity (Q_p) and the quantity of waste (Q_w) are two factors which are co-related and are variable with respect to increase or decrease in quantity. The rate of risk (R_r) may originate from a defined quantity of waste which itself is dependent on that quantity and hence on the production capacity. Therefore, $Q_p/Q_w = {}_rR_f$ (the risk factor rate). In the USA, for example, the quantities of risk which may be anticipated in relation to the annual production of four chemicals in million tonne quantities (250 day yr^{-1} = 100%) were reported for an annual rate of risk as: ethene 13.2%, propylene 6.4%, benzene 5.8%, and phenol 1.4%.

The real rate of risk for a complete operating programme including transport, handling, preparation, production, distribution, and disposal of residues, was estimated as an annual risk for one million tonnes of production yr^{-1} for 1980 as: transport 11%, storage 9.5%, production 31%, distribution 22%, disposal 27%, and treatment 3%, including the risk from residues as slag and sludge disposal.

4.3 The Estimation of Safety and Risk Factors

An example of the risk factor is provided by the estimation of the effective rate of risk, *e.g.* for titanium dioxide production. The influence of hazardous wastes on the environment can be calculated for wastes which arise from wastewater, polluted air, other emissions, *etc.*

The waste arising from the sulphate process in a titanium dioxide plant having an annual production capacity of 50,000 tonnes is:

Crystalline $FeSO_4.7H_2O$ 140,000, and in solution 83,000 tonnes yr^{-1}
Sulphuric acid (pure) 75,000, and 'weak' acid 40,000 tonnes yr^{-1}

The gaseous fractions from all processes are *ca.* 5000 m^3 hr^{-1} and the estimated annual gaseous emissions will be *ca.* 10,000,000 m^3.

- Energy for production 6000 kW h $tonne^{-1}$
- Energy for recycling 2000 kW h $tonne^{-1}$
- Energy for air purification 1200 kW h $tonne^{-1}$
- Energy for wastewaster treatment 1000 kW h $tonne^{-1}$

Total energy requirement for product 10,200 kW h $tonne^{-1}$, which corresponds to a cost of US$ 102 $tonne^{-1}$ of product.[8]

This leads to an economic–ecological relationship (E:E*) *i.e.* 1 tonne product = 7 tonnes of residues. The cost for minimization of risk based on the total energy requirement illustrates the essential rate of risk avoidance cost factor which is necessary to reduce the E:E* ratio from 1:7 to 1:–1. This will then equate to a 100% safety factor (SF). For an E:E* ratio of 1:–1, the ratio 1:–1 means an absolute minimization of risk by application of preventive measures, *i.e.* zero option. The economic:ecological relationship will normally tend to be 100%, when the normal safety condition (SF) equals 100% then E:E* = $1:1^{-1}$, in which case the economic:ecological risk ratio ($E:E_r$) will be zero.

The elimination of wastes generated by the production of titanium dioxide is possible by recycling technologies, but the disproportional ecological–economic relationships result in expensive pre–treatment and recycling of waste to ensure limitations in the risk factor. The cheapest way to reduce the risk potential is to avoid waste; to utilize a proportion (or all) of the intermediates and to recycle.

The energy requirement for pre–treatment and recycling processes is already high, but by the use of recycling, the E:E* relationship can be reduced from 7:1 to 1:1, especially by means of energy recovery.

The principle for the prime responsibility applied by industrial countries is the requirement that the cost of hazardous waste treatment and disposal is carried by the producer, *i.e.* the 'polluter pays' principle.

4.4 Hazardous Waste Management Criteria

The diversity of solid, semi–solid, and liquid wastes arising from the chemical industry requires a standardized systems analysis approach to identify specific methods for the collection, storage, pre–treatment on–site, or transport to off–site ultimate disposal. This is necessary because wastes arising from the chemical industry in many cases involves a wide range of scientific, technical, and industrial techniques.

4.5 Estimation of Risk Potential by Systems Analysis

By means of new systems analysis studies it is possible to classify which groups or classes of waste and residues may be used directly as chemicals or additives. In some cases a useful cost advantage can be gained in the manufacturing process itself or a suitable reuse as a raw material in other factories, *e.g.* a range of residues and by-products arising from titanium dioxide and phosphoric acid production are recyclable.

The reuse of ferric chloride, cuprous chloride, or ammonium and sodium persulphate to etch the copper surface of printed circuit boards is a classic example of the use of recycled residues.

Certain organic solvents can also be recycled. For example, in the engineering, metal, and engine industries, a large range of mechanical components and equipment are degreased with trichloroethene or 1,1,1-trichloroethane. In the plastic and rubber industry acetone, diethyl ether, trichloromethane, or benzene derivatives are used for cleaning but have an important function in other processing and extraction procedures. These processes will contaminate the solvents with inorganic residues, oils, and general dirt. However, the solvents can be recovered for reuse.

Nevertheless, the solvent residues have to be considered as hazardous fractions which are usually incinerated. In general, recovery of hydrocarbon solvents is the easiest, as chlorinated solvents, especially in the presence of copper (as catalyst) at a temperature of 380 °C, can lead to the formation of dioxins.[8]

It is important to perform an input-output analysis to obtain the appropriate disposal or treatment route. A complete physico-chemical analysis is required to determine the composition and properties of residues and to interpret the data to estimate a rate of risk potential for handling, treatment, and disposal. It should be remembered that classical mixing and dilution methods are no longer acceptable.

During the past decade, more than 80% of solvents have been recovered or recycled. Risk to the environment now only occurs from sludges and wastewaters which may contain heterogenous suspended solvents, chiefly from wastewater treatment plants which do not have solvent absorptive systems. An example is the emission pollution potential from a large dry cleaning factory using perchloroethene or 1,1,2-trichloro-1,2,2-trifluoro- ethane plus detergents and water. The solvent is collected on a continuous basis and then filtered and redistilled for reuse. The leads to a high solvent loss to the atmosphere and to wastewater. The sludge from the latter has to be collected for distillation. The input-output analysis showed that soluble fats, oils, filter-cake residues, including organic matter and dirt, accounted for 20-25%, although the solvent recovery could be as high as 60%. However, the atmosphere could be contaminated by as much as 15-20% by weight of the spent solvent and this could be considered as a risk factor to the surrounding environment. The improvement to such plants and the reduction of secondary discharges reduces the risk which is in any case proportional to the rate of waste minimization. The final residues for disposal are in the form of a dry, crumbly solid or oily liquid, which is only partially biodegradable in such media.

It should be remembered that the environmental health risk is, in such cases, concerned with pollution arising either during production or as condensates from solvents during plant cleaning. In the future, the principles of 'Best Practical Environmental Option' (BPEO) (see chapter by F. Feates) will demonstrate clear advantages.

4.6 Specific Management and Elimination of Hazardous Wastes

A specific management and elimination concept has been developed for a range of organic substances such as pharmaceuticals, fertilizers, pesticides, cosmetics, food and feed additives, and their waste residues. This is based on a bioenvironmental hazard profile which identifies and describes the most economic operations for management, minimization, and elimination of chemical compounds which are not economically and ecologically suitable for conventional disposal.

It is recognized that the biocidal effects/side-effects of chemical compounds and their residues on the environmental matrices is of great significance and that consideration has to be given to their minimization, the elimination of chemical waste, or at least reduction of the toxic hazard to the bioenvironmental matrix. This must be appraised by both the producer and the consumer, in view of the relationship between the producer and user benefit which must result in bilateral co-operation.

4.7 Bioaccumulation and Biosorption of Hydrocarbons

Under heteroautotrophic and hetero-organotrophic conditions a mixed population of aerobic micro-organisms and assimilative algae can yield a high rate activity for the oxidative degradation of hydrocarbons, biosorption of inorganic ions, and the accumulation of halogen and heavy metals. The symbiosis of blue-green algae (unicellular cyanobacteria) assimilates carbon dioxide and yields oxygen via photosynthesis at 30–35°C at 500–10,000 lux and is therefore ideal for use in biologically stressed wastewater treatment facilities. Hence, to avoid BOD-deficiency, to increase the purification ability, and to minimize the concentration of hydrocarbons in wastewater and sludges, it is recommended to integrate biological treatment systems with industrial air purification, wastewater, and sludge treatment plants. The average rate of minimization of the total dissolved organic substances, using biological treatment such as biosorption, transformation, conversion and utilization as carbon sources, and energy, has been calculated as 73.5–75% of the total dissolved organic carbon concentrates of the effluent.[8,9]

According to experimental data obtained from chemical wastewater treatment plants, an average rate of degradation of 89.5% of dissolved hydrocarbons under anaerobic conditions and a secondary rate of 62.5% of total hydrocarbon residues in an activated sludge by an aerobic system have been obtained. However, in a complex industrial wastewater with a heterogenous composition the rate of 17.0–27.5% for non-biological degradation was obtained, *i.e.* intermediary chemo-oxidative degradation.

The bioconcentration factors for micro-organisms and algae need to be considered as a critical step for the estimation of the overall bioaccumulation factor. The variable rate of biodegradation and bioconversion of organic wastes can be calculated, *viz.*

Bioaccumulation factor (BFn) =

$$\frac{\text{concentration of compounds in cells (cc) in } \mu g\ g^{-1}}{\text{concentration of compounds in media (CM) in } \mu g\ g^{-1}}$$

Bioabsorption rate (BA) in % =

$$\frac{\text{concentration of DOC5 x 100}}{\text{concentration of DOCt}_0} - 100$$

Biodegradation rate/day =

$$\frac{\text{concentration of DOCt}_0 - \text{DOCt}_5}{dt_5}$$

Where duration time dt_n = 5 days,
$DOCt_0$ = 1000 mg l^{-1}, $DOCt_5$ = 375 mg l^{-1}, dt = 5 days

This resulted in a rate of biodegradation of 125 mg l^{-1} day^{-1} or equivalent to 12.5% of total DOC reduction day^{-1} with a biodegradation ability of 62.5% within the 5 days of biological treatment.

4.8 Photochemical Degradation of Organic Substances

In addition to biodegradation photochemical oxidation of up to 98% has been reported.[10] The photolysis and photochemical oxidation of organic compounds in liquid media or in the gaseous phase, originating from chemical production or landfill disposal, has been described. The economically acceptable method is to degrade the hazardous waste to natural gases such as carbon dioxide, ozone and water.

The application of solar energy combined with photovoltaic systems and ozone/UV generators has been shown to be effective for the minimization of secondary wastes such as methane, carbon monoxide, nitrogen oxides (NO_x), and hydrogen sulphide to reduce the risk potential to both man and the environment.

In combination with a hot gas purifier,[8] a range of aromatic, aliphatic, and complex polychlorinated gases/liquid wastes can be degraded to mineral acids and carbon dioxide.

The conversion of volatile landfill gases *e.g.* nitrogen, carbon dioxide, methane, sulphur dioxide, ammonia, and hydrogen sulphide, by biosorption is not economic, but the application of photolysis/photochemical:UV/O_3

oxidation processes can yield several end products such as ammonium salts, carbon dioxide, water, calcium nitrate, hydrogen chloride, hydrogen fluoride, *etc*. Any hydrogen sulphide will be oxidized to sulphuric acid (calcium sulphite).

5 CONCLUSIONS

5.1 Risk Management by Waste Minimization Processes

The quantitative evaluation of risk of hazardous substances and the identification of their effects on different environmental matrices require interdisciplinary knowledge of the composition, properties, and behaviour of individual substances.

The risk to humans and to the biosphere may be classified by localized effects of production, storage, transportation, or utilization locations by:

i) Directly by emissions, contact, and entry through respiratory, dermal, or oral routes;

ii) Indirectly by absorptive contamination of environmental matrices, and secondary effects to biotic systems;

iii) By absorption, distribution, and conversion *in situ* in environmental media and tertiary transfer to biotic and ecosystems.

Detailed information and interdisciplinary knowledge including molecular interactions, biochemical interactions within cells of the absorbed substances, distribution, metabolism, conversion, and storage parameters, in addition to biochemical and cytotoxicological pathways and elimination, is necessary to develop a practicable risk management concept and in turn to accomplish the establishment of a waste management and minimization strategy plan.

The identification of organic substances is now a much easier task in light of the availability of new techniques and not least the increasing availability of data.

A model for the identification, selection, assessment, and classification of chemical substances has been originated by the United Nations Environment Programme/International Register for Potentially Toxic Chemicals (UNEP/IRPTC) which embodies a modern, informative, and interdisciplinary data profile system incorporating a broad spectrum network within UN member countries.[1] The file structure of the IRPTC data bank demonstrates in 17 categories a basis for the development of a sound waste management plan.

The International Programme for Chemical Safety (IPCS)[11] and the IRPTC co-operative venture on technical exchanges will provide for each type of hazardous waste fraction an estimation of safety and risk potential by production, transportation, storage, and elimination or recycling processes. In addition, there is a triangular relation between production, consumer, and disposal practices with respect to their interaction with the air/land/water environmental matrices - all aspects of the biosphere must be considered.

5.2 The Risk Potential of Secondary Wastestreams

Secondary wastes include a range of hazardous and toxic substances and elementary ions which are included in wastestreams or are formed during pre- or post-treatment of waste fractions. These can arise from both domestic and industrial waste.[5]

A practical risk management programme will need to take into account the risk assessment and mathematical estimation of risk potential to the environment including societal and industrial aspects and to include the following risk factors:

i) Waste sources including production and consumption statistics, *i.e.* collection, transport, storage, preparation, and disposal;

ii) Waste composition, characteristics, physico-chemical properties, quantity and quality of material prior to and during treatment;

iii) Waste avoidance, minimization, improvement in the quantitative and qualitative reduction of waste production entailing a relevant waste and risk management concept based on the application of clean technologies and cleaner production.

The execution of the foregoing will depend on the financial capacity of the waste producer, coupled with new regulatory requirements.

5.3 Systematic Waste Minimization and Elimination Processes

In some European countries a large proportion of recyclable materials, *e.g.* glass, paper, and aluminium cans, are sorted and separated from domestic waste. However, there are also large quantities of other recyclable materials, *e.g.* plastic, wood, paint, cardboard, batteries, *etc.* (see chapter by G. Vonkeman).

Waste system analysis on the collection and recycling of recyclable waste can show a saving of 1200 kWh $tonne^{-1}$ of waste plus a 50% saving in landfill capacity.

More than 2000 million tonnes of waste per annum are produced within the European Communities of which 30-59 million tonnes, *i.e.* 1.5-2.5%, is ajudged to be hazardous.

Waste elimination and recycling technologies offer a rate of reduction of up to 93.7% with a residual amount of 6.3%. For instance, the plasma ultra–high temperature reactor (Plasmox) offers new perspectives for destruction of highly toxic wastes, including chemical and biological weapons. The process involves technology for safe elimination of highly toxic and biocidal substances under environmentally sound and sealed operation systems. The rate of elimination of influent material can be 93.7–96.3%. It produces a very leach–resistant slag residue (to a maximum of 6.7%). This may require the construction of a subterranean hermetically sealed device. The Plasmox process corresponds with 'the state–of–the–art' technology, equipped with integral process air, wastewater, and slag treatment systems combined with energy recovery units and an integral on–line analysis and data–monitoring system which meets the requirements of all regulators.

Hence, in considering the reduction in transport and disposal costs additional economic advantages in cost–benefit equations can be expected. This can be accomplished by the manner in which technical waste minimization and the risk potential can be eliminated from discharges to every environmental medium.

A systematic adaptable waste classification, minimization, and management concept embodying specified description of integral waste recycling and elimination technologies is being developed to assist industrial efforts in waste management which depends on approaches to recycling and environmentally clean techniques. These processes are introduced as economical/ecological friendly environmental technologies for the requirement of industry and for effective control of discharges to the environmental matrices and therefore reduction of pollution sources.[1,2]

6 REFERENCES

1 UNEP/WHO, Health Aspects of Chemical Safety, Interim Report on Hazardous Waste Management, UNEP/WHO, Geneva, 1982.

2 J.W. Huismans, 'What is Waste? Definition and Classification', International Symposium on Special Waste, 1990, **1**, 67, International Environment Bureau, Geneva.

3 'Waste Economy and Waste Management Promotion Programme', 1990–95, Federal Ministry of Research and Technology (BMFT), FRG, Bd.1/1990.

4 R.A.J. Arthur, *Industrial Waste Management*, 1990, **1** (7), 17.

5 'Research into Environmental Pollution', WHO Technical Report No.406, Report of Five WHO Scientific Groups, 1968, pp. 61, (available from ICCI, Zürich).

6 'Chemistry, Agriculture and the Environment', ed. M.L. Richardson, The Royal Society of Chemistry, Cambridge, 1991.

7 D.R. Corning and R.L. Sykes, 'Waste Management in the Tannery Industry', Case Study, Industrial Waste Management, 1990, **1**, 11.

8 ICCI Report on Transnational Technology Transfer Programme for Environmental Technology, Interim Report to the Commission of the European Community, DGXI, 1989/1990.

9 D. Freitag, L. Ballhorn, H. Geyer, and F. Korte, *Chemosphere*, 1985, **14**, (10), 1589.

10 C. Darpin, ABB, 'Deodorization and Purification of Waste Air by Ozone', 'Chemicals, Technologies and Hazardous Waste, Chemical Trends in Environmental Protection Technologies', The International Society for Environmental Protection (ISEP), *Envirotech*, 1989, **2**, 79.

11 M. Mercier, 'Risk Assessment of Chemicals: A Global Approach', in 'Risk Assessment of Chemicals in the Environment', ed. M.L. Richardson, The Royal Society of Chemistry, London, 1988, pp. 73–91.

12 B. Samimi-Eidenbenz, 'The New Technology at the Service of the Chemical Industry', The International Society for Environmental Protection (ISEP), *Envirotech*, 1989, **1**, 7.

19

HMIP's Role in Integrated Pollution Regulation

Frank S. Feates

DIRECTOR, HER MAJESTY'S INSPECTORATE OF POLLUTION, 1989–1991,
ROMNEY HOUSE, 43 MARSHAM STREET, LONDON SW1P 3PY, UK

1 INTRODUCTION

At the time this chapter was prepared (April 1991) the UK Government was in the final stages of creating new institutional arrangements for the protection of the environment from pollution caused by operating industrial processes, in particular those with the potential to cause most harm. The Environmental Protection Act 1990[1] introduces, in Part 1, a new system of integrated pollution control (IPC). In Part 2 it increases controls over the disposals of waste on land, and in Part 5 it modifies the regulation of the handling and disposal of radioactive substances under the Radioactive Substances Act 1960. These provisions together with certain responsibilities under the Water Act 1989 provide the main legislative basis for Her Majesty's Inspectorate of Pollution's regulatory responsibilities (see Annex 1). To meet these responsibilities HMIP has developed a new regime called integrated pollution regulation (IPR) which began operation in April 1991. Preparations for IPR began in April 1987 when HMIP was formed by combining Her Majesty's Industrial Air Pollution Inspectorate, Her Majesty's Radiochemical Inspectorate, the Hazardous Waste Inspectorate, and a new Water Pollution Inspectorate. In October 1989 HMIP's field inspectorate was reorganized into three regional divisions, based in Bristol (West Division), Leeds (North Division), and London (East Division) subsequently located in Bedford. From April 1991 the provisions of Part 1 of the Environmental Protection Act began to be phased in. IPR is not just about new legislation; it is about a radical shift in the Inspectorate's thinking and approach.

There have been significant developments in 1991, including the start of implementation of IPC and the announcement of the plans to create an Environmental Protection Agency bringing together HMIP and the pollution regulation function of the National Rivers Authority (NRA). This chapter does not address these more recent developments.

2 INTEGRATED POLLUTION REGULATION – HMIP'S RAISON D'ETRE

Integrated Pollution Regulation can be summarized as follows:

i) Carrying out all its work within the framework of what HMIP has defined as **Integrated Pollution Regulation**;

ii) **A preventive approach**, which is designed to avoid problems arising, through rigorous assessment of plant design and waste management proposals, authorization and monitoring, backed by enforcement action and prosecution where necessary (see Section 2.1);

iii) A positive but **structured** relationship with operators;

iv) Provision of information and guidance to industry, through publication of **Chief Inspector's Guidance to Inspectors – 'IPR Notes'**; and

v) A systematic **targeting of resources** to the highest priorities.

From April 1991 the statutory framework of control for non-radioactive processes was reflected in a cross-media philosophy. IPC and HMIP's other main functions – regulation of premises under the Radioactive Substances Act 1960, audits of the new Waste Regulatory Authorities, and regulation of discharges of Red List substances to sewers – will be dealt with in an integrated manner, through **Integrated Pollution Regulation**.

2.1 The Preventive Approach

From an environmental standpoint, prevention is normally better than cure. For operators, it is more cost effective and less disruptive to design effective controls and operating procedures into a plant, rather than face later remedial action and reactive adaptation. It is also better to minimize the creation of waste at source and to encourage recycling wherever practicable and to dispose of remaining wastes in the most environmentally acceptable way. This thinking lies behind HMIP's preventive approach.

The three elements in the preventive approach are:

i) **Guidance** on process design and operation, in particular, through IPR Notes;

ii) **Avoidance** of pollution risk, by rigorous scrutiny and process design and operating arrangements, and **reduction of waste creation at source**, through the authorization process; and

iii) **Deterrence**, by using authorizations to set up monitoring regimes which will bring lapses in performance quickly and reliably to the attention of the operator, HMIP, and the public; and by effective monitoring, inspection, and enforcement regimes.

2.2 Relationship with Industry

The traditional approach to pollution regulation in Britain has been one of informal working together between operators and enforcing authorities. This is a legitimate approach, and one which has been productive and effective in achieving high standards. However, IPR marks a shift to a more structured approach to regulation. In line with this, HMIP's relationship with the individual operators whom it is charged with regulating will become more formal, and the provision of company and plant specific technical advice will become less necessary.

This does not mean that there will not be dialogue with industry. HMIP will continue to provide general technical guidance through publication of the Chief Inspector's IPR Notes and other formal guidance to inspectors. More than 180 such notes are planned for preparation over the next five years. HMIP also maintains close links with trade associations and bodies representing relevant sectors of industry such as the Confederation of British Industry and the Chemical Industries Association and in addition The Royal Society of Chemistry.

2.3 Technical Guidance Material – IPR Notes

Ministers have undertaken that the introduction of IPC will be accompanied by technical guidance on the operation of each process under IPC. Formally, these will have the status of guidance to inspectors, but will be published for the benefit of applicants.

Draft guidance on industry sectors has already been prepared for the following categories:

- fuel and power industry;[2]
- metal industry;[3]
- mineral industry;[4]
- chemical industry;[5]
- waste disposal industry.[6]

These will be followed over the next five years by full IPR notes covering all processes prescribed for IPC, beginning with the publication of an IPR note on large combustion processes.[7] Annex 1 gives the timetable for the preparation of Chief Inspector's Guidance to Inspectors for industries and processes.

2.4 Cost Recovery Charges

Reflecting the 'polluter pays' principle, the Environmental Protection Act contains provision for HMIP to charge fees for the determination of IPC applications and for the holding of an IPC authorization (subsistence charge), to meet its ongoing costs of oversight and enforcement. A fee will also be charged to cover the costs of considering an application for a substantial variation of an authorization. Processes which involve releases to controlled waters (estimated to be about one in six of IPC processes) are subject to monitoring by the National Rivers Authority. To cover the cost of this, the annual subsistence charge for such processes will additionally include the amount payable under the NRA's charging scheme.[8]

These charges do not represent a 'pollution tax': they are by statute limited to the recovery of HMIP's regulatory costs, although the recent White Paper on the Environment[9] states that the Government will be considering a form of incentive charging very soon.

To ensure that the level of charges on individual installations will fairly reflect the amount of regulatory effort involved in dealing with them, charges are linked to the number of specified 'components' which the installation contains. There is a lower application fee for existing installations, being converted from an existing statutory approval, to IPC. For example, a combustion process which utilizes three boilers of more than 20 megawatts, three gas turbines of more than 20 megawatts, and a waste gas treatment plant would contain 3+3+1=7 components for charging purposes.

Charges commenced from April 1991 – but are payable only when an application for authorization is required. HMIP has consulted industry on the details of the charging proposals, and will take comments into account in framing the final scheme.

The Act also introduces a charging scheme for premises regulated under the Radioactive Substances Act 1960 from April 1991. The scheme will apply to all premises holding existing registrations and authorizations.

3 PUBLIC REGISTER OF POLLUTION INFORMATION

The Environment Protection Act[1] requires that all information directly related to applications, and the issuing of new authorizations and compliance requirements be placed on a public register. This will be a major innovation and will require:

i) Applications for IPC authorizations to be advertised;

ii) Third parties to make representation to HMIP regarding these applications, and any changes to be made public;

iii) Authorizations and monitoring data;

iv) The information to be held on public registers, copies of which will be held by the local District Council and HMIP; and

v) Authorizations including discharges to water to be copied to the National Rivers Authority's registers so that they hold a complete record.

Part 1 of the Environmental Protection Act provides for the introduction of integrated pollution control, which covers processes responsible for the majority of releases to air, land, and water with the greatest pollution potential. These are known as 'Part A' processes because of the wording in the Environmental Protection (Prescribed Processes and Substances Regulations SI 1991 No. 471).[8] Prior authorization will be needed by those wishing to operate prescribed processes. The Secretary of State for the Environment (in England) specifies the processes to be prescribed for IPC: HMIP's responsibility is to operate the regulatory system.[10] At present it is envisaged that about 105 types of process will be prescribed, involving at least 5000 installations in England and Wales. IPC processes fall into three main categories:

i) Processes currently regulated for air emissions under the Health and Safety at Work *etc.* Act 1974;

ii) Processes giving rise to significant quantities of special waste;

iii) Processes giving rise to emissions to sewers or controlled waters of Red List substances.[11]

The Environmental Protection Act (Section 7) requires HMIP to have in regard the following proposals in setting IPC authorization conditions:

i) A general duty on the operator to use of **'best available techniques not entailing excessive cost' (BATNEEC)** to prevent or minimize releases of prescribed substances, and to render harmless substances which are released;

ii) Application of the **'best practicable environmental option' (BPEO)** to minimize pollution to the environment as a whole; and

iii) **Compliance with limits, plans, quality standards, and objectives** set by the Secretary of State for the Environment.

4 IPC AUTHORIZATIONS

An IPC authorization is required before a prescribed process may be operated.[12] The applicant has a right of appeal to the Secretary of State against refusal of an authorization, or against conditions attached to it. HMIP is required by the legislation to determine an application within a prescribed period of time, normally 4 months, or such longer period as may be agreed with the applicant. Otherwise, the applicant can appeal to the Secretary of State as if the application has been refused. HMIP must consult the Ministry of Agriculture, Fisheries and Food, sewerage undertakers (for discharges to sewers), the Health and Safety Executive, and the NRA on applications for authorizations involving releases to controlled waters, and must incorporate conditions as to the release not less stringent than those which the NRA may require. General guidance on IPC has been published separately.[13]

4.1 Implementation Procedures and Timetable

It is not feasible to issue IPC authorizations immediately for all existing installations and the Act allows for staged implementation. Implementation will be by class of process. Every firm operating a particular class of process will require an IPC authorization - and will start to pay IPC charges at the same time as operators of similar processes elsewhere in England and Wales.

However, all new installations and existing installations undergoing major change have required an IPC authorization from April 1991. Moreover, the Act, reflecting in part an EC Directive requirement, provides that IPC authorizations will be subject to review at least every four years. In practical terms this makes it desirable to implement IPC within four years from April 1992 so that HMIP will have completed all initial authorizations before having to start on the first round of reviews.

HMIP will aim to implement IPC on the following timetable:

i) The Environmental Protection (Prescribed Processes and Substances) Regulations SI 1991 No. 472[10] prescribes the IPC processes, and requires all new installations and substantially changed existing installations in Schedule A (see Annex 1) to be subject to authorization, as from 1 April 1991, unless approved under an existing statutory system by that date;

ii) From 1 April 1991, all large combustion plants (those over 50 MW) are subject to IPC. Operators are required within one month to submit an application for IPC authorization;

iii) IPC will be progressively extended to other existing installations from April 1992 with the target of completion by April 1996 (see Annex 1).

4.2 Local Authority Air Pollution Control

Part I of the Act also establishes a parallel local authority air pollution control regime. The controls apply to a range of what can best be termed 'medium-polluting' processes - or 'Part B' processes as they are commonly known as, because of the heading in the Regulations.[10] The new functions will be exercised by district or borough councils and by port health authorities.

Part B processes have a significant potential for air pollution, but are not so complex as to require the cross-media approach of IPC. It is estimated that the local authority controls will apply to some 12,000 such plants, together with between 10,000 and 15,000 small waste oil burners in garages and workshops.

The implementation of the local authority controls will be in three stages. The processes have been divided into three blocks. For those processes in the first block, new processes and substantial changes to existing processes were not to be carried on without an authorization after 1 April 1991. Operators of existing processes were required to apply for authorization between 1 April 1991 and 30 September 1991. For processes in block 2, the dates were 1 October 1991 for new processes and substantial changes, and 1 October 1991 to 31 March 1992 for existing processes. The three blocks are set out in Schedule 3 of the Regulations.[10]

It is intended that Secretary of State guidance on BATNEEC for each category of process will also be issued in three blocks. The first block of 25 notes was published by HMSO in February 1991. The other two blocks of notes were issued on 1 October 1991 or 1 April 1992, as appropriate.

Local authorities will be required to levy an application fee and an annual subsistence charge, as for IPC. The charging scheme is, however, different: notably, the Part B processes are not subdivided into components and the fees and charges are lower. The application fee will in most cases be £800 per process, and the annual charge will be £500.

General guidance notes on the working of the local authority air pollution control system were published by HMSO at the end of April 1991.

5.1 Waste Regulation Functions

HMIP's waste regulation functions are placed on a statutory basis by the Environmental Protection Act, which places a statutory duty on the Secretary of State to keep under review the regulation authorities' discharge of their functions. Waste Regulation Authorities (WRA) will be created as separate entities from the Waste Disposal Authorities under the provisions of the Act. There will be 63 Waste Regulation Authorities in England (at County and Metropolitan District Level) and 37 in Wales (at District level). HMIP will assist the Secretary of State to discharge this duty by reporting to him on their periodic audits of each authority and continuing general oversight of regulation standards in the country. HMIP will also comment as necessary on WRA Annual Reports.

HMIP's new role will come into effect three months after the relevant provisions for separating waste regulation authorities from waste disposal authorities are enacted. This occurred in the first quarter of 1991 and the authorities first annual reports covered in April 1991 to March 1992, to be published in Autumn 1992.

5.2 HMIP's Audit and Oversight of WRAs

The prime responsibility for standards at individual waste management facilities will rest with the Waste Regulation Authorities. HMIP will audit the authorities' capability, methods, systems, and performance. Inspection of a sample of sites will also be carried for background information. HMIP's audit reports on WRAs will be published and HMIP may also publish interim reports on any authority as necessary.

5.3 Co-ordination of Waste Regulation Authorities

HMIP will play an active part in the promotion of higher standards by working with voluntary regional groups set up by the waste regulation authorities to help to ensure consistency of licensing policy, procedures, and standards within their committee areas. HMIP will report to the Secretary of State on the operation of the groups. HMIP's role will include:

i) Discussion/presentation of new guidance, procedures, and initiatives;

ii) Provision of guidance on the interpretation of codes of practice and other guidance material and its application to individual cases;

iii) Advice on good practice, for example on waste management licence conditions, inspection regimes, enforcement, *etc.* and

iv) Disssemination of information on good and bad practice to authorities.

6 RADIOACTIVE SUBSTANCES ACT 1960 – AUTHORIZATION, REGISTRATION, INSPECTION, AND ENFORCEMENT

There are nearly 9000 premises regulated by the Radioactive Substances Act 1960 (RSA60). There has been an increase of approximately 30% over about the last 5 years. They can be split into four basic categories as shown in Table 1.

With the introduction of Integrated Pollution Regulation (IPR) it has been decided that disposal authorizations under RSA60 will be reviewed every 4 years. This is a substantial task and will be phased over 4 years starting in 1991.

Table 1 *Categories of RSA60 premises*

Site	*No. of Premises*
BNFL Sellafield	1
Other sites	30
CEGB, SSEB, and UKAEA sites licenced under the Nuclear Installations Act 1965	
Amersham International	
Other nuclear and UKAEA sites and sites authorized under RSA60 to dispose of radioactive wastes	1100
Registrations under RSA60 to keep and use radioactive materials	7000+

7 WATER ACT NOTICE: RED LIST DISCHARGES TO SEWERS

HMIP is already responsible on behalf of the Secretary of State for the Environment for the regulation of discharges of Red List substances[11] to public sewers under the Water Act 1989. HMIP estimates that around 2000 discharges are from processes which will remain outside IPC control, *i.e.* those not prescribed under IPC. Under EC requirements, Water Act consents, like IPC authorizations, are subject to review at least every 4 years.

8 SUPPORTING RESEARCH

HMIP is developing its portfolio of information about the techniques available for pollution abatement of industrial processes with a view to providing its inspectors with guidelines on internationally recognized standards of good practice. HMIP is also planning a research programme to establish methods of characterizing and quantifying the impact of pollutants on air, land, or water to allow comparative evaluation of cross–media process options. This is in addition to an extensive programme of research to assess the impact of alternative radioactive waste disposal systems.

9 FUTURE DEVELOPMENTS

The Environment White Paper,[9] published on 25 September 1990, announced two new initiatives of particular importance to HMIP. The first was that HMIP should be a candidate to become a 'Next Steps' agency (*i.e.* a separate agency within the Department of the Environment) as soon as possible. HMIP would thereby have a clearly separate identity in discharging its new regulatory responsibilities under the Environmental Protection Act. The second initiative involves the setting up of an independent committee to advise HMIP on all its responsibilities. Membership would be drawn from industry, public bodies, and independent members. It will ensure that views and advice from a wide range of informed opinion would be available to HMIP in implementing IPR.

10 CONCLUSIONS

In this chapter a broad overview is given of the future role of HMIP in regulating industrial pollution in England and Wales. Until each of HMIP's new responsibilities under the Environmental Protection Act is phased in during the next five years, existing single medium controls will remain in force. HMIP and industry face many challenges over this period, and there is much to learn but we believe that Integrated Pollution Regulation provides the basis for coherent and fully effective industrial pollution control.

11 REFERENCES

1 Environmental Protection Act, HMSO, London, 1990.

2 Chief Inspector's Guidance to Inspectors – Fuel and Power Sector Guidance Note (Consultation Paper), IPR1, HMIP, September 1990.

3 Chief Inspector's Guidance to Inspectors – Metal Industry Sector Guidance Note (Consultation Paper), IPR2, HMIP, September 1990.

4 Chief Inspector's Guidance to Inspectors – Mineral Industry Sector Guidance Note (Consultation Paper), IPR3, HMIP, September 1990.

5 Chief Inspector's Guidance to Inspectors – Chemical Industry Sector Guidance Note (Consultation Paper), IPR4, HMIP, September 1990.

6 Chief Inspector's Guidance to Inspectors – Waste Disposal Industry Sector Guidance Note (Consultation Paper), IPR5, HMIP, September 1990.

7 Chief Inspector's Guidance to Inspectors – Large Boilers and Furnaces 50 MW and our Process Guidance Note IPR1/1, HMIP, April 1991.

8 Fees and Charges for Integrated Pollution Control, HMIP, March 1991.

9 'This Common Inheritance, Britain's Environmental Strategy', HMSO, London, September 1990.

10 The Environmental Protection (Prescribed Processes and Substances) Regulations 1991, SI 472.

11 C.D. Byrne, 'Selection of Substances Requiring Priority Action', in 'Risk Assessment of Chemicals in the Environment', ed. M.L. Richardson, The Royal Society of Chemistry, London, 1988, pp. 414–434.

12 The Environmental Protection (Applications, Appeals and Registers) Regulations 1991, SI 507.

13 'Integrated Pollution Control, A Practical Guide', Department of the Environment, 1991.

ANNEX 1: PREPARATION OF CHIEF INSPECTOR'S GUIDANCE TO INSPECTORS FOR INDUSTRIES AND PROCESSES

Final guidance	*Issue with IPC*	*Initiated*	*Apply*
Interim Guidance for Industries			
Fuel and Power industry Metal industry Mineral industry Waste disposal industry	1.2.91	1.4.91*	

*All new and substantially changed processes require an IPC authorization from 1 April 1991.

Guidance on individual processes

Schedule Reference	*Process*			
Fuel and Power industry				
1.3	Combustion (>50 MW boilers and furnaces)	1.4.91	1.4.91	1.4.91 30.4.91
1.1	Gasification			
1.2	Carbonization	1.10.91	1.4.92	1.4.92
1.3	Combustion (remainder)			30.6.92
1.4	Petroleum			
Waste Disposal Industry				
5.1	Incineration	1.2.92	1.8.92	1.8.92 31.10.92
5.2	Chemical recovery			
5.3	Chemical waste treatment			
5.4	Waste derived fuel			
Mineral Industry				
3.1	Cement	1.6.92	1.12.92	1.12.92 28.12.93
3.2	Asbestos			
3.3	Fibre			
3.5	Glass			
3.6	Ceramic			
Chemical Industry				
4.1	Petrochemical	1.11.92	1.5.93	1.5.93
4.2	Organic			
4.7	Chemical pesticide			
4.8	Pharmaceutical			

4.3	Acid manufacturing	1.5.93	1.11.93	1.11.93 31.1.94
4.4	Halogen			
4.6	Chemical fertilizer		1.5.94	1.5.94
4.9	Bulk chemical storage			
4.5	Inorganic chemical	1.11.93		
Metal Industry				
2.1	Iron and steel	1.7.94	1.1.95	1.1.95
2.3	Smelting			31.7.94
2.2	Non-ferrous	1.11.94	1.5.95	1.5.95 31.7.95
Other Industry				
6.1	Paper manufacturing	1.5.95	1.11.95	1.11.95 31.1.96
6.2	Di-isocyanate			
6.3	Tar and bitumen			
6.4	Uranium			
6.5	Coating			
6.6	Coating manufacturing			
6.7	Timber			
6.9	Animal and plant treatment			
Full Implementation		1.4.96		

Risk Management During Chemical Use

20
Risk Minimization in the Use of Chemicals

Gerrit H. Vonkeman*

PO BOX 90740, 2509 LS, 'S-GRAVENHAGE, THE NETHERLANDS

1 INTRODUCTION

This chapter raises the question of how to minimize the risk in the use of dangerous chemicals. The minimization of personal risk for an individual user to a small amount of a chemical will not be considered here. This would lead to the difficult discussion of how to define personal risk, let alone how to cope with the perceived risk to individuals. Discussion on this matter will be found in other chapters. Some of the author's views have been expressed in an earlier publication.[1] Environmental risks will be covered in general terms: risk to plants, animals, ecosystems, and human beings can be considered collectively. It is felt that this approach is justified for those cases where chemicals having a high human and ecotoxicity and which are persistent and have a tendency to accumulate in the environment are involved. In general, these are substances on the so-called 'black lists'.[2] This has more justification when the knowledge of toxicity is limited, including the effects on human beings, and when even less is known about synergistic effects. Data on effects on animals, plants, and ecosystems are often rudimentary or even lacking completely.

In the case of such substances, the philosophy of risk minimization is quite simple: as many uses as possible and dispersion of the substances in the environment should be avoided. However, this is more easily said than done. It is evident that the strict application of such an approach would have far-going technical and economic consequences. Additionally, answers are needed to questions such as which uses to tackle, which effluents to prevent, what should be the role of reuse and recycling. These are often difficult questions to answer. In those cases where answers can be given, they often appear to be counter to common sense and seldom agree with existing policies.

*Foundation for European Environmental Policy; Dutch Committee for Long Term Environmental Policy.

2 THREE CASE STUDIES

Before considering such questions and attempting to formulate objectives of a more or less universally applicable approach, three different case studies relating to chemicals are given.

The first will deal with cadmium, which is a heavy metal with a long record of environmental pollution and detrimental and sometimes fatal effects on human populations. Cadmium is not produced directly from cadmium ore; it is a by-product of the production of other metals, particularly zinc. Another important source of cadmium is phosphate ore. It is processed in the production of fertilizer, phosphoric acid, and phosphates for detergents and other applications.

Mercury, the second example, has been used for thousands of years, again having a long record of disease and death. The British expression 'mad as a hatter' stems from the fact that hat makers who used mercury in the treatment of felt used to become mentally ill and finally died. As distinct from cadmium, mercury is produced directly from mercury ore. As is the case for cadmium, there are also important diffuse sources, including the use of fossil fuels such as carbon, oil, and natural gas.

Whereas in the cases of cadmium and mercury consideration has been given to substances with a relatively low annual production, the third example, chlorine, is one of the most important bulk chemicals in use in the world. Some of its compounds, mainly inorganic, can be seen as environmentally innocuous. Organo-chlorine compounds, on the contrary, form a family of persistent substances which do not occur naturally and which can have detrimental effects on humans and other organisms; in addition they can play an important role in smog formation, acidification, ozone depletion, and effects on the climate. Chlorine is produced from rock salt, which implies that its production is linked directly with another bulk chemical - sodium hydroxide. (See chapter by A. Mottershead.)

The above will be discussed in greater detail, including reference to mass balances or materials flow charts[3,4] for the European Community (EC) (see Figure 1). The bold line marks the boundary of the area under investigation, within which there are two subsystems.

In the economy section the inflow comes from imports (1) and/or primary production or extraction (2). The outflow that does not enter the environmental section consists of the economic output, or exports (5), and an item called 'responsible disposal' (6). This may relate to complete incineration of an organic substance, immobilization in products or waste, controlled storage, or return to the substrate (for example storage in mineshafts). Some of the substance will accumulate in the economy (11). This accumulation can be subdivided into materials (ma) products (pr), and waste for processing (wa).

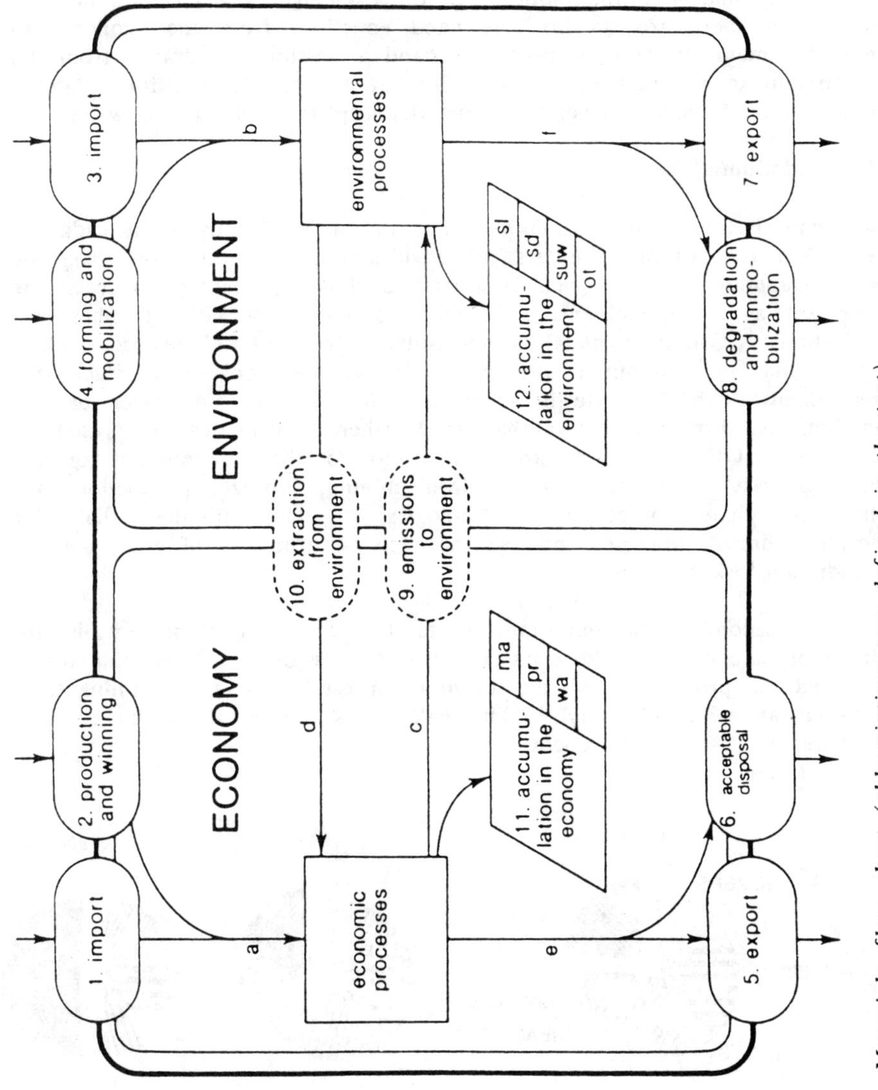

Figure 1 *Materials flow chart (abbreviations are defined in the text)*

In addition, the environment section has a transboundary input from the environment outside the studied area (3), an item called formation (*e.g.* nitrogen fixation), and mobilization (*e.g.* due to weathering) (4). The outflow from the environment also comprises transboundary output or export (7) and an item called breakdown and immobilization (8). Accumulation in the environment (12) is divided into ground-local (bl), ground-diffuse (bd), waterbase (wb), and others (ot). The latter relates largely to groundwater and biotics. In the exchange between economy and environment there are on the one hand emissions from the economy to the environment (9) and on the other hand a possible withdrawal from the environment to the economy (10). The latter may include, for example, the dredging of harbour sludge and the dephosphating of surface water.

2.1 Cadmium

As previously indicated, cadmium is not produced directly from cadmium ore. Primarily, cadmium enters the environment via mining of phosphate rock. Depending on its origin a greater or lesser percentage of cadmium is present as a contaminant. Cadmium is associated with all phosphate applications. Cadmium metal is not isolated from phosphate rock, which implies that all cadmium present in phosphate rock will reach the environment or be a waste in some form or another. A major part of cadmium is dispersed on arable land when phosphates are used as fertilizers. Other amounts are transferred to the environment or are found in waste deposits: in wastes from mining, fertilizer production, and other phosphate processing technologies. This implies that the phosphate-linked processes and uses have to be a matter of great concern in cadmium risk reduction.

The second most important route for cadmium dispersion is the mining of zinc ore (and to a lesser extent copper ore). When zinc ore is processed, a proportion of the cadmium present, as a contaminant, is produced as cadmium metal. The balance is emitted as effluvia or as wastestreams of the zinc process.

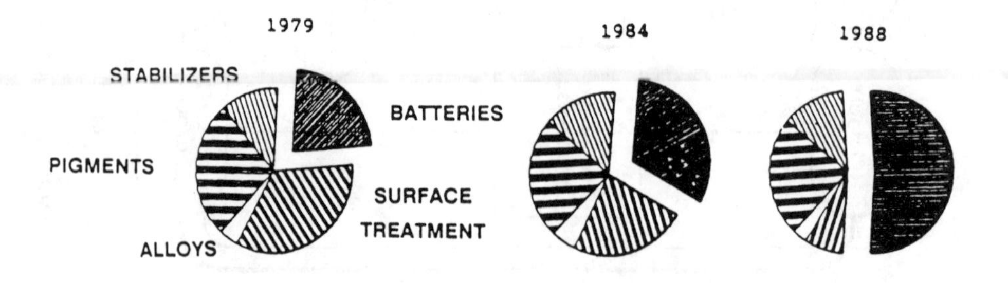

Figure 2 *Distribution of the end uses of cadmium from various applications for the years 1979, 1984, and 1988*[5]

Cadmium has many applications, but substantial changes have occurred over the past ten years because of environmental restrictions and for economical reasons (see Figure 2). The cadmium transfer in the economic system of the EC is summarized in Figure 3;[6] the flow chart is shown in Figure 4.

Many applications of the use of cadmium are such that it is almost inevitable that it will reach the environment at the end of its life cycle. These uses have been forbidden or are under serious threat of being banned. Such applications include: its use in anti-corrosion coatings, in anti-fouling paints, as a pigment, and as a plasticizer in polyvinyl chloride (PVC). In particular its use in pigments and plasticizer has been extended and these have caused significant problems. Large amounts were used in the PVC industry, a substantial part of which was discarded packaging in household waste.

This has led to serious cadmium contamination from waste incinerators and wastewater treatment plants. Despite restrictions, there has been a steady growth in the use of cadmium in accumulators and batteries. Rechargeable nickel-cadmium cells are present in an ever-increasing number in small apparatus used in large consumer markets, ranging from electric screwdrivers, drills, and garden tools, to portable electronic apparatus such as audio-visual apparatus, computers, telephones, and faxes.

2.1.1 Minimum Risk Policy for Cadmium. In considering the phosphate route, it is evident that only two approaches are possible. One, in the extreme, would be to abandon the use of phosphate rock; the second to remove cadmium from phosphate fertilizer and associated products, thus preventing cadmium from entering the environment in a diffuse and uncontrolled manner. Unfortunately, removal of cadmium from fertilizer is a difficult and expensive process because of the huge amounts of phosphate rock involved and the low cadmium concentration. Nevertheless, there is no choice in the longer term because of the problems currently encountered and the accumulative nature of cadmium contamination. Abandoning phosphate rock would be an extreme step, because of its availability and the consequences upon the national incomes of Morocco, Togo, Senegal, and a number of other Third World countries. However, the releases of cadmium and the overall phosphate releases are a problem in several countries, and drastic reductions in primary phosphate use are necessary and possible, *inter alia* by using animal manure and by application of phosphates that can be produced from the tertiary treatment of wastewaters. The latter will become obligatory in the EC in the forthcoming decades.[7]

With respect to cadmium produced from zinc, and applying the approach that diffuse uses and emissions should be avoided, there are two rather drastic options. The first and more drastic option would be to ban all zinc production. The second, less drastic option, would be to store

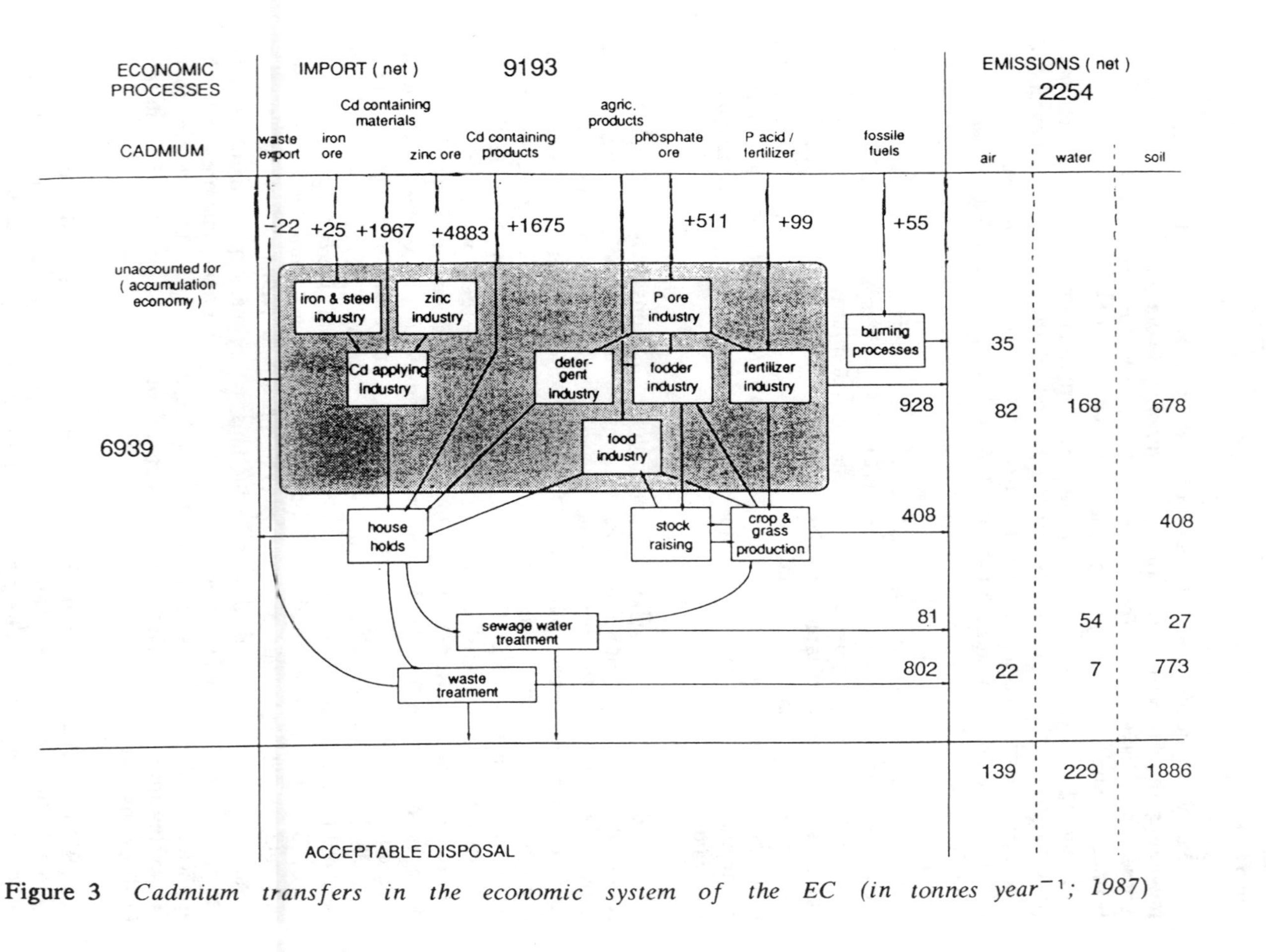

Figure 3 *Cadmium transfers in the economic system of the EC (in tonnes year^{-1}; 1987)*

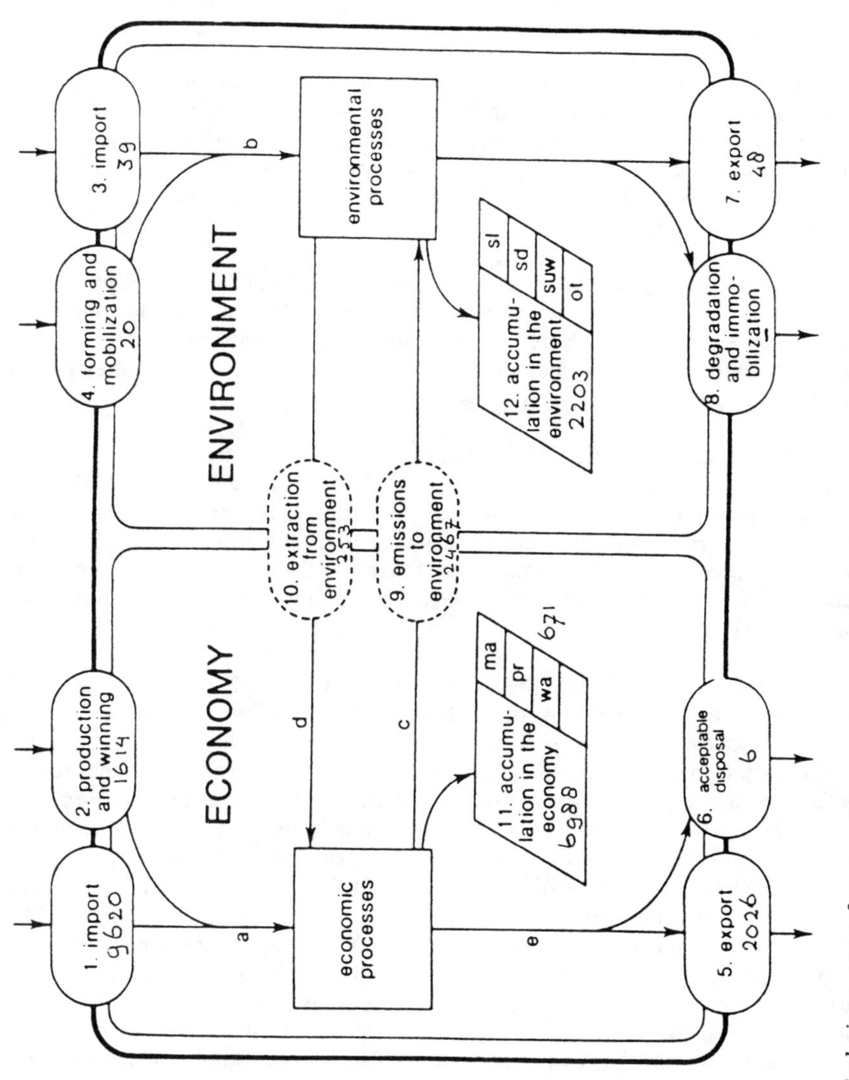

Figure 4 *Cadmium transfer in the economy and the environment of the EC (in tonnes year^{-1}; 1987)*

the cadmium produced from zinc and abandon all cadmium uses. Both solutions have merits that are not immediately evident.

Zinc is a metal that is becoming increasingly rare and its current production rate cannot continue for much more than two decades. Hence, immediate reduction in zinc production would retain a quantity of zinc ore for future generations, which is in line with the principle of sustainable development.

One consequence of reducing zinc production would be that the principal source of cadmium is eliminated. On one hand this could be an incentive to produce cadmium from phosphate rock, and on the other hand to store cadmium or cadmium waste for future generations. However, there is a third choice, the use of cadmium in batteries and recycling of used cadmium products.

Examining the principal uses of cadmium as indicated in Figure 2, it is noted that, apart from its use in batteries, all other applications are diffuse and recovery would be almost impossible. Batteries are currently disposed of into the environment and this situation can be redressed by installation of good collection systems.

As it is not realistic to abandon all cadmium uses at present, a positive policy for transferring cadmium to battery production and then to install a system in which virtually all batteries are recovered would be advantageous. However, the question would then be what to do with the recovered batteries? The most instinctive answer would be recycling, which would have two possible effects pertaining to the situation (see Figure 5).

The upper part shows the situation when no recycling occurs (a). Box E represents the amount of material present in the economic system. Recycling is shown as the best case situation (b). The recycled product replaces virgin material and the primary production is reduced accordingly. This is not the case, however, for cadmium. As previously outlined, the primary production of cadmium is determined by the zinc production. Since recycling cadmium would not result in an increased amount of zinc, it has no effect on the zinc market. The amount of zinc produced remains constant and hence the production of primary cadmium. This implies that the recycled cadmium would lead to an increased availability of cadmium on the market, in turn resulting in lower prices and a pressure to increase its use. This situation is shown in diagram (c). Hence, the ultimate result of recycling batteries is an increased discharge of cadmium to the environment via diffuse emissions, which is the converse of that which is required.

Therefore, the minimum risk policy for cadmium is:

i) Reduction of zinc and phosphate production;

ii) Cadmium present in phosphate rock should be recovered;

iii) The application of cadmium, other than for battery production, has to be minimized;

iv) The batteries should be recovered but not recycled. Batteries would need to be stored safely until the circumstances are such that recycling becomes a viable and environmentally beneficial option.

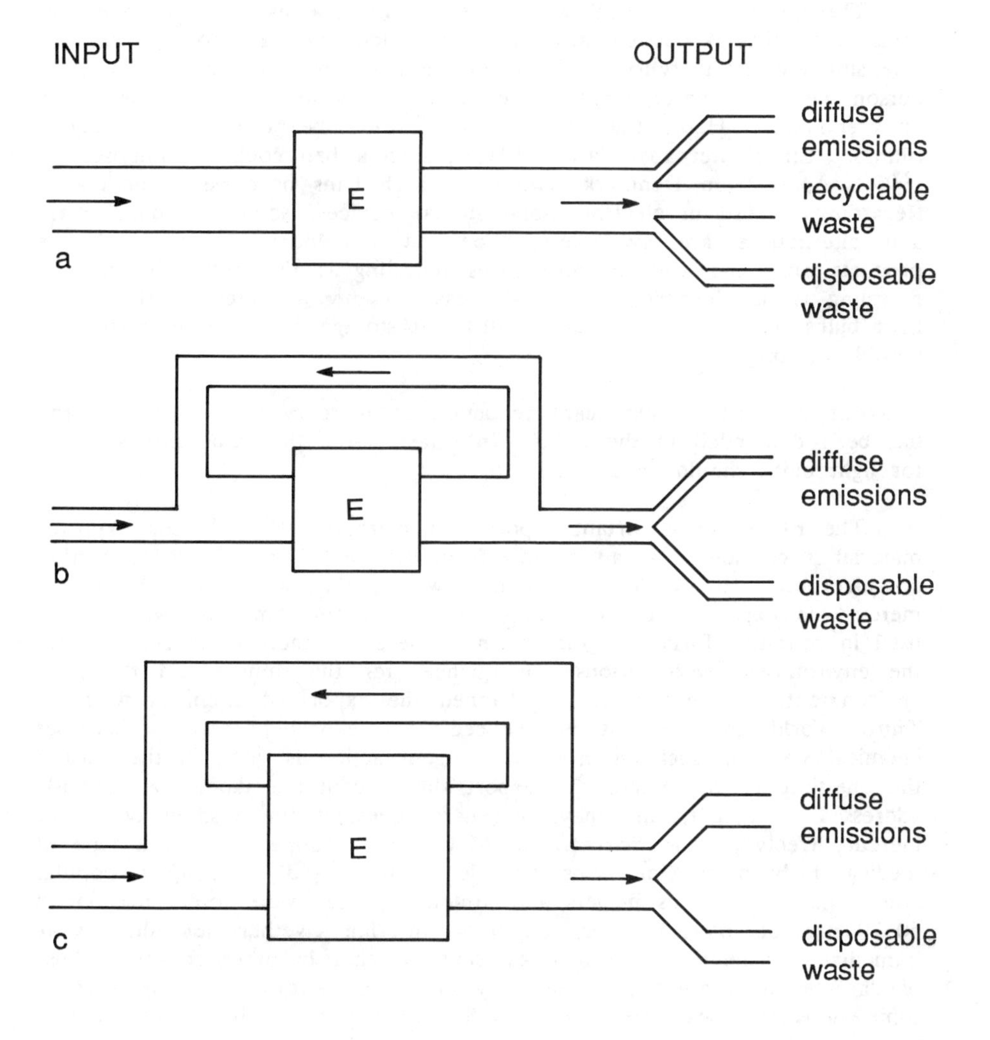

Figure 5 *Influence of recycling on production processes*

2.2 Mercury

As can be seen in Table 1, the predominant diffuse primary sources are fossil fuels, which implies that action is urgent.

For the use of mercury in chlorine production, alternative processes are available (diaphragm and membrane electrolysis); hence the amalgam process should be banned. Within the EC, legislation to achieve this goal has recently been passed, but the phasing out will take a (too) long period.[8]

The use in dentistry likewise leads to diffuse emissions. Recovery of spills at dentists is possible and should be undertaken as soon as possible. The situation for mercury in the filled teeth is more problematic. When a person dies, this mercury enters the environment when the body is buried or cremated. Here, the only solution would be a ban on amalgam fillings. Since alternatives are available, such a ban could be imposed by 1995. Apart from Denmark (2000), no such bans have been announced. Regarding its use in electronic and similar devices, some are unnecessary and alternatives are available. For the remainder, recovery of the products after use and in most cases recycling is the best solution. If recycling is not feasible, *e.g.* in the case of mercury used in fluorescent light bulbs, recovery of the used bulbs and storage as chemical waste is a feasible option.

For batteries, a significant reduction of mercury use is possible and has been demanded in the EC.[9] In other cases, the same procedure as for light bulbs should be enacted.

The EC is an important exporter of mercury. Usually the exported material is considered as an outflow from the system that needs no further consideration. It should be noted, however, that a major part of the mercury is exported to the Philippines and Latin America, where it is used in primitive forms of gold mining, where the mercury is emitted into the environment with serious consequences for the miners and the local environment. Since the EC has banned the export of chemical waste to Third World countries, it is deduced that exported mercury becomes chemical waste in such countries and urgent action is vital; in the interim the question as to whether this export should continue should be seriously addressed. In fact, one may in general question the wisdom of selling mercury freely on the market. In May 1991, it appeared that a type of medicated bandage (for arms and legs) was rapidly becoming popular among golf and tennis players and athletes. They were found to contain 60–100 g of mercury per length. In The Netherlands they were immediately banned,[10] but it is not unlikely that in other countries these bandages contain amounts of mercury in the same order of magnitude as some industrial releases – all these will ultimately reach the environment.

Table 1 *Mercury emissions in the EEC in 1989 (preliminary data;* tonnes year^{-1})

Source	*Destination of immediate emissions*		
	Air	*Water*	*Soil*
Natural sources	75	150	–
Atmospheric fallout transferred to	–	100	200
Water transferred to:	100	–	75
Mercury mining	15	1	3
Non-ferrous metal production and refining	42	–	–
Base metal production	–	20	–
Coal combustion	121	?	6
Natural gas combustion	3	–	–
Industrial production:			
i) chlorine, caustic soda	38	50	100*
industrial catalysts	–	13	–
cement	?	–	–
ii) batteries; dental amalgams; electrical equipment; paints; pharmaceuticals; thermometers; laboratory products; phosphate fertilizer	5	10	36
Product use:			
chlorine, caustic soda	?	?	*
cement	–	–	–
batteries	?	?	316
dental amalgams	–	–	24 (dental amalgams to laboratory products)
electrical equipment	–	–	
paints; pharmaceuticals	–	–	
thermometers			
laboratory products	–	–	
phosphate fertilizer	?	?	?
Solid waste disposal municipal; industrial	33	–	6
Sludge waste disposal agricultural municipal, inclusive domestic wastewater industrial	3	35	28
Totals (without natural sources)	**260**	**129**	**699***

*180 metric tonnes unaccounted for by the chloralkali industry

Apart from banning as many mercury uses as possible, recovery and recycling of used mercury must be a key requirement in a minimum risk policy for mercury. Contrary to the case of cadmium, recycling mercury will result in a decrease of primary production and hence is creditable.

2.3 Chlorine

The case of chlorine is extremely complicated.

Chlorine is largely produced from sodium chloride, and to a much lesser extent from potassium chloride. The world production is over 40 million tonnes $year^{-1}$; 32% is produced in the USA, 27% in Western Europe, and 20% in Eastern Europe. (Due to the reunification of Germany, the last two figures may need to be slightly adjusted.) The contribution of different countries to total EC-production capacity are:

	%
Belgium	4
Netherlands	8
FRG	38
France	15.5
Italy	12
Spain + Portugal	8
UK	14.5

The electrolytic decomposition of sodium chloride solution (brine) produces hydrogen and caustic soda as co-products. Every 1000 tonnes of sodium chloride produce theoretically 610 tonnes of chlorine, 680 tonnes of caustic soda, and 20 tonnes of hydrogen gas.

Both the volume and the nature of the production make it a questionable process from an environmental viewpoint. Three different electrochemical processes are in use. All are energy intensive and produce wastestreams, although there are important differences. The amalgam process has a very high energy consumption and is associated with serious mercury losses, in the order of 2–10 g of mercury per tonne of chlorine. However, it produces very pure sodium hydroxide.

The diaphragm process is almost as energy intensive. Although mercury electrodes are absent, a new environmental disadvantage is that it usually requires diaphragms made of asbestos. A practical and economic problem is that the sodium hydroxide produced is mixed with sodium chloride and is only purified to a certain degree.

More recently, a membrane process having less environmental impact and energy consumption has been introduced. Like the amalgam process, it produces chlorine and sodium hydroxide of high purity.

In particular, the mercury process has been the subject of EC environmental policy since the early 1970s, when such policies began to be formulated. Legislation that envisages a phase-out of the mercury based process is now in place[8] and replacement of asbestos by other diaphragm material is also envisaged.[11]

The North Sea Conference decided upon a complete ban on mercury emissions by 2010. Companies that invest in new installations often opt for the membrane process. In the Netherlands, Akzo now uses such techniques for 55% of its capacity. The diaphragm process is used for 30%, whereas only 15% is produced by the amalgam process.[12] In spite of these developments, the figures for the EC as a whole are about the reverse: 70% of the present chlorine production still uses the amalgam process and another 25% the diaphragm process. The change to the membrane process will probably take far more than ten years.

Caustic soda and chlorine rank among the most important chemicals in the world. Their main uses are given later in this chapter. Their complementary requirements imply that their markets are directly linked and that it is important that there is a chlorine–caustic soda balance. Chlorine is used as a bulk chemical as a gas and as hydrogen chloride or its solution in water, hydrochloric acid. Both chlorine and hydrogen chloride are difficult to store. They can easily be converted from one to the other and industry uses this 'Cl_2–HCl balance' to cater for their respective market requirements. Measures that interfere with the chlorine market, *e.g.* for environmental reasons, can have important economic consequences.

Chlorine and hydrogen chloride are highly aggressive gases. Elemental chlorine was used as a war gas in the first world war. This implies that chlorine production and its transport and the use of chlorine and hydrogen chloride present a hazard to workers, the community in the vicinity, and the environment, when a leak or accident occurs. For other reasons, similar hazards apply for caustic soda and hydrogen.

According to Akzo[12] about 5% of the annual chlorine production of Western Europe (EC + Scandinavia) is used in its elemental form, and another 15% to prepare inorganic compounds. Slightly more than 80% is used to produce halogenated organic chemicals by reacting the chlorine with hydrocarbons. This figure of 80% can be subdivided into 35% for polyvinyl chloride (PVC), 20% for more common solvents: chloro–methane, –ethane, and –ethene, and 25% for other compounds. A more detailed breakdown is given below.

In the USA, the use pattern differs, with relatively more uses in elemental and inorganic forms.[13] Undoubtedly this means that differences will also exist between the member states of the EC. Predictions of the USA usage in 1994 indicate a further shift to solvents and PVC despite environmental pressures for their reduction. These data are summarized in Table 2.

Table 2 *US chlorine applications in 1989 and 1994 (%)*

Application	*1989*	*1994*
Water treatment	5	5
Inorganic compounds	8	8
Pulp and paper	14	8
Titanium dioxide	3	3
Epichlorohydrin	5	5
Propene oxide	8	9
Chlorinated methanes + ethanes	14	15
Dichloroethane/VCM/PVC	28	33
Other organic compounds	12	11
Miscellaneous	3	3

2.3.1 Inorganic Uses.* Within the inorganic chemical area chlorine and inorganic compounds are used as disinfectants or anti-foulants for drinking water and industrial water production, bleaching of paper pulp, textile, and other materials, and similar applications. Pulp and paper and, to a lesser extent, textile industries are well known for the difficult wastestreams they generate, not only within the EC but also in EFTA countries, *e.g.* Austria, Finland, Norway, and Sweden. Additionally, the use of chlorine for industrial, or indeed even domestic water disinfection, is questionable. In all cases where chlorine is used as a bleaching and disinfecting agent it will react with living organisms or organic substances, forming organochlorine compounds. In view of their toxicity, accumulative nature, and the significant quantities so formed, these should be added to the products manufactured for organic chemistry, hence forming one combined threat to the environment which is discussed later in this chapter.

Hydrogen chloride is used to neutralize alkaline streams in chemical processes: in steel production, petroleum drilling, and in the food and pharmaceutical industries.

After simultaneous production from sodium chloride, part of the chlorine and caustic soda will be combined into salts like sodium chlorite and sodium hypochlorite (NaOCl). Aqueous solutions of these products are less dangerous to handle and transport than chlorine. Likewise, they are used for bleaching and disinfection (a dilute solution is the well known Javelle-water) and can be used to release chlorine *in situ* by adding hydrochloric acid. The food and metal industries and sugar production are important users of these substances, as is the case for potassium and calcium (hypo)chlorites. An intermediate in production processes, chlorine

*Most data have been derived from the Foundation of European Environmental Policies files from the literature mentioned at the end of this chapter. Notably references 20 and 28 present a useful, general picture. The figures indicated illustrate the problem, rather than presenting the most recent and complete data.

dioxide, is similarly an effective water disinfectant, but is much more dangerous to handle.

The drawbacks of using chlorine-based disinfectants and bleaching agents are well known and the search for (equally cheap) substitutes continues. Partial replacements have appeared in ozone and hydrogen peroxide but problems still exist for which considerable production capacity is being developed.

Potassium chlorate is used in fireworks, matches, and explosives, and sodium chlorate is used as a weed killer. The potassium and ammonium perchlorates are used in fuels for rockets and jet propulsion.

Small amounts of chlorine are used to produce metal chlorides like aluminium chloride (a well known catalyst), titanium chloride (a smoke screen agent), ferrous chloride (a flocculating agent, etching agent, and an agent for phosphate removal), metal oxides (*e.g.* titanium dioxide; an important white pigment in the paint and paper industries), and pure metals like titanium (airplanes and ships), iron, magnesium, aluminium, silicon, and boron.

Relatively small amounts of chlorine are used to produce compounds of phosphorus and sulphur that are used as catalysts and chlorinating agents in the chemical industry.

2.3.2 Organic Uses; Organochlorine Compounds. Replacement of some (or all) of the hydrogen atoms in hydrocarbons by chlorine atoms, using chlorine or hydrogen chloride, has been used to synthesize tens of thousands of new compounds. The vast majority of these do not occur naturally, are persistent, and can harm the environment to a greater or lesser extent. A major proportion of the organohalogen compounds produced may well remain in use for shorter or longer periods but these will all finally appear in the environment.

It sometimes can take up to 100 years for these compounds to decompose and release chlorine in some chemical form and produce (directly or indirectly) carbon dioxide from the hydrocarbon moiety. Since this carbon dioxide is of fossil origin, it contributes to the greenhouse effect, although this is a relatively small amount compared with carbon dioxide from the direct combustion of fossil fuel.

Whilst it is difficult to prioritize the enormous number of compounds, a distinction between four production categories can be made:

i) Production in which chlorine or hydrogen chloride acts as an intermediate, but which yields chlorine-free end-products;

ii) Production which yields chlorine-containing products intended to be released to the environment;

iii) Production which yields chlorine–containing products that are not explicitly intended to be released to the environment but, nevertheless, will contaminate the environment;

iv) Production which yields chlorine–containing products that will probably not be released to the environment, or can be prevented from so doing.

All manufacturing with chlorine and hydrogen chloride is hazardous leading to risks, as emissions to air, water, and soil will always occur, extremely poisonous by–products or intermediates may be formed, and noxious chemical waste may result. Most risks are located in one place: the production site. This implies that the responsibility is clear and that, in principle, very strict preventive measures can be demanded and enforced. Hence, industry representatives often argue that this sector of chlorine chemistry is no problem (see below).

2.3.2.1 Category 1: Chlorine as an intermediate

Processes in which chlorine is used as an intermediate but not present in the end product involve about 50% of the chlorine use in organic chemistry.[12] This results in most of the chlorine leaving the process as an (inorganic) chloride, or as hydrogen chloride that may be reused. For example, methyl chloride (monochloromethane) is an intermediate for many substances including silicones, methylcellulose, tetramethyl–lead and higher chlorinated methanes. Allyl chloride/epichlorohydrin is used to produce epoxy–resins that are used in paints, electric insulations, laminates in electronics, adhesives, and composite materials for airplanes. Oxidized chlorine products are used as intermediates in the production of *inter alia* polypropyleneglycol, polyethers, and poly(alcohols) which finally result in polyurethanes (car industry, furniture, shoes, textile, packaging materials, thermal insulation in buildings) and carboxymethylcellulose (food, cosmetics, glues, oil industry). Benzyl chloride is a reactant in the pharmaceutical industry for the production of amphetamines, phenobarbital, and many other medicinal compounds. Phosgene is reacted with diamines for the preparation of di–isocyanates which are then reacted with glycol via another route to form polyurethanes. Phosgene is also used as an intermediate in the production of polycarbonates, including important engineering plastics in the car industry, production of household equipment, artificial glass, *etc.*

Industrial representatives often state that these types of processes have little relevance to the chlorine pollution problem. However, this may not be the case as is evident from the view taken by environmentalists and by part of the scientific community.[15–21]

Firstly, some of the processes involve extremely dangerous intermediates, *e.g.* vinyl chloride monomer (carcinogenic) and phosgene (a well known war gas).

Additionally, the losses and by-products will always occur and chlorine-containing emissions and waste are unavoidable. Even small percentages can be significant, bearing in mind the large volume of production and the nature of these substances. Furthermore, whilst it may be possible and perhaps probable that any escape of material can be prevented ultimately, the present situation remains that these productions cause real and substantial problems and much progress is necessary to close all the gaps.

2.3.2.2 Category 2: Products intended to enter the environment

Among the chlorine-containing products that are designed for release to the environment, pesticides and similar compounds predominate, with pharmaceuticals and cosmetics also being significant. High-volume pesticides include substances such as aldrin, dieldrin, DDT, chlordane, lindane, and heptachlor. In the Netherlands alone, some 10,000 tonnes of halogen containing pesticides are used annually.[22]

Consideration has to apply equally to production and use. Even legitimate use will always generate local environmental effects. Whether these are acceptable or otherwise is a matter for political decision; some products have been banned, others restricted and linked with strict conditions. Nevertheless, practice has shown that there remain illegal use and misuse, and soil and (ground) water are sometimes polluted. Moreover, banning products nationally or within the EC does not necessarily mean banning production. Several banned chlorinated pesticides are still produced in relevant quantities and exported to countries outside the EC, including the Third World. Similar remarks apply to pharmaceuticals and cosmetics, although both their quality and their quantities are much less of a problem than in the case of pesticides.

2.3.2.3 Category 3: Products that enter the environment unintentionally

The major components of this group of products are the volatile organohalogen compounds and those which will probably eventually reach the environment. About 20% of the chlorine production of Western Europe is used to produce chloro-methanes, -ethanes and -ethenes. The vast majority of this production will evaporate ultimately into the atmosphere, at an estimated rate for Western Europe of 500,000 tonnes annually.[23,24] These compounds play a key role in atmospheric chemistry, including smog formation, ozone depletion, and the greenhouse effect.

The 1983 production of dichloromethane (ME) in (Western) Europe was some 210,000 tonnes, which was 40% of global production. In 1986, FRG alone produced 155,000 tonnes. About 45% of the production is used as cleaning agents and paint removers, another 25% as propellants in aerosol cans. Of the remainder, the application in extraction processes is important and these substances are also used for soil disinfection.

1,1,1-Trichloroethane (1,1,1-Tri) is also an important solvent used as cleaning or degreasing agent. In 1984 the European production was 150,000 tonnes, 33% of the world production. No data are available for the FRG production in 1986, but the use was estimated at some 45,000 tonnes and the FRG is a major exporter of solvents.

Trichloroethene (Tri) is the third most important solvent. The European production in 1984 was 200,000 tonnes or 55% of the world production. The use in the FRG in 1986 was about 30,000 tonnes from which is estimated a production of about 50,000 tonnes.

Tetrachloroethene or perchloroethylene (PER) had a European production capacity in 1985 of about 450,000 tonnes, or 45% of the world capacity. The real production is unknown. In the FRG 157,500 tonnes were produced in 1986, of which 45-50,000 tonnes were used in the FRG. Trichloroethene and tetrachloroethane are widely used solvents in dry cleaning, metal cleaning, and degreasing.

In addition to dry cleaning, solvents are used by *inter alia* the electronic industry, the paint and adhesives industry, car production, the plating industry, and the printing industry.

In Figure 6, the consumption of chlorinated solvents in Western Europe is indicated. The graph was taken from reference 24 which claims that it is 'based on information from CEFIC and other producer estimates'. However, a more recent publication from CEFIC on solvent use for degreasing operations suggests that the amounts should be higher.[25]

At present, the bulk of the organohalogen solvents evaporate, *i.e.* about 500,000 tonnes year^{-1} for Western Europe. Reference 24 indicates a total of 460,000 tonnes for 1984, and 100-120,000 tonnes per year for Germany (FRG) in 1987. It further states that 86% of ME, 70% of 1,1,1-Tri, 65% of Tri, and 50% of the PER ultimately evaporate; this means a weighted average of 68%. The emissions to air can be partly prevented by limiting evaporation or absorbing the vapours. Considering applications, the cost of adequate preventive measures, and the widespread and diffuse uses, a major percentage of the produced solvents cannot be prevented from evaporation in practice.

Most of the solvent residues becomes contaminated with (organic) compounds and frequently these chemicals can be harmful for man and the environment. Depending upon the application, the concentrations may range from a few percent of 'dirt' in the solvent to a few percent of solvent in the 'dirt'. The latter can often be the case in industrial cleaning or degreasing operations. Such substances tended to form a substantial part of the chemical waste destined for incineration at sea. It is likely that 25% of the used solvents, *i.e.* in the order of 170,000 tonnes, will be present finally in such wastes. Obviously the total weight of such wastes will be a multiple of this figure.[26]

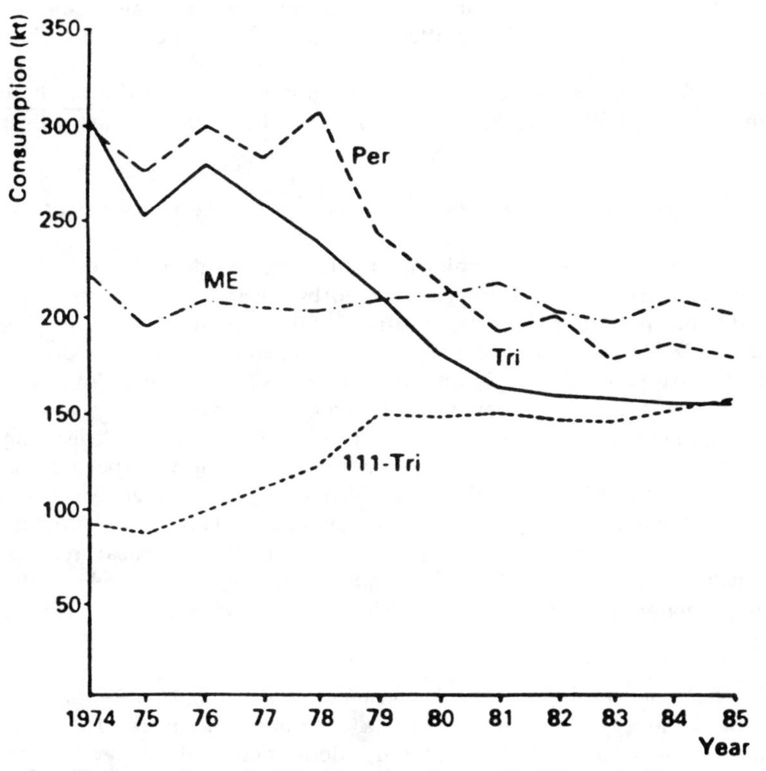

Figure 6 *Consumption of chlorinated solvents in Western Europe*

From the moment that incineration at sea was virtually banned, remarkably little additional organohalogen waste has been presented to other treatment or storage facilities. For these and other reasons, the waste generated by the use of organohalogen solvents is a matter for immediate concern. As earlier stated, significant pressures and efforts exist to reduce the evaporative losses of solvents. However, there must be an awareness that such reduction will hardly reduce the amount of residual liquid and (semi-)solid wastes.

A special category is the chlorofluorocarbons – CFCs. They are used as solvents, propellants in spray cans, blowing agents in foam plastic production, and coolants in refrigerators and air conditioning. The CFCs form a large family. All members have more or less severe capabilities for ozone depletion and greenhouse effects. Some of them are subject to the Montreal Protocol (an international agreement to reduce their use) and similar regulations that envisage a phase-out of several products and/or applications for a number of countries (including the EC) within the next one or two decades.

Nevertheless, not all compounds are covered, and production will continue for some time and shift to other CFCs in parallel. The effects of their (still extensive) production, including the associated wastes, will remain for many years. This matter is outside the scope of this chapter.

A third category consists of plastic products destined for individual use, and in particular packaging materials. These aspects are considered later.

2.3.2.4 Category 4: Products that need not reach the environment

By far the most important chlorine-containing product in this group is polyvinyl chloride. In Europe it absorbs over 35% of the chlorine produced, in past decades more than 100 million tonnes in Western Europe. PVC has been under serious discussion due to its presence in municipal waste and the emissions it generates if this waste is incinerated. Part of the concern is associated with cadmium, which is usually present in PVC pigments and plasticizers. Although the recent EC legislation [27] is neither in force, nor bans cadmium use completely, it is expected that this aspect will be phased out in the next decades. [4] The other concern is the formation of dioxins and furans in incineration plants. [28] Although the role of PVC in this process is uncertain, public and political reaction has been such that there exists a strong pressure to abandon PVC in packaging material. It is hard to predict how this will (or should) develop.

As such, packaging material consumes only about 10% of the PVC produced. More than 60% has a lifetime of over 15 years, in applications such as building material, floor and wall covering, artificial leather, insulation for wires and cables, pipes for drinking water, and sewage effluent transport. The table below shows a breakdown of PVC used in the Benelux countries. [22]

Hard PVC (in %)	
Pipes and appliances	29
Film/foil	18
Door and window frames, *etc.*	9
Bottles	2
Other	3
Flexible PVC	
Wall and floor cover	12
film/foil (packaging)	10
Cable insulation	7
Textile and paper lining	4
Pastes (artificial leather, tarpaulins)	3
Hoses and tubing	2
Other	1

In many of these applications, PVC would be difficult to replace. It has excellent properties for these purposes and it is relatively cheap. (Remember that 55% of its weight consists of chlorine, hence chlorine-free plastics may demand a double quantity of hydrocarbon compounds.) Provided that the use of cadmium in PVC products is abandoned, a thorough environmental impact assessment is required before the abandonment of PVC is considered. Some 100 million tonnes of PVC, most of it containing cadmium, is still in use but threatens to appear as waste in the future.

Currently, only an amount equivalent to 1% of the annual PVC production is (uncontaminated) waste. Although recycling of PVC has commenced, it encompasses only a minute fraction of the annual production. If the use of PVC in long lasting applications remains a serious option, attempts to offer recycling facilities with a capacity of the same order of magnitude as the production are needed.

In addition to PVC, several elastomers or rubbers are made which incorporate chlorine in the product. Neoprene and hypalon are two examples. In volume, their production is low as compared with PVC.

As previously stated, enormous amounts of PVC are still in use and these will appear as waste in the future. It remains to be seen whether they will be recycled. Currently, the capacity is simply not there and, in addition, recycling is not without problems even where it exists. In Austria, a recycling plant for PVC building material may be (or has been) closed for environmental reasons. Industry usually argues that incineration of PVC, with recovery of heat and hydrogen chloride, is also a form of recycling. The environmental organizations and some governments see this as unacceptable which, in part, is due to the problems associated with cadmium, dioxins, and furans from incineration plants.

The pressure on PVC has intensified the search for new products that may replace PVC as a bulk polymer. In this area, results may come from enterprises that do not produce chlorine themselves and envy the dominant role of PVC in the market. Rumours that Shell will soon produce a competitive product polyketone (PECO) have been heard in chemical circles.[29] Given the chemical formula, the hydrocarbon demand could be in the same order as that of PVC.

Industry tends to state that chlorine chemistry is so important and the use of chlorine-containing (organic) products so beneficial that society cannot live without them. Emissions and wastestreams are important and can have serious effects in some cases, but they can be kept under control. For the chemical industry itself, it is claimed that the process of reducing and preventing emissions and waste production is in hand and does not really need further (external) attention. However, for products, the suggestion is that here 'the gaps should be closed', and that some applications, where this is possible, may have to be abandoned (CFCs, open use of solvents), but generally, all (organic) chlorine products will be recollected and recycled or incinerated in the future.

Industry representatives, *e.g.* Europe–wide organizations such as CEFIC and EuroChlor, also predict that as a consequence of these developments, chlorine demand will become lower and industry will react by producing less chlorine from rock salt.

It must be added here that some individual industries take much more aggressive leads than their international associations, from which a 'common denominator', which is often close to the position of the least environment–conscious members in the most lenient countries, has to be found. The Dutch Akzo company recently published a balanced and detailed position on the role and impact of chlorine production and use as a whole, revealing its views and intentions regarding production, transport, products, recycling, and waste management and inviting and promoting an open discussion with all interested parties.[11]

There is a fundamental difference between the approach of industry to the 'chlorine problem' and that of the environment movement (and some political parties).[15–21] The most extreme view is that chlorine chemistry is outdated and environmentally unacceptable and should be completely abandoned. Slightly more lenient views perceive a demand to abandon at least the production of most organic chlorine compounds, notably solvents, pesticides, and PVC. Similar views are occasionally heard from scientists.[15]

2.3.4 Synthesis, Comments, and Conclusions Regarding Chlorine. In considering the present situation within the EC, it can be seen that each year some 500,000 tonnes of chlorine–containing solvents evaporate into the atmosphere or form chemical waste. Of the annual chlorine production of 10 million tonnes, 50% is released into the economy and sooner or later the environment; the other 50% will also reach the environment, but in the form of much less harmful inorganic chlorides.

Some 100 million tonnes of PVC have been produced already and most of it is still 'somewhere'. Capacity to recover and process this by appropriate technology is virtually absent.

Chlorine compounds have been major components of chemical waste. There is little environmentally acceptable chemical waste treatment capacity and also the storage is often questionable. In the long term storage is not really a solution. Also, the problem with chemical waste is that its nature and scope are not really known. Even a country such as the Netherlands, which has well developed legislation and endeavours to keep stock of the chemical wastestreams, must nevertheless admit that tens, if not hundreds of thousands of tonnes or more are 'missing'. A serious and skillful attempt on behalf of the German Environmental Ministry to draft a 'chlorine mass balance' for Germany suggests forcibly that most of such 'missing' quantities find their way into the environment.

Another question is related to the effect of recycling. Will it really result in a reduction of primary chlorine production? On studying the figures that are available for the United States,[13] note must be taken that it is expected that the chlorine market will grow from 11 million tonnes in 1989 to 11.5 million tonnes in 1994, which implies that there is a diminishing growth rate. Chlorine prices have dropped from US$ 140 per tonne in 1986 to US$ 75 in 1990. At the same time the demand for sodium hydroxide continued to grow, with an accompanying price rise from US$ 120 to US$ 300 per tonne and demand exceeding the supply. In Table 3 the major applications of sodium hydroxide are summarized for 1989 and 1994. The production is expected to grow over this period from 11.2 to 12.0 million tonnes. The nature of the uses is such that further growth in demand looks likely.

Table 3 *USA sodium hydroxide applications 1989 and 1984 (%)*

Application	*1989*	*1994*
Water Treatment	8	8
Inorganic compounds	12	12
Pulp and paper	24	23
Aluminium	3	3
Textile	5	5
Petroleum/petrochemical	8	9
Soaps and detergents	8	8
Other organic compounds	12	12
Miscellaneous	10	10

This implies that sodium hydroxide is the current market leader, which will exert strong pressures to produce more sodium hydroxide. If this continues to be produced by electrolysis of sodium chloride this may result in an even greater excess of cheap, primary chlorine, at the same time reducing the attractiveness of recovering chlorine or hydrogen chloride from chlorine-containing products or wastes.

Another feature is that the extra chlorine that will be produced in the USA is expected to be used by additional production of solvents and PVC. These products will also absorb the chlorine that will become available from the expected diminishing use in the pulp and paper industry. (In recent years, a reduction by 50% has already taken place in the USA). Naturally processes are known that produce sodium hydroxide without the simultaneous production of chlorine. However, it is outside the scope of this chapter to speculate on their possible future role, except to note that this role will increase little until the market price of chlorine falls considerably further.

Finally, it must be kept in mind that although some saturation may be observed in the EC and the USA, the market for solvents, PVC, and other organohalogen products is considered as an important growth market elsewhere and it is not likely that limitation will be in place within the next decade. European producers will not hesitate to obtain their share of this market, for which the environmental effects of production will be felt inside the EC.

What does this all mean in terms of a minimum risk policy for chlorine? In fact we have a terrible dilemma. Unlike the cases of cadmium and mercury, we are faced with a major category of chemical industry, spread all over the world and producing bulk chemicals in quantities of millions of tonnes. As is the case for cadmium, the production of chlorine does not stand on its own, but is linked with another bulk chemical: sodium hydroxide. As long as this link is not broken, any real intervention in chlorine production for reasons of environmental protection and risk minimization will cause tremendous economic effects. Even if the link is broken, the case of chlorine chemistry itself remains difficult.

Recovery and recycling can have different effects: Individual compounds can often be recycled to produce the same compound again, but almost every chlorine compound can also be transformed into elementary chlorine and/or hydrogen chloride, which can be used anywhere in the chlorine chemistry network. Recycling will seldom result in a reduction of primary production. Given this situation, a minimum risk policy for chlorine must be based on a realistic approach and then rather optimize than minimize risks.

i) Evidently serious attempts should be made to break the link between chlorine and sodium hydroxide production.

ii) Diffuse emissions of (particularly organo-) chlorine compounds into the environment must be prevented. This will imply a ban on a number of uses.

iii) Recovery of used products is necessary, but their recycling is not always feasible. Where recycling is applied, one must decide on a case by case basis whether recycling into the same product or into elementary chlorine or hydrogen chloride is the best option.

iv) Given the high volume of chlorine production and the economic importance of chlorine chemistry globally, it is not realistic to expect a rapid fall in production. This implies that 'outlets' must remain available for a long time in the future. These have to be found in products that have relatively low environmental impacts and/or can be prevented from entering the environment. Production of PVC could provide such an 'outlet', under the condition that only long-lasting uses must be allowed, under circumstances wherein recovery can easily be realized. At a certain stage, the recovered products must be recycled, but it remains to be seen whether this should be done immediately or after a lengthy storage period. Under all circumstances, the application of cadmium in pigments and plasticizers must be forbidden.

3 CONCLUSIONS

There is no single minimum risk policy that can be applied to the use of any substance or product. It is essential that a mass balance or mass flow chart is drafted before judgements on the most appropriate policy are made. Both intentional and unintentional sources must be taken into account.

Linked sources, which yield other substances simultaneously, can cause serious complications, which may determine the minimum risk policy. Such linked sources can be contaminants in ores, but also co-products from the main component of the ore itself, as in the case of sodium chloride.

Recycling is not always a useful tool in risk reduction. A careful study of the mass flow diagram is necessary before decisions regarding recycling are taken. Complications and adverse effects may be caused by physico-chemical factors, but also by factors that are determined by the market. The latter implies that the application of so-called financial or economic instruments can be a useful tool in minimum risk policy.[6] This is the more important where substances as discussed in this chapter cause problems world wide and, therefore, are difficult to regulate by classical means of 'command and control'.

The perception of risk is determined by both objective and subjective factors. This chapter has concentrated on objective factors, but in real life subjective factors may be the most decisive at the end of the day.

4 ACKNOWLEDGEMENTS

The ideas proposed in this chapter were developed from research projects for the Commission of the European Communities, the Dutch Committee on Long Term Environmental Policy, and the Dutch Ministry of the Environment. The author thanks Professir Iain Thornton from the Global Environment Research Centre of Imperial College, London and Professor Helias Udo de Haes from the Centre for Environmental Research of the University of Leyden and their staff, who were partners in some of these projects. Special thanks are also due to Peter Maxson, Senior Researcher at the Brussels Office of the Foundation for European Environmental Policy.

5 REFERENCES

1 G.H. Vonkeman, 'Risk Assessment and Risk Communication: Attitudes of Interested Parties and Their Consequences for Action at Community Level', in 'Communicating with the Public about Major Accident Hazards', ed. H.B.F. Gow and H. Otway, Elsevier Applied Science, 1990, pp. 461–469.

2 C.D. Byrne, 'Selection of Substances Requiring Priority Action', in 'Risk Assessment of Chemicals in the Environment', ed. M.L. Richardson, The Royal Society of Chemistry, London, 1988, pp. 414–434.

3 H.A. Udo de Haes, G. Huppes, and J. Guinee, 'Stofbalansen en Stroomschema's: de Accumulatie van Stoffen in Economie en Milieu' (Mass Balances and Flow Diagrams: Accumulation of Substances in the Economy and the Environment), Milieu: *Tijdschrift voor Milieukunde*, 1988, **3** (2), pp. 21–55.

4 H.A. Udo de Haes, G.H. Vonkeman, and G. Huppes, 'Cadmium Policy: from Prohibition to Control', Commissie Lange Termijn Milieubeleid, 's–Gravenhage, March 1990, pp. 34.

5 Anon., 'Lead, Zinc and Cadmium into the 90s', Lectures ILZIC Silver Jubilee Conference, Brussels, 1988.

6 G. Huppes, E. van der Voet, W.G.H. van der Naald, G.H. Vonkeman, and P. Maxson, 'Financial Instruments for European Environmental Policy', Study for European Commission, DG XI, Leyden/Brussels, 1990, pp. 170.

7 'EC Directive on the Treatment of Urban Waste Water', 91/271/EEC, *Off. J. Eur. Comm.*, L135, 8 May 1991.

8 'EC Directive on Limit Values and Quality Objectives for Mercury Discharges by the Chloralkali Electrolysis Industry', 82/176/EEC, *Off. J. Eur. Comm.*, L81, 27 March 1982.

9 'EC Directive on Batteries and Accumulators Containing Dangerous Substances', 91/157/EEC, *Off. J. Eur. Comm.*, L78, 26 March 1991.

10 'Regeling Kwik–bevattende Arm– en Beenbanden', (Dutch Directive on Mercury–containing Bandages), *Staatscourant*, **143**, 26 July 1991.

11 'EC Directive on the Prevention and Reduction of Environmental Pollution by Asbestos', 87/217/EEC, *Off. J. Eur. Comm.*, L85, 28 March 1987.

12 Akzo N.V., 'Chloor en de Samenleving – de Rol van Akzo' (Chlorine and Society – the Role of Akzo), Brochure, Arnhem, 1990, pp. 40.

13 Anon., *Chem. Eng. News*, 21 May 1990, p. 18, and 8 October 1990, p.18.

14 M. van Opstal and J. van den Broek, 'Chloor in Beweging', (Chlorine on the Move), *Chemisch Magazine*, February 1991, p. 68–81.

15 Tweede-Kamerfractie Groen Links, (Parliamentary Faction Dutch 'Green Left'), 'Grenzen aan de Chloor – Met een Chloortax op weg naar een Duurzame Chemie', (Putting Limits to Chlorine: Towards Sustainable Chemistry by Means of a Chlorine Tax), November 1990.

16 Die Grünen im Bundestag (Parliamentary Faction FRG 'Greens'), 'Verzicht auf PVC und Chlorchemie, Fakten und Argumente für den Ausstieg', Bonn, December 1989.

17 W. Berends and D. Stoppelenburg 'Van Keukenzout tot Gifcocktail – Chloor: Toepassingen, Milieu-effecten en Alternatieven', (From Table Salt to Deadly Elixir – Chlorine: Uses, Environmental Impacts and Alternatives), Association for the Preservation of the Waddensea and Milieudefensie (Dutch Friends of the Earth), Amsterdam, September 1990.

18 Association for the Preservation of the Waddensea and Milieudefensie 'Chloorchemie in Nederland – Verslag van de Studiedag op 2 Oktober 1990, (Chlorine Chemistry in the Netherlands – Report of a Seminar), Amsterdam, November 1990.

19 Non-governmental Conference on Environment and Development (March 1990), 'Bridging the Gap – An Agenda for Action', London, 1990.

20 S. Winteler and A. Ahrens, 'No Future for Chlorine', ed. Oekopol, Hamburg, 1990.

21 Aktionskonferenz Nordsee e.V. (Hg.) 'Ausstieg aus der PVC-Nutzung', ed. Aktionskonferenz Nordsee e.V., Bremen, 1990.

22 R. Kleijn, 'Stromen van Organochloorverbindingen in de Nederlandse Ekonomie', (Streams of Organohalogencompounds in the Dutch Economy), Centre for Environmental Studies, State University of Leyden, October 1990.

23 G.H. Vonkeman, 'The Management of Liquid Organohalogenated Wastes after the Prohibition of Incineration at Sea', in 'Report of the 1989 Seminar Management of Liquid Organohalogenated Waste', ed. European Commission, Brussels, 1990, pp. 55–67.

24 Bundes-Länder-Arbeitskreises Umweltchemikalien (BLAU) 'Chlorierte Kohlen-Wasserstoffe', 1989.

25 ECS (European Chlorinated Solvent Association) 'Comparaison des différentes Méthodes de Dégraissage des Métaux - Procédés Aqueux et Procédés aux Solvants', ed. CEFIC, Brussels, 1990.

26 Verband der Chemischen Industrie e.V., 'CKW-Lösemittel - Entwicklung, 1986-1990/1995', January 1988.

27 'Proposal for an EC Directive Relating to the Tenth Amendment of Directive 76/769/EEC Concerning the Harmonization of the Legal and Public Administration and Use of Certain Hazardous Substances and Preparations', COM (89) **548**, *Off. J. Eur. Comm.*, C8, p. 8-11, 13 January 1990.

28 U. Lahl and B. Zeschmar-Lahl, 'Facette der Chlorchemie', *Müllmagazin*, (3), 1990, pp. 52-59.

29 F. Valk, 'Polyketon, Vervanger van PVC?' (Polyketone, an Alternative for PVC?), *Chemisch Weekblad*, 1991, **87**, 53.

30 V. Norberg-Bohm, J. Yanovitz, and J. Prince, 'Materials Balance for Bromine, Chlorine, Sulphur, and Nitrogen in Europe', Working Paper WP-88-073, IIASA, Laxenburg, 1988, pp. 34-42.

21

Pesticides and Risk Perception

David M. Foulkes (deceased)

SCHERING AGROCHEMICALS LTD., CHESTERFORD PARK, SAFFRON WALDEN, ESSEX CB10 1XL, UK

'There is no point in getting into a panic about risks of life until you have compared the risks which worry you with those that do not - but perhaps should'

Lord Rothschild[1]

1 INTRODUCTION

We live with risks. To avoid it entirely would require us not to live. Consciously and unconsciously we continually assess risk in our everyday lives. In so doing each of us establishes personal risk criteria which we believe will permit us to conduct our affairs without harmful outcome. Objectivity may not be part of this process; indeed, the so-called 'facts' may only confuse an issue or, if unacceptable, be rejected.

In his review of perception of risk Slovic[2] shows how four principal factors can govern the risk-ranking an individual may place on a particular activity. Firstly, is the activity known and familiar? For example driving, smoking, or consuming alcoholic drinks. Conversely, are the risks unfamiliar, novel, or unquantifiable? For many people the health risks of synthetic chemicals, including pesticides, might fall into this area.

Thirdly, is it within our control? We have probably heard that excessive caffeine intake is bad for us and hence we can control this by drinking less coffee. Most of us believe we are safely in control of our car whilst driving. We can give up smoking at any time - or so we might claim.

Alternatively, there are risks due to influences beyond our control as individuals. Examples of this might be earthquake, nuclear reactor fall-out, and major pollution incidents.

Activities falling within these four main factors will be weighted according to a person's experience and knowledge but research indicates that the size and acceptance of a risk is, overall, based upon emotional and not rational judgement. Most people are more concerned over the airline flight they are about to make than over the drive to the airport despite the fact that, statistically, the latter is more likely to lead to death or injury.

A risk is more readily accepted if the activity in question is seen to provide benefits which override its potential for harm. Starr,[3] employing psychometric scales in developing methods for weighing technological risks against benefits, concluded that the acceptance of a risk is approximately proportional to the third power of the benefits of an activity. He also established that we will accept risks from voluntary activities such as swimming or skiing that are about a thousand times higher than we would tolerate from involuntarily experienced hazards which provide benefits of a more practical nature. In this last class fall pesticides whose benefits are often incompletely appreciated and whose hazard may be deemed to be real as they are regularly viewed as 'toxic'.

This opinion is reflected in a study reported by Upton[4] of the ranking of perceived sources of health risks in some sectors of the US population. These perceptions were contrasted with the known actuarial data relating to death and injury for the activities listed. Housewives and students ranked pesticides within the top ten and business professionals ranked them as the fifteenth most likely hazard. In reality pesticides ranked as number 28 out of 30 on the actuarial listing, with no fatalities linked to them during the 12 month survey period.

In summary, pesticides can present several aspects of risk: they can be viewed as without apparent benefit, unfamiliar, unquantifiable, inherently toxic, and imposed without choice. However, the evidence from toxicology, from safety testing, from quantitative estimates of exposure, and from epidemiology provides a relatively positive overall picture.[5]

The benefits of pesticides to mankind are multiple and do not just cover food production. It is a surprise to some to be reminded that rodents, cockroaches, disease vector carrying insects such as mosquitoes, locusts, and other insect species which harbour parasites harmful to animals and man are controlled by the use of pesticides. The chlorination of water supplies is, in effect, a pesticidal operation and is a benefit which is not yet fully available in global terms. DDT, restricted in use because of its harm to birds and environmental persistence, nevertheless continues to save lives in parts of the world where malaria is endemic and overall has probably provided greater human benefit than any other chemical substance invented by man. A recent estimate by Berry[6] suggests that DDT has saved over fifteen million lives.

Turning to crop production, it is generally estimated that up to 30% losses occur during growth or subsequent storage as a result of the competitive action of weeds or destruction by fungi or insects.[7] The role of pesticides in ameliorating these effects is of profound significance in economic and human terms. In contrast to fifty years ago, before the introduction of any significant chemical means of pest control, the proportion of income spent on food in developed countries was twice what it is today. Here the range and quality of fresh foods available have been extended in a way which could not be imagined in the post-war period. In the developing world the availability of food remains a daily priority. Without the use of materials to control crop predators the extent of shortages in these countries would be even greater. It is easy, in an affluent society, to overlook the pressing challenges which are faced elsewhere. A world population which is increasing by 200,000 people per day requires to be fed. Reversion to the non-use of pesticides would open the door to the devastation of potato crops by blight as was witnessed in Ireland in the last century and to the wholesale spoilage and loss of cereals and vines in many countries by fungal infection.

Pesticides, because they are 'meant to kill' are viewed by the layman as a class of chemicals that must be toxic. All substances are chemical in constitution and all chemicals toxic. The key lies in the quantitative dose which needs to be administered to induce a toxic response. This observation, often overlooked, was made some centuries ago by the German philosopher Paracelsus.[8] It is the role of the toxicologist to determine the type of toxic effects revealed by a particular substance and the level of dose that produces these effects. To know that a substance possesses 'toxic properties' is not enough. We need also to know whether or not this substance will present a risk, a risk being the probability, over a defined time period, of a hazard manifesting itself.

2 PESTICIDE TOXICOLOGY

As indicated above it is vital to determine, for any pesticide, the extent of its toxic properties and the circumstances under which it may pose a hazard. This must take into account the manufacturer, the user, the consumer, and the environment. For the purpose of this discussion the potential effects on human health will be given primary consideration.

Any potential for harmful effects of a pesticide on the environment are investigated to considerable breadth and depth. Rates of breakdown and persistence in the environment and the likelihood of there being any short- or long-term effects on important sectors of animal, bird, fish, and insect populations have to be established as part of the international registration process.

In order to build up a picture of the likely effect upon man the toxicologist sets out to explore a series of determinants. These, when taken together, will lead him to conclude whether or not the chemical in question, albeit possessing certain toxic properties, is likely to present a risk in use. The overall process of deriving toxicity data is one which has close parallels between pesticides and pharmaceuticals. Without listing all the many combinations of effect which can be studied, the overall areas which are pursued are as follows: carcinogenicity; neurotoxicity; reproductive effects; mutagenicity; specific organ effects; pharmacology and behavioural parameters; metabolic effects.

Although the studies which are undertaken are performed in accordance with the internationally agreed protocols which are required by regulatory agencies, findings may emerge that require further study and clarification in order to put them into perspective. For example, there could be an influence on hormonal levels which may precipitate tumour formation. An understanding of the role of such an hormonal alteration would be important in judging the carcinogenic potential of the substance in question. Such additional work would be undertaken beyond the basic requirements laid down for registration.

The studies which are performed in a range of species take a minimum of five years to complete at a cost of several million pounds to the agrochemical company. At all times the toxicologist is seeking to establish the relevance of his findings to possible effects in man.

Without entering into the physiological debate over the use of animal models for predicting effects in humans, animal studies conducted to the protocols of today are probably over-cautious in their prediction of a likely effect in man. This is clearly to the good.

Protocols are often deliberately set up to weigh against the chemical under test. In long-term toxicity studies it is a requirement to include a high dose level which must be shown to reach the Maximum Tolerated Dose (MTD). The use of such high doses can result in artificially induced chronic cell death and thereby increase further cell division of neighbouring cells. This process can lead to tumour formation and consequently the use of the MTD approach is open to question. Results from tests run at the MTD have shown a similar percentage of chemicals to be carcinogenic whether or not they are synthetic or of natural origin.[9] Although it is intended to weigh studies in favour of reduced risk the interpretation of studies conducted at such high dose levels can lead to confusion in the public mind. Concern results when a substance, administered at doses excessively higher than any likely human exposure, produces cancer in animals. The emotive reality of cancer formation takes on a major importance and the quantitative, overdosing aspect is overlooked or dismissed. The manner in which the growth regulator Alar came to be withdrawn is a classical example of this. In wishing to bring the Alar issue before the American public, the National Research Defense Council (NRDC) avoided reference to a quantitative perspective on the risk involved. Sectors of the public became seriously concerned that the consumption of apples and apple produce linked to Alar usage could cause cancer. What was completely overlooked was the fact that it would be necessary for a child to consume 18,000 litres a day of apple juice for a lifetime in order to receive a dose approximate to that which produced a cancerous effect in rodents. Clearly any interpretation of risk must take into account all the facts of a scientific investigation. This was certainly not the case in the Alar debate, as a fascinating review published by the American Council on Science and Health concluded.[10]

Public concern over risk presented by pesticides is often directed to the issue of residue in foods as in the case of Alar. In order to estimate the likelihood of risk from this quarter, three disciplines are brought to bear in conjunction with each other. Firstly, the biologist seeks to establish what amount of chemical needs to be sprayed onto a crop in order to control either the weed, insect, or fungus. Secondly, the analyst takes samples from the harvested crop and determines how much, if any, of the chemical remains in it before and after processing. Finally, the toxicologist links his data with that of the others to see the relationship between the amount which in his studies produces an effect, in comparison with the amount found in the crop. The toxicologist applies a number of quantitative safety factors and generally imposes a hundred-fold factor to bias his findings against the chemical. If the calculations and results show any indication of risk then the product is not pursued for use on that particular crop.

Confusion and concern often arise when the so called Maximum Residue Level (MRL) has been found to have been exceeded. The MRL is derived by determining the amount of residue present when the product is used in accordance with the label - so called 'Good Agricultural Practice'. For this reason the MRL is purely a measure of the latter and in no way in itself is an indicator of safety or hazard. The MRL is expressed in mg of pesticide per kg of crop. In practice it may be possible to exceed the MRL by several orders of magnitude before the consumption of such a crop might pose a risk to health. Thus consideration must be given to the **actual amount** of pesticide consumed. This is clearly dependent upon factors such as the amount of the foodstuff which is likely to be eaten on a daily basis. The toxicologist takes into account his findings and the built-in safety factors and arrives at a value termed the Acceptable Daily Intake (ADI). This is simply the quantity in mg of a substance which can be consumed daily for a lifetime with the practical certainty, on the basis of all the known facts, that no harm will result. For example, an ADI for a particular pesticide may be 60 mg. The MRL for the same substance when used on tomatoes may, for sake of argument, be 1 mg of the pesticide per kg of tomatoes. Suppose that a farmer has, contrary to the label (and hence the law in many countries), applied more of a pesticide than required and the resulting amount is 2 mg of pesticide per kg of tomatoes. Although the MRL had been exceeded by a factor of two the consumer would nevertheless have to eat, for a lifetime, 30 kg of tomatoes over-treated in this manner every day before reaching the amount which the toxicologist has deemed an Acceptable Daily Intake. Within the ADI there are additionally numerous built-in safety factors which make it improbable that these particular tomatoes, containing a pesticide at levels exceeding the MRL, are likely to be harmful.

Opponents of pesticides often gain media attention by drawing attention to analyses of crop samples where the MRL has been exceeded and imply that some health barrier has been breached. An appreciation of these concepts is fundamental to determining whether or not a risk is likely to ensue from the use of a particular pesticide. Because of the methods employed and the mathematics involved it is not always easy to convey the appropriate information to the consumer to enable him to make a judgement.

The principles of international registration are set up on the assumption that a pesticide may be harmful. Before registration can be granted unequivocal evidence of the lack of harmful potential must be presented and accepted.

In setting a perspective upon the likely risk of harm resulting from pesticide residues in food it is valid to note that, as knowledge progresses it is becoming clear that most of the crops we consume contain significant levels of natural compounds which are in themselves toxic.[11] Given the expense and the resource required for long-term toxicology studies, it is not surprising that the health effects of these natural toxins are as yet largely undetermined. Where studies have been undertaken the results show that Nature does not differentiate between natural and synthetic chemicals. An equal proportion of either of them (*ca.* 50%) shows a potential for cancer formation in rodent studies.[9] The conclusion has to be that Nature does not differentiate between the chemicals she makes herself and those which are contrived by man.

3 THE EPIDEMIOLOGICAL APPROACH

Can epidemiological studies in man provide a ready answer to the question of whether pesticides are producing adverse effects which could not have been foreseen from animal models? Unfortunately epidemiology is a blunt rather than a precise instrument whose ability to measure is often confounded by a variety of factors. For example, it is essentially impossible to establish what pesticides a consumer will ingest in his diet at any particular point in time, if they are there at all. It is also equally difficult to determine precisely what other factors in a lifestyle might be present which in themselves could lead to the manifestation of ill health. Smoking, drinking, and other dietary habits could have an influence far beyond that of a minute amount of pesticide in a diet. Surveys in the UK and the US continue to show that most foodstuffs, especially after processing, do not contain any measurable residues of pesticides whatsoever. When residues are detected they are not found at levels which represent a health risk. Furthermore, in developed countries persistent pesticides are no longer used. Only products which break down quickly in the environment or are rapidly metabolized in mammalian systems are permitted to be registered.

Evidence is rapidly accumulating that natural toxins are present in our food to an extent much greater than was previously thought.[11] In a quantitative sense Ames concludes[9] that 99.99% of the toxins we ingest from our food are of natural origin and it is certainly true to say that in the case of many of them their complete potential for harmful effect is not yet fully established.

It should be remembered that the toxins produced by plants provide a defence mechanism against specific predators. When exposed to attack, *e.g.* by fungal infection, the levels of these toxins may be raised to levels which could pose a risk to the consumer. In this example the use of synthetic fungicides eliminates this source of risk.

Against this background it is difficult to see how epidemiological research alone could ever demonstrate an injurious effect from the presence of pesticides in foodstuffs where the crops have been grown and treated at the proper rate.

An alternative approach is to examine health trends over the last fifty years in those countries where statistics are particularly sound. In the US, for instance, mortality information has been available in a detailed form for many years. Taking cancer as an end-point it is clear that the age-adjusted mortality rate for cancer in the US has been declining since 1950. The exception is lung cancer which, in contrast to the reduction in cancers of other organs, has increased by nearly 250% as a result of smoking. The time span upon which these statistics is based is also coincidental with the introduction of modern synthetic pesticides in developed countries. Although it is not possible to prove a negative, it is nevertheless reassuring to observe that, in these countries, life expectancy has steadily increased alongside a reduction in disease and a mortality rate which could be linked to chemical involvement.

The sector of the population contained within the manufacturing and user groups is most likely to provide useful data linking pesticides to human health. Usually persons involved in the manufacturing, formulation, and use of pesticides will have a more consistent contact with them and this contact is potentially at higher rates than that which the consumer of food would encounter. In many manufacturing environments, particularly in developed countries, there are occupational health services whose records allow an assessment to be made regarding links between work activity and health. A considerable body of literature has been published which has investigated work force health effects in which pesticides can be implicated.

A forthcoming review[12] of all the relevant published literature concerning cancer mortality amongst pesticide manufacturers and applicators concludes that workers in the pesticide industry do not appear to be at more of a health risk than workers in other industries involving the handling of chemicals. The evidence for association between pesticides and certain types of cancer being contracted by farmers remains equivocal, as there are contradictory data from one study to another. At present conclusions in this area must remain open. Overall the scientific evidence derived from mortality studies of these two groups, the ones with the highest potential for exposure, do not support a high risk for pesticides in general. Despite the inadequacies of the epidemiological approach it nevertheless remains an important additional complement to data derived from other sources.[13]

The mis-use of pesticides, particularly in developing countries where control, training, and education as to use is often less advanced, is a cause for concern. Economics often direct such countries towards the use of older, less safe, and cheaper products. Pesticides have regrettably been used as an agent of suicide, which has increased the apparent risk factor they present. In practice, evidence shows that providing pesticides are applied in accordance with the directions provided by the manufacturer, no ill health will result. It is certainly true to say that no evidence has yet emerged to demonstrate that consumers have suffered any ill effects from eating crops which have been likewise treated in accordance with the label instructions. Some instances have occurred when pesticides have been illegally applied at excessive rates or to crops for which no approvals have been given. Short-term ill health has resulted for those who subsequently ate the crops. From the point of view of risk perception such incidents naturally lead those who are not conversant with the facts to believe that this is further proof that pesticides pose an inherent and serious risk.

4 CONCLUSIONS

Pesticides, principally in the form of synthetic chemicals, will continue to be used in the foreseeable future as a means of improving the quality and quantity of food production as well as playing an important role in many sectors of human and animal health. The products used in these areas today represent a significant magnitude advance in terms of safety compared with nicotine, arsenic, and sulphuric acid which were among the principal agents available to the farmer in the early part of this century. As understanding of biochemical processes continues to be enlarged so it will be possible to produce pesticides which are highly targeted to effect a very specific function within the plant, fungus, or insect without having impact upon other organisms. The significance of this trend has not yet received full recognition and public perceptions, supported from time to time by the media, often reflect the very real but outdated issues raised by Rachael Carson in her seminal book 'Silent Spring'. It is not proper to assume that the problems she listed over twenty-five years ago are still represented in the face of current technologies.

Added to this base of improved technology are a number of important consequential factors which, taken together, considerably reduce the basis of apparent risk presented by the use of pesticides. Firstly the quantities present in food and the environment are increasingly small and the residence time of the newer pesticides is becoming shorter and shorter. Given the extensive safety testing to which a pesticide is now subjected (a level of testing akin to that used for pharmaceuticals), it is possible to be increasingly confident that the health threat from pesticides is vanishingly small. Placed in context with the fact that Nature is not benign, it is not entirely logical to place concern in the direction of a pesticide residue without taking into account the dictum of Lord Rothschild and considering the toxins which are already there in the first place.

Society in developed countries is for the most part the most affluent, the most well fed and certainly the longest lived in the history of mankind. To go one step further into a hypothetical zero–risk society would involve political and economic change which would be untenable. The elimination of pesticides by no longer using them may be of superficial attraction to those seeking the immortality implied in a zero–risk.[8] Such a step would actually lead to food becoming increasingly expensive. Paradoxically it could also increase the presence of natural toxins within food that is the consequence of attack by pests.

For many the risks associated with pesticides will remain unfamiliar, disassociated from everyday activity and beyond direct control and management. Clearly it remains for the scientific community and the farmer to ensure that the consumer is aware of the safety of modern pesticides and the benefits that they convey. With this information the consumer will be better placed to put pesticides in perspective and perhaps not to regard them as a source of risk.

Finally a commonsense approach can provide healthy reassurance. Daniel Koshland[14] in his editorial in the journal *Science* has elegantly and humourously presented an antidote to which I would direct anyone seeking a balanced illumination of risk.

5 REFERENCES

1 Lord Rothschild 'Risk', The Dimbleby Lecture, *The Listener*, 1978, **100**, 715–718.

2 P. Slovic, *Science*, 1987, **236**, 280–285.

3 C. Starr, *Science,* 1969, **165**, 1232

4 A.C. Upton, *Sci. Amer.*, 1982, **246**, 41–49.

5 D.M. Foulkes, 'Agrochemicals in the Environment', in 'Chemistry, Agriculture and the Environment', ed. M.L. Richardson, The Royal Society of Chemistry, Cambridge, 1991, pp. 3–10.

6 C.L. Berry, Proceedings of the Brighton Crop Protection Conference, 1990, **1**, pp. 3–13.

7 Food and Agriculture Organization (FAO), 'Agriculture towards 2000', Rome, 1981.

8 'Risk Assessment of Chemicals in the Environment', ed. M.L. Richardson, The Royal Society of Chemistry, London, 1988, pp. xxi, 532–541.

9 B.M. Ames and L.S. Gold, *Angew. Chem. Int. Ed. Engl.*, 1990, **29**, 1197–1208.

10 American Council on Science and Health, 'Alar one year later', Washington, 1990.

11 R.C. Beier, *Rev. Env. Contam. Toxicol.*, 1990, **113**, 47–137.

12 L.P. Brown, 1991, A Review of the Cancer Mortality of Pesticide Manufacturers and Applicators, *Hum. Exp. Toxicol.*, in press.

13 D. Coggon, Proceedings of the Brighton Crop Protection Conference, 1990, **2**, pp. 657–663.

14 D.E. Koshland, *Science*, 1989, **244**, 1529.

22

Industrial Liability – The Future in Europe

J.-M. Devos

GENERAL SECRETARY, EUROPEAN CHEMICAL INDUSTRIAL COUNCIL (CEFIC), AVENUE E. VAN NIEUWENHUYSE 4, BTE 1, 1160 BRUSSELS, BELGIUM

'All losses and all damages caused by the deeds of any person, through negligence, thoughtlessness, ignorance of what one should know or similar shortcomings, however trivial they may be, must be compensated by the person whose fault or negligence has created the damage'

Domat

1 INTRODUCTION

The hallmarks of the European chemical industry are its considerable size as an economic sector and its huge diversity. A significant number of research, production, and processing activities are carried out by a host of companies. Once manufactured, the most diverse products – from the simplest to the most technologically complex – are placed on the market, and are subject, among other matters, to transport, packaging, labelling, and storage operations. All these activities may cause damage either to persons, to goods, or to the environment. It is thus easy to comprehend the importance of 'civil liability law' or 'third party liability' and the economic impact on industry in general and the chemical sector in particular.

'Manufacturer's liability' must be considered in the framework of general legal systems of third party liability, or 'responsibilité ou quasi-délictuelle' in French civil code systems.

Faced with growing political and legal constraints aggravated by the situation of chemical producers through the progressive introduction of 'strict liability' regimes, CEFIC has developed a consistent and coherent approach based on the principle 'Legal liability must be linked to operational responsibility'.

The chemical industry has called for fair and economically viable systems of liability where the interests of claimants and those of the industry would be balanced. To various degrees, these international instruments of civil liability are gradually discarding the traditional basis, fault, or negligence, and moving towards 'strict liability' regimes.

This chapter reviews various major international legal instruments recently adopted or under discussion in this area.

2 THE EC REGIME ON LIABILITY FOR DEFECTIVE PRODUCTS

2.1 Background and Legal Basis

The directive was approved by the Council on 25 July 1985 and published in the official journal of the European Communities on 7 August 1985 (O.J. No. L 210). Member States had to implement the directive by July 1988. However, only a few EC Member States did actually meet this deadline.

The declared objective of approximation of the laws is partially attained, as the text allows the Member States to introduce certain variations to three specific provisions.

Article 100 of the Treaty of Rome provides the legal basis of the directive. Since the adoption of the Single European Act amending the EC Treaty, a new Article 100 (a) was introduced to enhance the achievement of the internal market. Article 100 lays down the following conditions:

i) Unanimous council decision, based on a proposal from the Commission.

ii) Compulsory consultation of the Assembly (European Parliament) and the economic and social committee if there is a need to amend national laws. Both institutions were consulted and expressed their opinion.

iii) Grounds for the directive: according to (i), the approximation of the laws of the Member States concerning the liability of the producer for damage caused by the defectiveness of his products is necessary because the existing divergencies may distort competition, affect the free movement of goods within the Common Market, and lead to differing degrees of protection of the consumer.

iv) Social–political justification: it is assumed in the preamble that 'liability without fault on the part of the producer is the sole means of adequately solving the problem, peculiar to our age of increasing technicality, of a fair apportionment of the risks inherent in modern technological productions'.

The industry has challenged strict liability and expressed preference for liability regimes based on fault or negligence in accordance with traditional principles of the law of the torts. This battle was lost, however, when even Italy, once a strong opponent of strict liability, accepted this regime in the political compromise that led to the adoption of the directive in 1985.

Was this a real source of concern to industry and should strict liability of *a priori* be perceived as hostile by industry? My personal analysis is that the answer very much depends on three interrelated factors:

i) The content of actual provisions of the directive;

ii) The content of the national laws;

iii) The case law, including that of the EC Court of Justice, which could play an important role in clarifying ambiguities.

More important than the 'labels' or the 'names' used to qualify liability are their specific features. Other elements are important too: the practice of the courts, the judicial organization and its cost, and the general attitude of the public towards the liability and compensation system.

In this connection, the directive, though far from perfect from the legal certainty view point, was probably a reasonable compromise acceptable to governments, industry, and consumers (though some would not share this view, both in consumer circles and in industry). As will be seen, the question of recognition of the State of the Art Defence was an important aspect of this 'package'.

2.2 Nature of the Liability Regime

2.2.1 Principle: no fault liability. The directive provides for the following system: the injured party has to prove the damage, the defect of a product and the causal relationship between the defect and the damage (Art. 4). Thus, negligence and fault are no longer relevant as sources of producer's liability.

The suppression of the 'fault' standard should ease the task of the victim. However, it should be stressed that the case law of most countries had already introduced a softening of the position of the victim, for example in reversing the burden of proof or in extending considerably the concept of fault or 'negligence' of the producer. In a way, we could say that the duty of care is now very strict indeed, with the important exception of the State of the Art Defence.

2.2.2 Lines of defence and exemptions from liability (Art. 7 of the directive). It is for the producer to prove that his is one of the cases of exemption provided for, *viz.*

i) That he did not put the product into circulation;

ii) That the defect did not exist when the product was put into circulation or that it came into being afterwards;

iii) That the product was neither manufactured for sale or any other form of distribution, nor manufactured or distributed by him in the course of his business;

iv) That the defect is due to compliance of the product with mandatory regulations issued by the public authorities;

v) State of the Art Defence ('development risk'). This fifth defence will be dealt with in the following section because it is of special importance;

vi) In the case of the manufacturer of a component, that the defect is attributable to the design of the product in which the component has been fitted or to the instructions given by the manufacturer of the product.

This last point is very important and was stressed by the chemical industry. All sectors of the chemical industry should be fully aware of this defence. It will be more and more important to have clear agreements between purveyors and buyers about the purity and characteristics of component products.

2.3 The Three Options of the Directive

The harmonization dilemma. A fundamental question that had to be addressed by the authors of the directive was the so-called 'harmonization' question, *i.e.* should the EEC directive provide for strict harmonization ('maximum directive') or should it create a 'minimum system' allowing the Member States to establish different and additional standards, depending on their concepts of the 'protection of consumers'? Real harmonization is one of the 'raisons d'être' of this text and therefore was preferable. The EEC Member States delegations recognized that a 'maximum directive' would be the best solution from the Treaty of Rome point of view, but at the level of details, compromises became necessary and the 'maximum' approach turned into a limited approximation of laws.

2.3.1 The State of the Art Defence. A highly important question is, should the manufacturer of a defective product be held liable if the state of scientific and technological development did not enable him to detect the existence of a defect at the time the product was put into circulation?

It was strongly argued that the inclusion of development risks in the scope of liability would interfere with technological innovation. Until now, in the vast majority of Western legal systems, no party has had to accept the risk of the unknown and unforeseeable. As CEFIC said in a position paper addressed to the EEC Council:

> 'While it is perfectly understandable that a system of liability should oblige producers to take all possible precautions against any possible injury to the consumer, it seems totally unreasonable to require that these same producers be liable for the unknown, and for risks over which they have no control whatsoever. Injury caused by a defect in

a product cannot be a source of liability if the defect was undetectable and unforeseen at the time of its being put into circulation because of the state of the art'.

After long and difficult debates, the EC Council finally introduced in the directive what is known as the 'State of the Art Defence'.

Accordingly, the producer has the possibility to prove that the state of scientific and technical knowledge when he placed the product into circulation was not such as to enable the existence of the defect to be discovered. This defence may therefore be introduced in the legislation of Member States and was effectively introduced in most laws including that of the UK. However, the community regime allows individual Member States to maintain the liability of the producer for 'development risk' if it already existed in their internal laws when the directive was approved (*i.e.* July 1985). Only Germany was affected by this provision, as its 'Pharmaceutical Law' did not allow producers to use the State of the Art Defence.

The community regime also allows Member States to introduce derogations to that principle in new legislation, though this possibility is tempered by a clause of community 'stand-still' and the obligation to follow a specific procedure under the supervision of the Commission (Art. 15(1) of the directive).

The importance of recognizing a State of the Art Defence is clear if one considers its impact on R & D efforts, which in certain sensitive areas might be discouraged otherwise. There is already evidence, especially in the USA, showing that the liability regime based on absolute liability and disregarding important factors like the state of technology and plaintiff's behaviour has had adverse effects on new products.

If we take the USA experience, the impact of certain Court cases based on 'absolute producer liability' has been enormous insurance costs and (non)availability of cover. In West Germany, the Pharmaceutical Laws have not – so far – generated similar experiences. However, nobody can predict what would be the consequences for the German pharma pool of a serious development risk case similar to the Thalidomide Affair (which, in the strict sense, was not a development risk case).

Wisdom in the debate over the State of the Art Defence inclines to moderation. I personally believe that the compromise of the directive preserves fair balance and efficiency.

<u>2.3.2 Derogation in respect of financial limitation of liability</u>. The directive makes no provision for a community system of financial ceilings to liability. It does, however, allow the Member States to introduce on an individual basis a liability 'ceiling' for personal injury, which may then not be less than 70 MECU. The German and Belgian laws have, for example, introduced such ceilings.

The Commission is instructed to submit a report to the Council 10 years after the date of notification of the directive (1995). Acting in the light of this report the Council is to decide whether or not to cancel the option of introducing ceilings. As with the 'State of the Art' issue, such a provision mirrors the Council's political embarrassment at the time of the adoption. The Council will, however, be free to decide to maintain the State of the Art and the ceilings option if a political majority so wishes.

2.3.3 Derogation in respect of agricultural products. The directive normally excludes agricultural products from its scope. However, Member States are authorized to include them within their laws and to apply the regime of the directive.

2.4 Other Important Aspects of the Directive

The directive covers not only personal injury but damage to property (with a 'lower threshold' of 500 ECU in the second stage). This provision has led to some divergent interpretations.

2.4.1 Definition of producer. 'Producer' is extensively defined in the directive as:

The manufacturer of a finished product;

The producer of any raw material;

The manufacturer of a component part;

Any person who, by putting his name, trademark, or other distinguishing feature on the product presents himself as its producer.

Clearly, the concern of the directive's drafters has been to avoid the possibility of an injured person finding himself in a situation where there was no responsible person. Hence the care taken in defining the producer. So true is this that the directive provides for treating the importer into the community as a producer within the meaning of the directive and, should it prove impossible to identify the producer, treating the supplier as the producer.

2.4.2 Product defectiveness. For the purposes of the directive, a product is defective not according to objective or material criteria but with reference to the 'safety which a person is entitled to expect'.

Paragraph 6 of the recital states that it is not the product's fitness for its intended use but the lack of the safety which the public at large is entitled to expect. Such a reference creates difficulties of interpretation. In reality, it will be for the judge to assess. However, the directive provides usefully that all circumstances should be taken into account in assessing the safety that the product should provide. Three criteria are mentioned, without the list being exhaustive: (a) the presentation of the product; (b) the use to which it could reasonably be expected that the product would be put; and (c) the time when the product was put into circulation are just some of the circumstances to be considered in assessing the product's safety.

Safety is to be assessed by taking into account the expected use of the product, the marketing, the given information and user instructions, and other similar factors related to the product.

2.5 Conclusion

The new community regime on liability for defective products contains both 'pros' and 'cons' and seems generally balanced. Despite its imperfections and weaknesses, it genuinely contributes to the establishment of similar legislations in Europe. Its practical impact, particularly on insurance costs, has so far not been dramatic but is still uncertain and will depend on the content of national laws and their application by national courts under the control of the European Court of Justice.

3 THE RECOGNITION OF ECOLOGICAL DAMAGE IN INTERNATIONAL AND COMMUNITY LAW OF CIVIL LIABILITY

One of the most significant aspects of civil liability law and current work in the international sphere is the extension of the concept of damage (death, personal injury, damage to property) to 'ecological damage'. This same ecological damage is defined in various ways depending on the instruments analysed. Some common features can, however, be identified:

Environmental damage concerns the community, not the individual;

Environmental damage is difficult to quantify.

As will be seen, harmonization of international laws in this field is a particularly complicated task.

The proposal for a Council directive on civil liability for damage caused by waste[1] defines 'environmental damage' as 'an important and persistent interference in the environment caused by a modification of the physical, chemical, or biological conditions of water, soil, and/or air ...'. The criterion of persistence could be abandoned according to Commission sources. This definition would appear to be more restrictive than the one envisaged in the draft Council of Europe Convention,[2] which refers to 'any loss or damage resulting from impairment of the environment, provided that compensation ... is limited to the costs of measures of reinstatement actually undertaken or to be undertaken'. This definition includes the definition of the 'CRTD Geneva Convention' of 10 October

1989.[3] The word environment is itself defined very broadly in the Council of Europe draft, as it covers natural abiotic and biotic resources, such as air, water, soil, fauna, and flora, and the interaction between the same factors, assets making up the cultural heritage and the characteristic aspects of the landscape. These last two elements of the definition serve to illustrate admirably the erosion of the traditional concept of damage, as it is now a matter of covering damage to collective property by civil liability mechanisms. Such damage is particularly difficult to quantify. It is worth noting that the Commission of the European Communities has avoided giving a definition of the word 'environment' in the 1989 draft of its proposed directive on liability for waste.

In response to these initiatives concerning the definition of new legal regimes that take into consideration ecological damage, CEFIC has advocated multilateral solutions, not just because of the transfrontier effects of pollution but also on account of distortions of competition that might result from the application of divergent national legal regimes.

4 COUNCIL OF EUROPE WORK ON COMPENSATION OF DAMAGE CAUSED DURING THE EXERCISE OF DANGEROUS ACTIVITIES

4.1 Background to and Status of Work

In 1987, following the Conference of European Justice Ministers (Oslo 1986), the European Council of Ministers Committee set up a Committee of Experts on compensation for damage to the environment. The programme is based on the belief that issues relating to civil liability and compensation for damage resulting from dangerous activities need to be addressed at international level. The main aim of the programme is to draft a set of rules designed to ensure prompt and effective compensation of damage to people, to property, and to the environment during the exercise of dangerous activities carried out in a professional capacity in installations or on sites.

After its sixth meeting, the Committee of Experts instructed the Secretariat General and the Legal Affairs Directorate of the Council of Europe to prepare an interim activity report. This report was to be submitted to the European Committee on Legal Co-operation (CDCJ), a senior Council of Europe body in the legal field, and to Member States of the Council. Containing a description of the Committee's work, the report encloses preliminary draft rules on compensation for damage resulting from dangerous activities. The report refers explicitly to the CEFIC memorandum on civil liability for damage linked to an industrial activity, sent to the Council of Europe in January 1988.

When adopting its report, the Council of Europe Committee of Experts decided to inform and officially consult CEFIC, the European Insurance Committee, the European Environment Bureau, and the

International Union for the Conservation of Nature and Natural Resources.

In January 1991, the Committee of Ministers agreed to the publication of a revised version of the draft Convention, urging that interested circles be consulted. The draft rules are expected to lead to an international convention.

4.2 A Brief Look at the Draft Council of Europe Convention

4.2.1 Liability regime. The draft provides for a regime involving 'strict' civil liability of 'the operator of a dangerous activity'. The operator's responsibility is not based on the concept of fault, but on the causal link existing between the damage caused and the incident occurring at the time or during the period when the operator was in charge of a dangerous activity. It will be noted that this general approach is in line with the one advocated by CEFIC, which has always wanted to see a strict link maintained between civil liability for damage and operational responsibility.

The Council of Europe report rightly points out that one of the major reasons for making the person in charge of the activity from which damage resulted bear the cost of damage is that this person is best placed to prevent its occurrence or to limit its extent.

4.2.2 Incident. The word incident is defined as any instantaneous or fully continuous occurrence, such as an explosion, a fire, a leak, an emission, or any sequence of events having the same origin ...

4.2.3 Concept of operator. The operator is defined as being 'the person in actual control of a dangerous activity'. The concept 'actual control' is not, however, defined in the draft as it stands in early 1991.

4.2.4 Concept of dangerous activity. Dangerous activity is defined as being: 'One or more of the following activities, provided that it is performed professionally, including activities conducted by public authorities:

i) The handling, storage, production, or discharge of one or more dangerous substances or any operation of a similar nature dealing with such substances;

ii) The handling, storage, production, or discharge of one or more dangerous genetically modified organisms or micro-organisms or any other operation of a similar nature dealing with dangerous genetically modified organisms or micro-organisms;

iii) Activities involving technologies producing dangerous radiations (like those mentioned in the planned Annex II of the Convention);

iv) The operation of an installation or a site for the incineration, the treatment, the handling, or the recycling of waste like the installations or sites mentioned in the planned Annex III of the Convention, in so far as the quantities involved cause a significant risk to man, the environment, or property;

v) The operation of a waste disposal installation or site.

The definition of dangerous activities under point (i) refers to lists of dangerous substances to be attached to the text of the rules. It should moreover be noted that the operators of a waste installation or site are explicitly subject to the new system.

4.2.5 Concept of damage. The Council of Europe text mentions legal categories traditionally included under the concept of 'damage' in civil liability matters, *i.e.*

Damage arising from death or personal injury caused by the dangerous substance or waste;

Any loss of or damage to property caused by the hazardous substances or waste.

As was seen previously, the text also introduces the concept of loss or damage to the environment, apparently going beyond the precedents established in the 'UNIDROIT' Convention and in the European Commission draft on liability for waste.

4.2.6 Right of action for private groups. The text to some extent authorizes actions by private associations whilst restricting the scope of such actions before the courts:

Requests for prohibition of any unlawful activity posing a serious threat of injury to the environment;

Requests for an order to the operator to prevent an incident or damage;

Requests for an order to reinstate or clean up the environment.

4.2.7 Causality link and administration of proof. While maintaining the requirement of a causal relationship between the incident giving rise to damage and the activity of the operator, rule 5 specifies that the court 'shall take due account of the increased danger of causing such damage inherent in the dangerous activity'.

4.2.8 Other points. The draft rules also mention:

The possibility of introducing compulsory insurance (or other equivalent financial security) to be maintained by the operator;

A system of access to information held by authorities or operators.

The same system explicitly recognizes the right to protect the confidentiality of data. The system is broadly based on the provisions of community law governing the subject.

These two points will doubtless be much debated within the Council of Europe.

5 CIVIL LIABILITY FOR DAMAGE CAUSED DURING THE CARRIAGE OF DANGEROUS GOODS BY LAND: THE CONVENTION OF 10 OCTOBER 1989[4]

The International Institute for the Unification of Private Law and the United Nations Economic Commission for Europe have actively co-operated in framing an international instrument on liability and compensation for damage caused during the carriage of dangerous goods by road, rail, and inland waterway vessel. Prepared within the Institute of Rome, the draft was subsequently forwarded to the UN Economic Commission for Europe. CEFIC, along with other international organizations representing transport and insurance interests, participated as observer in the work of UNIDROIT and the United Nations.

The draft convention was finalized at the extraordinary session of the Inland Transport Committee of the UN Economic Commission for Europe, held in Geneva from 2nd to 10th October 1989. CEFIC was represented at this session, reacting favourably to the main thrust of the Convention, which enshrines the linkage between civil liability and operational responsibility. This last, dated 10th October 1989, has been open for signature by the States since February 1990. Five ratifications are required for it to come into force. Among the countries which have expressed their active support, one should mention EEC countries (especially the Netherlands, Denmark, Germany, France with some reservations), EFTA countries (notably Finland and Austria), and central and eastern European countries as well as the United States.

The salient features of the Convention are as follows:

'Strict' liability regime;

For the purposes of the Convention, carriage comprises the period from the beginning of loading of goods to the completion of unloading operations;

Channelling of liability on to the carrier (except in the case of loading or unloading when such operations are carried out under the control of another person);

Reference to the 'ADR' list for the definition of substances to which the Convention will apply (including hazardous wastes);

Principle of compulsory insurance (or any system of equivalent financial security);

Limiting of total financial liability of the carrier (road – rail) to a ceiling of 30 million ECU per vehicle (18 million per personal injury; 12 million for damage to property and the environment). In the event of the 'personal injury' ceiling being used up, the amounts of unmet claims for personal injury could compete with claims for damage to property within the 'damage to property' ceiling. The liability ceiling specified for inland waterway transport is 15 million ECU (8 million for personal injury, 7 million for other damage);

Definition of the carrier (road, waterway) as being the person having use of the vehicle. The latter is presumed to be the person on behalf of whom the vehicle was registered. This presumption may be reversed;

The rail carrier is defined as being the person(s) operating the railway on which the incident giving rise to damage occurred;

Grounds for exemption from liability based on:

> The lack of information by the shipper or any other person as to the nature of the goods;
>
> The fact that a third party acted or omitted to act with intent to cause damage;
>
> The deliberate fault of the injured party;
>
> The occurrence of certain exceptional events: act of war, civil war, or natural phenomenon of an exceptional and irresistible character;

Extension of the definition of 'damage' which covers ecological damage in addition to the more traditional categories 'death or personal injury' and 'damage to property'. Ecological damage is defined as 'any loss or damage by contamination of the environment caused by dangerous substances, provided that the compensation for impairment of the environment ... is limited to costs of reasonable measures of reinstatement actually undertaken and to be undertaken'.

The States will, however, be able to introduce some reservations if they so wish (particularly concerning defences and the limiting of liability). From the economic and practical standpoints, it does not look as though the Convention will bring upheaval in land and rail transport. The Counseil des Bureaux (international insurers' body designed to facilitate the settlement of damage caused by motor car traffic – 'green card system') has stated that the 'green card system' will adapt fairly easily to the introduction of the Convention into Member States' laws. As far as rail is concerned, the vast majority of railway companies have their own systems of insurance cover. The Convention is, by contrast, a novelty for waterway transport, for which there has been no compulsory insurance regime up to now.

The authors of the draft have forwarded a resolution to the International Maritime Organization for a similar convention to be adopted for maritime transport, thereby complementing the work carried out in regard to land transport.

6 COMMUNITY DRAFT ON LIABILITY FOR WASTE

Under the framework-directive (EC/84/631 of 6th December 1984) on the control of transfrontier movements of waste, the Commission has submitted a proposal for a directive on a liability regime.[5*] CEFIC has and will continue to plead for there to be coherence between international instruments. The Community directive on liability for hazardous wastes should be compatible with the UNIDROIT Convention and, on a broader level, with the Council of Europe approach. With the content of the current 1989 Commission draft, this compatibility does not seem ensured. This last is, in effect, based on a strict channelling of liability on to the 'producers' of waste, departing from liability of the operator or the economic operator in control of the waste.

Many questions raised by the Proposal for a Directive will need to be addressed and resolved in full consideration of the Council of Europe's work on industrial liability.

The following sections describe the main aspects of the directive.

6.1 The 'Polluter Pays' Principle and the Channelling of Liability

Article 3 of the 1989 proposal poses the principle of the liability of the waste producer for damage and injury to the environment 'caused by such waste' irrespective of a fault on the part of the producer. The proposal is intended to implement the 'polluter pays' principle. It should be noted, however, that this principle – the legal ambit of which remains ill-defined, does not necessarily imply a strict liability regime. A liability regime based on the polluter's fault could have been compatible with this principle, if the regime ensures that the polluter actually suffers the consequences of the pollution. The Commission's choice of a strict liability

*<u>Note</u>: At the time of writing this report (beginning of 1991), the Commission was considering amending its proposal on a number of points, thereby going some way to meeting the opinion of the European Parliament. This Institution has advocated a regime based on liability of the operator (producer or disposer) for damage caused by his waste, provided this same operator has control of the waste (report of the session on 22nd November 1990, PV 43, EP 146, 824).

regime 'channelled' on to the producer is explained by the concern to use civil liability as an instrument to eliminate or reduce waste. It seems to us that there is a distortion of the very function of civil liability and – more seriously perhaps – a contradiction with the goal of responsible management sought by the Commission.

One positive feature of the Community text is article 2.2.C., providing for a transfer of liability whenever the producer delivers waste to a licensed or approved treatment or disposal undertaking. Applying the same logic, it is hard to see how a similar transfer of liability would not occur upon delivery to a licensed carrier.

6.2 Definition of the Producer

The Commission proposal refers to any natural or legal person whose *professional activity creates waste* and/or anyone who carries out *pre-processing, mixing*, or other operations resulting in a change in the nature or composition of the waste, until the moment when the damage or injury to the environment is caused.

Article 2(2), however, sets out that the following are to be deemed to be the producer of the waste:

The person who imports the waste into the Community;

The person who had *actual control of the waste* when the incident giving rise to the damage or injury to the environment occurred –

If the producer is not identified;

If the waste is in transit in the Community;

The person responsible for the installation, establishment, or undertaking having obtained a licence, particularly under the directives 75/442/EEC on waste (being amended), 78/319/EEC on hazardous wastes (being amended), and 75/439/EEC on the disposal of waste oils and 76/403/EEC on the disposal of polychlorinated biphenyls and polychlorinated terphenyls.

The method chosen by the Commission leads to a complex definition based on a system of references to previous directives, all of them providing for treatment storage, and disposal operations to be subject to licensing.

This leads to a contradiction in the Commission's philosophy when it 'channels' liability on to the producer (as defined in Article 1(1)) whilst acknowledging the responsibility of the operator or the person having 'actual control' of the waste (Article 25(2)), but only in the case where the producer was not identified.

6.3 The Concept of Waste

The Commission proposal refers to the definition of the directive 75/442/EEC, in the process of being amended.

The concept of waste has been broadly defined (see Appendix 1) but excludes any substances the holder of which is required to dispose of:

Gaseous effluents emitted to the atmosphere;

Radioactive waste;

Waste arising from mining operations;

Animal waste and certain kinds of waste from agricultural activities;

Polluted water, with the exception of waste in liquid form;

Decommissioned explosives.

6.4 Insurance and Compensation Fund

The 1989 proposal did not lay down any general obligation to insure. This position does seem to have changed somewhat in that the Commission is seemingly ready to meet the European Parliament's concern by asking for the producer/operator to be covered by a financial security.

Moreover, the directive would authorize the Commission to explore the possibility of setting up a fund to compensate for damage when the person responsible is unidentifiable or insolvent.

7 LIABILITY IN CONNECTION WITH THE CARRIAGE OF DANGEROUS SUBSTANCES BY SEA (HNS) (International Maritime Organization – IMO)

The matter of the adoption of an international draft convention on civil liability has for some years been included on the agenda for IMO Legal Committee work. An initial draft based on a 'two-tier' system of liability – both complicated and costly – failed to secure adoption in 1984. In spite of this setback, the International Maritime Organization still sees the issue as a priority.

The current work of the IMO Legal Committee revolves round several hypotheses:

Alternative 1: Sole liability of the shipowner coupled with a general raising of amounts of the total limitation stipulated in the 1976 Convention on the Limitation of Liability for Maritime Claims (CLLMC).

Alternative 2: Sole liability of the shipowner coupled with an additional specific amount provided for in the CLLMC Convention for damage caused by HNS substances.

Alternative 3: Shipowner liability limited in accordance with the provisions of the CLLMC and complemented by compulsory insurance, the cost of which would be borne by the shipper.

Alternative 4: Shipowner liability limited in accordance with provisions of the CLLMC and complemented by a fund financed by the shipper.

Other options remain possible, including the adoption, envisaged at the beginning of 1991, of a maritime carrier liability regime complemented by the setting up of international compensation scheme, to come into operation beyond certain amounts or where the person responsible is not solvent.

CEFIC has come out in favour of Alternative 1, being concerned to maintain the principle of the necessary linkage between civil liability and operational control. The increase in upper limits to liability set in the 1976 Convention on the limitation of liability should ensure adequate compensation, even for exceptional damage.

The issue of the adoption by the IMO of an international convention on civil liability is bound to remain an important priority in years to come.

8 QUESTIONS BEARING ON THE INSURANCE OF THE INDUSTRIAL RISK

In the past few years, industry has encountered growing difficulties in renewing its 'civil liability operation' insurance policies, both in terms of the cost of premiums and the extent of risks covered.

Insurance companies, as well as the re-insurance market, have been very hesitant, chiefly for the following reasons:

The impact of the USA liability crisis – characterized by the number and magnitude of claims for damages;

The fear of a similar situation developing in Europe, particularly in view of the introduction of new liability laws and more and more stringent technical regulations;

The potential liability of businesses involved in actions understood to be based, among other things, on damage to the environment whose cause is uncertain;

The anticipated development at national and international level of new laws which would deal specifically with liability for pollution.

This situation has prompted industry to study additional and new methods of risk financing. In the United States, chemical and insurance companies alike have created insurance pools to increase available capacity.

In Europe, industry has found it increasingly difficult to find appropriate cover, especially for 'gradual pollution'. Studies are therefore being undertaken to assess the 'feasibility' of and arrangements for setting up a European Industrial Insurance Pool designed to provide added capacity and better conditions. The covering of major industrial accident risks and the risk of environmental pollution are reportedly being envisaged in particular. However, the actual adoption and the entry into operation of a pool largely depend on the development of the traditional insurance market. In many cases, conditions surrounding the renewal of policies can be said to have improved in recent years. One reason for this favourable trend could be that the envisaged creation of an insurance pool put under direct industry control has served to make the insurance market more buoyant. The concept of a pool has – rightly so – been perceived as mirroring the grave concern of industry and as offering an alternative to insurance mechanisms that no longer look to be meeting industry's demand adequately.

The insurance crisis has not disappeared though. Its acuteness has merely been mitigated a little. Major incidents such as the Exxon Valdez maritime disaster or legislative initiatives unreasonably extending the scope of industrial civil liability could adversely affect the world insurance and re-insurance market.

9 CONCLUSIONS

To seek valid solutions for all European Community Member countries seems logical.

The issue could, however, be addressed in an even wider context given that other economic or geographic areas might be affected by the transfrontier effects of pollution, as illustrated by the accident at Chernobyl. As regards the European Community, a harmonization of national laws is to be supported, if leading to an elimination of distortions of competition and a simplification of existing provisions. A harmonization of this kind could, moreover, be achieved in a broader framework than that of the Community. It is interesting to note in this connection the declaration by a European Commission representative that the Community intends negotiating and signing the draft Council of Europe

Convention (Declaration made to the 9th meeting of the CJ–EN–Strasburg, 17th to 20th November 1990).

The international legislator should draw inspiration from the basic principles governing the law of civil liability as recognized in most modern states.

In defining coherent international regimes, account must be taken of the following points:

A strict link must be maintained between civil liability and operational responsibility. The draft CE rules on this, as well as the 1989 Geneva Convention (CRTD) have thus been welcomed by industry with some satisfaction;

That industry accepts its responsibilities in carrying out its activities does not mean that it is prepared to accept any criminal or civil liability when no fault or negligence has been committed or again – under a 'strict' liability regime – where it has nothing to do with the cause of damage. The chemical industry can accept the relatively ambiguous 'polluter pays' principle as long as the real polluter is referred to. Individual responsibility should remain the rule, as provided for in modern laws;

Causality has always been a key criterion in determining liability. The causal relationship between the damage and the incident generating liability should be unequivocally established. Such a criterion should, in any event, be retained irrespective of the liability regime under consideration;

The introduction of civil liability for environmental damage calls for defining the words 'environment' and 'damage'. These definitions should provide proper legal certainty and technical and economic foreseeability;

Interferences in the environment may be ascribed to a whole string of factors, which need to be analysed in concrete terms; whatever the basis of liability and the proof system chosen, the analysis of the facts and the establishment of the causal chain must remain priority instruments for seeking the truth in any civil action. In this process, the judge must continue to play a central role;

The concept of damage must be strictly limited to damage resulting directly and indisputedly from the incident giving rise to liability;

Reservations are called for on the recognition of a right of legal action for private organizations, especially where members of such groups have not suffered damage directly. Were such a right to be acknowledged, there would be a great danger of such groups acting on the basis of considerations unrelated to the law and without democratic representativeness or legitimacy. For damage caused to a community, the public authorities alone should be entitled to bring a civil liability claim against any polluters.

Industry has, moreover, constantly underscored the need to bear the economic aspects of the problem in mind. In the quest for new liability regimes, the cost of insurance and risk cover cannot be ignored.

10 REFERENCES

1 *Off. J. Eur. Commun.*, No. C 251, 4.10.89.

2 Draft Convention on damage resulting from environmentally dangerous activities (Council of Europe DIR/JUR/91 (1)).

3 Convention on civil liability for damage caused during the carriage of dangerous goods by road, rail, and inland waterway (CRTD) Geneva, 10 October 1989, ECE/TRANS/79.

4 CRTD Convention, Geneva, 10 October 1989, ECE/TRANS/79.

5 *Off. J. Eur. Commun.*, No. C 251, 4.10.89.

Appendix 1 *Directive 75/442/EEC was amended by the Council Directive (91/156/EEC) of 18 March 1991.*

Waste is defined under Article 1 (a) of the March 1991 Directive as 'any substance or object in the categories set out in Annex 1 which the holder discards or intends or is required to discard'. According to Annex I, categories of wastes are:

1 Production or consumption residues not otherwise specified below.

2 Off-specification products.

3 Products whose date for appropriate use has expired.

4 Materials spilled, lost or having undergone other mishap, including any materials, equipment, *etc.*, contaminated as a result of the mishap.

5 Materials contaminated or soiled as a result of planned actions (*e.g.* residues from cleaning operations, packing materials, containers, *etc.*).

6 Unusable parts (*e.g.* reject batteries, exhausted catalysts, *etc.*).

7 Substances which no longer perform satisfactorily (*e.g.* contaminated acids, contaminated solvents, exhausted tempering salts, *etc.*).

8 Residues of industrial process (*e.g.* slags, still bottoms, *etc.*).

9 Residues from pollution abatement processes (*e.g.* scrubber sludges, baghouse dusts, spent filters, *etc.*).

10 Machine/finishing residues (*e.g.* lathe turnings, mill scales, *etc.*).

11 Residues from raw materials extraction and processing (*e.g.* mining residues, oil field slops, *etc.*).

12 Adulterated materials (*e.g.* oils contaminated with PCBs, *etc.*).

13 Any materials, substances or products whose use has been banned by law.

14 Products for which the holder has no further use (*e.g.* agricultural, household, office, commercial and shop discards, *etc.*).

15 Contaminated materials, substances, or products resulting from remedial action with respect to land.

16 Any materials, substances, or products which are not contained in the above categories.

23

Knowledge, Responsibility, and the Safe Use of Chemicals

Sheila Jasanoff

DEPARTMENT OF SCIENCE AND TECHNOLOGY STUDIES, CORNELL UNIVERSITY, ITHACA, NEW YORK 14853, USA

1 INTRODUCTION

Among the two or three biggest problems confronting the chemical industry today, public opinion is often cited as the most critical. Indeed, it has become something of a central dogma among industry insiders that the public irrationally disagrees with scientific experts in its responses to the question addressed by the conference on which this book was based: 'can chemicals be used safely? Whether the conflict is over the safety of an individual substance (*e.g.* dioxin, Alar, formaldehyde) or over an entire class of products (*e.g.* pesticides), the public seems often to insist that safety is not obtainable at any price, even though the available science points in the opposite direction. This chapter will explore the reasons why public assessments of chemical safety so often diverge from what risk analysts would characterize as the 'actual' extent of hazard. In the process, I shall draw upon the social studies of science to challenge some of the preconceptions with which risk analysts have commonly approached the issue of public risk perception.

Researchers interested in the psychology of risk perception have long recognized that lay estimates of the magnitude of technological risk do not always accord with corresponding judgments by technical experts. An article by Paul Slovic, Baruch Fischhoff, and Sarah Lichtenstein, all well-known researchers in the field of risk perception, supplies a widely accepted explanation for this phenomenon: 'Risk' means different things to different people. When experts judge risk, their responses correlate highly with technical estimates of annual fatalities. Lay people can assess annual fatalities if they are asked to... However, their judgments of risk are sensitive to other factors as well (*e.g.* catastrophic potential, threat to future generations) and as a result are not closely related to... estimates of annual fatalities.[1]

Embedded in this analysis are at least three propositions that run counter to a significant body of research in the social studies of science and technology: (i) that given enough data experts will generally agree with each other in their risk assessment; (ii) that the only scientific way to think about risk is essentially in actuarial terms, in terms of expected annual fatalities or annual injuries from the activity in question; and (iii) following implicitly from the other two, that any other way of thinking about risk is possibly wrong, certainly unscientific, perhaps even antiscientific. To be sure, these propositions represent the biases of analysts trained to think in terms of engineering risks, which are especially amenable to quantitative measurement. Scientists concerned with biological effects have generally been more willing to recognize the need for qualitative judgments, and hence the possibility of expert disagreements, in risk assessment. Nevertheless, many risk analysts subscribe in principle to views very similar to those articulated by cognitive psychologists such as Slovic.

That these propositions need to be challenged, or at the very least re-examined, is apparent from even a cursory examination of technological controversies. Thus, the *New York Times* reported on November 3, 1987 that there was some disagreement within the scientific community about what should be done with the last remaining stocks of smallpox virus left in the world. Following the analysis of Slovic, Fischhoff, and Lichtenstein, one might have expected all research scientists to assess the risks of these materials in much the same way and to recommend a uniform course of action. But the *Times* reported that, although a large majority of some 61 scientists in 22 countries favoured destroying the virus, five had spoken out against this option. These five said that the virus supplies should be stored indefinitely, in spite of any potential risks. Experts clearly are not always in agreement about their responses to risk, even when the risk is as familiar and as statistically measurable as smallpox pathogenicity.

The *Times* article supplies some of the reasons why the experts who wanted to maintain the virus stocks felt the way they did; their fears apparently had little to do with immediate risks of injury to laboratory professionals. The two concerns mentioned most prominently were unforeseen future research needs and a philosophical objection to the deliberate extinction of a species, even if it was a dangerous virus. The second concern confirms our intuitive suspicion that willingness to tolerate risk is not simply a matter of recognizing actuarial probabilities. As the virus case indicates, scientists can be swayed by moral and ethical considerations when they evaluate risks related to research. I shall argue below that the public's view of chemical risk is shaped by similar ethical concerns, prompted in part by disagreements within science and, in part, by doubts concerning the responsible social control of science and technology.

2 SCIENTIFIC UNCERTAINTY AND EXPERT DISAGREEMENT

Disagreement among experts is clearly an important common ingredient in disputes about the safety of chemicals. Competing risk assessment methodologies can yield estimates varying by several orders of magnitude.

While experts seek to establish the relative credibility of different estimates, the public seem more inclined to insist that the worst-case scenario should be used as the basis for regulatory action. Bowing to the popular will in such cases may be politically expedient for regulators, but it can also result in serious economic inefficiency if limited societal resources are expended in protecting the public against negligible or nonexistent dangers.[2]

To forge a more reasonable decision rule under conditions of scientific uncertainty, experts have to persuade the public that risk assessments which are not based on worst-case assumptions may be worthy of respect even if absolute consensus cannot be reached. This effort at communication will in turn be effective only if both experts and the public become more sophisticated about the nature of scientific uncertainty and the causes of expert conflict.

3 THE CONTINGENCY OF KNOWLEDGE

Scientific uncertainty is popularly regarded as the primary reason for dissension among experts. When the facts are unknown or impossible to determine at reasonable cost, how can we expect science to arrive at a consensus? The potential for uncertainty, moreover, seems to rise in proportion to science's proximity to the policy process. As Alvin Weinberg pointed out nearly twenty years ago in his classic essay on 'trans-science',[3] the questions that policy makers ask of science cannot always be answered by science. Weinberg identified several factors that may make an issue trans-scientific, such as the impracticality of further research, the variability of behaviour in large populations, and the moral or ethical values affecting the selection of research priorities. These observations led Weinberg to conclude in a later essay that regulatory agencies should settle for 'a new branch of science, called regulatory science, in which the norms of proof are less demanding than are the norms in ordinary science'.[4]

More recently, researchers in the social studies of science have begun to document that the potential for expert conflict arises not merely from the trans-scientific character of questions relevant to policy but from the nature of science itself. According to this body of scholarship, scientific claims seldom if ever mirror nature purely. Rather, they are produced within the context of a prevailing theoretical paradigm[5] and in accordance with experimental and interpretative conventions that are negotiated among communities of experts and, in some cases, between scientists and society at large. Scientific knowledge thus is contingent on social factors lying outside nature, including the stakes, interests, professional affiliations, and ethical predispositions of the scientific community.

Debates about the risks of releasing genetically engineered organisms into the environment attest to the influence of disciplinary paradigms on the scientific evaluation of safety. In a 1987 issue of *Science* magazine, for example, there were two juxtaposed articles by Dr. Frances Sharples, an ecologist, and Dr. Bernard Davis, a geneticist, which illustrated one of the key problems the public faces in dealing with risk.[6] The articles were presented as visually parallel and, implicitly, as scientifically parallel explorations of the risks related to the environmental release of genetically engineered organisms. Yet in each article, the author more or less openly criticized the conceptual standpoint adopted by the author of the parallel piece. Sharples and Davis selected different models of real world phenomena in trying to explain whether risks would arise or not. Their choice of model, in turn, was coloured by their own scientific training and background. Sharples chose to think in terms of populations of exotic organisms that are transplanted to new environments and then multiply explosively. Davis preferred to think in terms of adaptation and suggested that the scenario envisaged by Sharples only makes sense when organisms are well adapted to the natural environment, which is not the case for laboratory engineered organisms. Therefore, Sharples's analogy was in Davis's view scientifically unsound.

My purpose here is not to take sides in this debate. The point I want to make is that such a debate in a prestigious scientific journal illustrates both the fragmented state of knowledge about risk and an important basis for public unease about science. Given two conceptually valid, yet hotly contested, models for assessing risk, the lay public understandably does not know which to believe. Public reaction often takes the form of increased scepticism about the entire scientific enterprise. Science, then, needs to project a truer image of its internal divisions and limitations before it can expect the public to arrive at a coherent and realistic accounting of risk.

Risk studies have shown as well that expert assessments of evidence sometimes contain embedded social assumptions that are never made explicit, that remain unreviewed and untested, and that may well be wrong. Brian Wynne called attention to a number of such cases in an essay describing the 'naive sociology' practised by scientists and technical experts.[7] For example, in investigating the fatal methane explosion at Abbeystead in 1984, the Health and Safety Executive (HSE) unquestioningly assumed that its public credibility would suffer unless it proved its case according to a standard of criminal negligence (that is, 'beyond reasonable doubt'). Yet this approach reflected more an institutional habit than a reasoned and rational choice, and it may have been inconsistent with HSE's actual statutory functions. More to the point, HSE's seemingly technical determination – the strength of the evidence of negligence – was dependent on a set of prior, and untested, legal and social presumptions. In this instance the agency's faulty assessment of the social needs undermined the apparent scientific objectivity of its conclusions. These observations have led Wynne and others to conclude that non-experts sometimes differ from experts not because they have failed to 'get the science right', but because, as a result of knowing the social context of the hazardous activity, they have actually 'got the sociology right'.

If scientific knowledge and its varying interpretations are 'socially constructed', then such claims can under appropriate circumstances be 'deconstructed' into their constituent elements; put differently, a seeming scientific consensus can dissolve back into conflicting interpretations of observation and data.[8] Empirical research bears out these theoretical predictions. The literature on risk controversies reveals many instances in which such deconstruction of scientific claims has occurred under pressure of politics or litigation.[9] As can be seen below, both the extent of deconstruction and the possibility of productive *reconstruction* are deeply influenced by the procedures through which scientific evidence is evaluated for policy purposes.

The fraying of scientific consensus in the context of risk management tends to follow certain characteristic patterns with important implications for public risk perception. Controversy exposes not only the areas of uncertainty – or 'interpretative flexibility' – in science, but also the interests that prompt experts to select among competing interpretations of the evidence. Thus, in most important debates about technological risk, it is possible to get reputable scientific opinion on both sides of the matter; in other words, the positions adopted by experts frequently pattern in the same way as basic interest groupings in society. Given a choice between alternative assessments of risk, 'anti-regulation' experts representing industry generally opt for the estimate that establishes a product or activity as safe. By contrast, 'pro-regulation' experts from environmental, labour, and consumer groups, and sometimes from health and safety agencies, tend to support higher estimations of risk.[10] Safety evaluations accordingly acquire a political cast that precludes dispassionate debate or good faith negotiation over policy alternatives.

Given the contingency of knowledge and the value-laden character of risk assessments, how can competing societal interests ever hope to reach consensus on the safe use of chemicals? Science by itself is rarely a strong enough force for convergence, for as we have seen science frequently falls captive to special interests in controversies over safety. Some policy analysts have even suggested that more science only leads to more contention in areas of high uncertainty and political divisiveness.[11] According to a more optimistic view, however, where knowledge fails process can sometimes provide an answer. Indeed, recent studies of chemical regulation suggest that an open, representative, and conciliatory process of communication among experts, regulators, and the public can promote the closure of controversies that could not have been resolved by scientific research alone.[12,13] Such a process in turn requires the existence of appropriate legal and political forums, as described below.

4 ACHIEVING CLOSURE

The US regulatory system provides an excellent research site for testing these observations because of the richly varied procedural approaches it employs in assessing the risks of chemicals. The initial assessment of risk, for example, may be carried out either by the agency staff, with or without the aid of external consultants, or by a panel of independent experts. Subsequently, the validity of this preliminary assessment can be tested through procedures that are either consultative and conciliatory or confrontational and trial-like. Comparisons among US regulatory agencies suggest that consensus about chemical safety can be achieved most readily when the risk assessment is initially carried out by independent scientists and thereafter is reviewed in a relatively non-adversarial procedural format that allows for significant public input. The regulation of sulphites by the Food and Drug Administration (FDA) and of Alar by the EPA provide an instructive contrast.

4.1 Sulphites

Sulphiting agents are widely used in the food preparation and processing industry to prevent discoloration and preserve the fresh appearance of foods. In the United States, FDA's most recent review of these substances was triggered by reports in the medical literature of acute allergic responses to sulphites, including several fatalities. Following earlier practice, FDA officials contracted with a respected professional society, the Federation of American Societies for Experimental Biology (FASEB), to analyse the medical reports and to determine whether sulphites posed a risk to public health. To carry out these tasks, FASEB convened an *ad hoc* panel of experts, almost all of whom had advised FDA on an earlier review of the safety of sulphites.

The panel concluded its initial evaluation of the literature with a draft report stating that there was indeed cause for concern about sensitive individuals, such as asthmatics, who could be exposed to sulphites through food. These concerns, the panel felt, could best be addressed by means of warning labels in restaurants and markets that offered salads and fresh produce treated with sulphites. The panel then held a public meeting and took evidence from a variety of sources including consumer groups, representatives of the food industry, and scientists working on sulphite sensitivity. This testimony led the panel to reaffirm its preliminary conclusion that sulphites were safe for the general population at allowed doses, but that they posed a risk of 'unpredictable severity' to specially sensitive individuals. With respect to warning labels, however, the panel reversed its original conclusion and advised instead that labelling alone would not adequately protect the sulphite-sensitive population. Certain uses of sulphites, the panel recommended, should be banned, most notably the preserving of fresh fruits and vegetables at salad bars.

The open public meeting held by the panel thus performed two important functions: it gave representatives of varied interests and affiliations the opportunity to participate meaningfully in the policy process, and it changed the panel's views with respect to a politically crucial issue – the adequacy of warning labels as a control technique for sulphites. In its final regulatory package, FDA went along with the panel's basic recommendations. The expertise of the independent advisers and the political consensus produced by the public hearing strengthened FDA's conclusion that sulphites posed a health threat deserving of attention, even though scientific evidence about the nature and magnitude of risk was by no means conclusive. In contrast with many other stringent agency actions, FDA's imposition of a ban on certain sulphite uses aroused no serious opposition or criticism.

4.2 Alar (Daminozide) (mono(2,2-dimethylhydrazide) butanedioic acid, hydrazide RN:1596-84-5)

The Alar case at EPA illustrates the relative inutility of confrontational procedures as a method of reaching closure on disputed issues of chemical safety. Beginning in the mid-1980s, EPA's Office of Pesticide Programs (OPP) carried out a review of the plant growth regulator daminozide (trade named Alar) and its breakdown product unsymmetrical dimethylhydrazine (UDMH) to determine whether the compounds were safe for use. Based on then available bioassay results, OPP concluded that Alar presented a significant risk of human cancer and should promptly be withdrawn from the market.

Asked to review EPA's risk assessment, the agency's Scientific Advisory Panel (SAP), however, came to quite a different conclusion. In the SAP's judgment, the animal studies relied upon by the pesticides office were badly flawed and should not have been used for quantitative risk assessment. This conclusion appeared consistent with forceful assertions by Uniroyal, Alar's manufacturer, that the methodologically problematic Alar studies should be dismissed as 'not science'. SAP's negative review essentially ruled out any possibility of immediate regulatory action on Alar, and EPA asked Uniroyal to carry out additional studies on the product's carcinogenicity. Environmental groups, led by the experienced and active Natural Resources Defence Council (NRDC), took EPA to court, claiming that the agency should not have bowed to the scientific panel's conclusions, but their plea was dismissed on procedural grounds.

EPA, however, encountered much negative publicity for its handling of the case when Uniroyal's tests apparently confirmed the earlier findings of carcinogenicity for UDMH. By this time, the adversarial flavour of the Alar review had aroused public suspicion that the SAP members had been allied with ('captured' by) industry.[14] Two US Senators characterized the panel as being 'riddled with pro-industry bias' and an ethics investigation was launched against two panelists who had represented Uniroyal before EPA following their tour of duty on the SAP.[13] Although these experts were eventually exonerated, the episode seriously undermined the credibility of EPA's pesticide advisory process.

In this polarized political environment, an effort by the NRDC to seize the scientific initiative and to project its own risk assessment to the public proved highly successful. A perturbing estimate of Alar's risk to children caught national attention after it was broadcast on the popular news programme '60 Minutes', even though NRDC's conclusions were based in part on the studies discredited by the SAP.[16] Something approaching a consumer panic ensued as large segments of the US public foreswore apples and apple products treated with Alar. Under pressure from congressional investigators as well as consumers, Uniroyal decided 'voluntarily' to withdraw Alar from the US market and later to stop shipments abroad.

Commentators on the Alar controversy have called attention to questionable elements in NRDC's risk assessment methods, including the reliance on the early cancer studies and the use of a time-dependent mathematical model to compute the risk to children.[17] NRDC's report, moreover, never underwent rigorous scientific peer review, despite claims to the contrary, and was ultimately disseminated to the public by a television reporter, whose expertise in manipulating visual images clearly exceeded his understanding of the intricacies of cancer risk assessment. From the standpoint of public perceptions, however, the central issue turned out to be not whether NRDC was right or wrong, but whether it commanded more credibility than either EPA or its scientific advisers. The environmental group in this case easily won the contest of credibility.

Process, we must conclude, was a major part of the reason. EPA's regulatory approach encouraged a tight coupling between industry and science, with little or no real opportunity for public involvement. SAP hearings seldom attract much public notice and the Alar review was no exception – a fact that enhanced the impression of an illegitimate alliance between the advisory panel and the chemical company. Once the legitimacy of its advisers came under siege, EPA could no longer seek refuge in its superior scientific expertise. NRDC was able to shoulder the regulatory agency aside in the public eye because trust at this stage counted for more than scientific expertise. NRDC's public interest credentials were of course impeccable, and its claims to scientific competence were substantial enough to persuade many that it, and not the EPA, was telling the true story with respect to Alar's safety.

5 THE CONTROL OF KNOWLEDGE

Expert disagreement and the contingency of scientific knowledge are not the only causes of dissension in the area of risk perception. Other recent controversies suggest that debates about safety as often as not are masks for underlying concerns about the social and political control of knowledge. Some examples may help illuminate the issues of control that are frequently embedded in discussions of chemical safety.

One instructive incident that occurred in 1982 concerned a decision by Arthur D. Little, a consulting company based in Cambridge, Massachusetts, to renovate a laboratory according to Department of Defense (DOD) specifications for work with chemical warfare agents. Because of its early involvement in recombinant DNA research,[18] the city of Cambridge already had some experience with citizen evaluation of potentially dangerous projects. This experience had demonstrated that citizens could sufficiently inform themselves about technical issues to play a productive part in policy making for new technologies. The city accordingly established a special advisory committee, with both scientific and citizen participation, to look at the risks posed by the A.D. Little initiative.

The committee concluded that the DOD project was too risky to go forward. For our purposes, the most interesting element in the story was the basis for that conclusion. Sheldon Krimsky, a policy analyst who both participated in and described the incident, says that 'a substantial majority of the committee did not feel there was justification for a single lethal exposure of the public to these chemicals in the name of chemical weapons research.[19] Having served on the committee and talked to its members, Krimsky found that their judgments about whether the risk ought to be tolerated were much influenced by their perceptions that the research was 'not unambiguously humanitarian'. If the research had been humanitarian, Krimsky suggests, the committee might have been much more prepared to find the relatively low risks of the project acceptable.

Similar concerns about the appropriate use of knowledge have previously arisen in the context of biomedical research, producing disparities in risk perceptions among scientists and between scientists and the general public. More than 15 years ago, for example, a psychiatrist and a geneticist at Harvard Medical School were forced against their will to shut down an experiment screening all newborn infants in a Boston-area hospital for the presence of a genetic abnormality known as the XYY chromosome.[20] The screening programme had been underway since 1968, and in 1975 the study received a lopsided endorsement - about 200 people in favour, 30 against - from the Harvard Medical School faculty, who concluded that it should be permitted to continue. The researchers nevertheless ended the project because they felt that it could not productively be continued in the face of anti-screening pressures, including protests by a 'science for the people' group.

The primary concern of the study opponents was the possible stigmatization of the newborn subjects. Although most geneticists now agree that the theory was incorrect or exaggerated, the XYY chromosomal pattern had been identified as the 'criminal chromosome,' and any child diagnosed as having it stood at risk of being labelled from birth as a potentially criminal personality. Those who opposed the study were afraid that this label would become for many newborns a self-fulfilling prophecy, because the parents would be notified that their child had a propensity to develop along antisocial lines. They felt that this was too much of a burden to impose on newborn infants.

The dispute in this case reflected a deep social ambivalence about the desirability of basic scientific knowledge. Are there certain categories of 'dangerous knowledge', things about ourselves and nature that it is better not to find out? Such questions are likely to arise with even greater force in connection with technology, which historically has enjoyed less autonomy and public trust than basic scientific research. Thus, as the chemical industry moves increasingly in the direction of biotechnology, it can expect to encounter a new kind of public questioning, focused not merely on the physical safety of products, but also on their broader social impact. Are there products that are socially unsafe, that we should not make and use because they manipulate nature or society in ways that are ethically or morally unacceptable?

The intense and still unresolved controversy about the safety of bovine growth hormone (BGH), a commercially promising product of agricultural biotechnology, can be seen as an early warning signal of this trend. Several multinational chemical companies, including Eli Lilly and Monsanto, invested in the development of BGH, claiming that the genetically engineered compound would generate immense economic benefits by boosting milk production in cows. Yet, their arguments concerning safety were conspicuously ineffective in winning the confidence of European and American consumer groups. Indeed, the BGH controversy aroused so much public and congressional concern in the United States that FDA in 1990 took the highly unusual step of publishing data on the safety of BGH a year before the agency expected to rule on the compound's commercial availability.[21,22]

FDA's action, however, may fall short of reassuring the anti-BGH interests because it addressed only the surface of the conflict and not its underlying causes. The dispute over safety in this case has rightly been seen as a surrogate for a more profound debate about the social control of biotechnology. Since the earliest days of recombinant DNA research, biotechnology has been dogged by two recurrent questions: what are the appropriate uses of genetic engineering and who should decide which uses to pursue? These are politically wrenching questions because they tap into culturally conditioned and politically divisive preconceptions about the appropriate relationship between human beings and the natural world.[23] The BGH controversy suggests that, more than fifteen years down the road from the Asilomar meetings, these questions and their implications have not yet been fully internalized by industry.

While deploring the public turmoil about BGH's safety, industry spokesmen concede in private that they may have been seduced into selecting this particular product more because it was possible to make than because it was needed. We are reminded of J. Robert Oppenheimer's poignant confession that he stopped opposing the hydrogen bomb because the Teller-Ulam design was 'technically so sweet'.[24] In the rush to commercialize biotechnology, the manufacturers of BGH also overlooked the possible structural and social impacts of their product, such as the disfavouring of family farms at the expense of large agribusiness. This blindness is reminiscent of the technocratic single-mindedness that drove the architects of the Green Revolution to engineer a product that increased yield at the price of exacerbating regional economic disparities.[25]

The failure of vision with respect to BGH in our day may well ensure that considerations of safety will henceforth be coupled in the public mind with greater sensitivity to the 'fourth hurdle' – that is, the requirement that industry should demonstrate not just the safety but the social and economic utility of proposed new products.

The preoccupation with physical risks that currently dominates the literature on risk perception shows signs of becoming increasingly irrelevant in the face of products and technologies whose impacts on society range beyond the purely physical. As the concept of 'safety' expands to include factors outside the domain of conventional risk assessment, products previously considered to be harmless may be subjected to closer scrutiny. The development of diagnostic kits, for example, is regarded by many in industry and the scientific community as a social good. The kits permit individuals to determine their own or their off-spring's risk of contracting certain kinds of diseases and to take early preventive action through selective abortions or other therapeutic means. But lawyers and social scientists have begun to worry about the possibilities for social engineering and other negative externalities lurking in the development of such technologies.[26] Once a cheap diagnostic kit exists, will it not be possible for insurance companies or employers to require that it be used to produce certain kinds of information? Will we not be labelling classes of people as unsuitable for insurance or employment, and thereby exposing them to possible discriminatory treatment, on the basis of such information? If kits are given to people who have no access to treatment or counselling, will knowledge about their condition simply lead them to despair? Until our legal and social institutions learn to deal with these types of questions, the chemical industry, including its biotechnological branch, may find itself enmeshed in even more acrimonious disputes over product safety than those it confronted in the opening years of the environmental movement.

6 RESPONSIBILITY IN TESTING

I have suggested to this point that public perceptions of chemical, and increasingly biotechnological, risk and safety are intimately connected with issues of credibility and control. In an increasingly sceptical world, the chemical industry is likely to find itself under increased pressure not only to guarantee the physical and social safety of its products but to reform the methods by which it establishes their safety.

Chemical manufacturers in the past enjoyed considerable autonomy in deciding what products to test and by what means, as is evident from an interesting dispute that erupted at the Ortho Pharmaceutical Company in the late 1970s. An Ortho researcher in charge of a project to test a drug decided that she could not, in conscience, go forward with the project, because the formulation in question included saccharin in doses that she felt were inappropriate for testing on infants and old people. An alternative formulation containing less saccharin had been considered by the company but had been rejected, presumably on economic grounds. As a result of her objections, the researcher was deprived of some of her responsibilities and eventually resigned from her job. In due course, she sued Ortho on the ground that she had been unfairly dismissed.

What the courts decided is not especially germane to the present discussion; in fact, the researcher lost her appeal.[27] Much more interesting are the reasons she herself gave for why she felt unable to test the drug clinically in the formulation that Ortho had developed. Questioned closely in depositions, the researcher said that she was not convinced the formulation was unsafe. What troubled her was the possibility that saccharin was a carcinogen and the continued uncertainty about its effects. She felt that it was inappropriate for the burden of that uncertainty to be placed on a test population consisting of infants and, potentially, old people, even though the drug to be tested complied with applicable FDA requirements.

The essential disagreement between the scientist and the company in this case was about where the burden of proof should lie in cases of uncertainty. The researcher implicitly balanced the risks and benefits of the project and determined that for this drug and this patient population it was better to wait for a formulation that used less saccharin than to go forward with the clinical trial as originally designed. The company, whose views prevailed, insisted that, as long as all legally applicable standards were met, safety considerations did not oblige it to delay or foreclose a profitable opportunity for product development.

In the future, however, growing public interest in supervising the applications of scientific knowledge may impose greater burdens of accountability during product testing. Evidence of such a tendency can already be seen in demands for more responsible use of animals in research and testing laboratories. The issue of animal rights appears to extend earlier concerns with consent in the scientific research setting. Since the 1960s, most industrial nations have adopted carefully structured rules of informed consent for research with sentient human subjects. Scientists, however, had devoted relatively little attention to the problem of consent when the experimental subjects could not speak for themselves. The vigour of the animal rights movement attests, in part, to the fact that the scientific community was not able, from within its own walls, to generate policies for protecting the welfare of such 'silent subjects'. The tighter guidelines for animal use and welfare adopted by US research-funding agencies in recent years represent an effort to bring science's system of values more in line with the public's on this explosive issue.

The episodes of fraud and misconduct that plagued the US research community during the 1980s[28] are also likely to intensify public demands for more oversight of safety testing. The adoption of 'good laboratory practices' by FDA and EPA and similar initiatives by the Organization for Economic Cooperation and Development responded partly to the need for international harmonization of test data, but partly also to specific instances of misconduct in testing laboratories.[29] With allegations of misconduct now penetrating to some of the highest pinnacles of scientific research, one can expect a substantial increase in the regulatory supervision of science, including the production of data related to chemical safety.

7 CONCLUSIONS

The history of public concern with chemical safety through the twentieth century is one of steadily rising expectations. In the early part of the century, during and after the so-called chemists' war, people were primarily worried about the traumatic or acute injuries inflicted by chemicals, but not about chronic illnesses and psychic or behavioural changes. In the 1970s, with the infusion of new scientific knowledge, the focus began to shift to include chronic disease, environmental pollution, and long-term ecological damage, such as ozone depletion or the loss of biodiversity. As biotechnological research develops, public concern is likely to mount over the still less tangible social risks of products made by the chemical industry.

How can the safety of chemicals best be established in an era of heightened environmental and social consciousness? I have tried in this chapter to provide a two-tiered answer. With regard to the impacts of chemicals on humans, ecosystems, and the physical environment, the basic procedures for safety evaluation are already in place, but much more attention should be paid to the methods by which we resolve differences among technical adversaries. The procedural structure of decision-making should take into account the contingent and socially constructed character of policy-relevant science, and regulators should strive to create an environment where differences that are at bottom as much ideological as scientific can be fruitfully negotiated.

In the wider scheme of scientific and technological change, however, chemistry and its sister sciences will have to confront the public's growing sophistication about the uses and abuses of scientific knowledge. Chemical 'safety', in a world that is both economically and ecologically interdependent, will increasingly entail the protection of human values and lifestyles along with the more familiar categories of health, safety, and the environment. Chemical products, in particular, may be rejected not because they are 'unsafe' in any conventional sense, but because the public is insufficiently persuaded that they serve a legitimate social need. By contrast, chemicals that pose some unavoidable degree of risk, such as pharmaceuticals and pesticides, may be adjudged 'safe' if the public accepts their benefits and recognizes that they can be controlled. It is toward satisfying these demands for accountability and control that chemical safety experts should turn their attention in the next decade.

8 REFERENCES

1 P. Slovic, B. Fischhoff, and S. Lichtenstein, *Risk Analysis*, 1982, **2**, 85.

2 R.J. Zeckhauser and W.K. Viscusi, *Science*, 1990, **248**, 559.

3 A.M. Weinberg, *Minerva*, 1972, **10**, 209.

4 A.M. Weinberg, *Issues in Science and Technology*, Fall 1985, 68.

5 T. Kuhn, 'The Structure of Scientific Revolutions', 2nd Edn., University of Chicago Press, Chicago, 1970.

6 F.S. Sharples and B. Davis, *Science*, 1987, **235**, 1329.

7 'Environmental Threats: Social Sciences Approaches to Public Risk Decisions', ed. J. Brown, Belhaven, London, 1989, Ch. 3, p. 33.

8 B. Latour and S. Woolgar, 'Laboratory Life', 2d Edn., Princeton University Press, Princeton, N.J., 1986.

9 S. Jasanoff, 'The Fifth Branch: Science Advisers as Policymakers', Harvard University Press, Cambridge, Mass., 1990.

10 B. Gillespie, D. Eva, and R. Johnston, *Social Studies of Science.*, 1979, **9**, 265.

11 D. Collingridge and C. Reeve, 'Science Speaks to Power', St. Martin's Press, New York, 1986.

12 S. Jasanoff, 'Risk Management and Political Culture', Russell Sage Foundation, New York, 1986.

13 J.D. Graham, L.C. Green, and M.J. Roberts, 'In Search of Safety: Chemicals and Cancer Risk', Harvard University Press, Cambridge, Mass., 1988.

14 S. Jasanoff, *Science, Technology, and Human Values*, 1987, **12**, 116.

15 E. Marshall, *Science*, 1989, **245**, 20.

16 NRDC, 'Intolerable Risk: Pesticides in Our Children's Food', 1989.

17 J.D. Rosen, *Issues in Science and Technology*, Fall 1990, 85.

18 S. Krimsky, 'Genetic Alchemy', MIT Press, Cambridge, Mass., 1982.

19 'Governing Science and Technology in a Democracy', ed. M.L. Goggin, University of Tennessee Press, Knoxville, Tenn., 1986, 207.

20 B.J. Culliton, *Science*, 1975, **188**, 1285.

21 A. Gibbons, *Science*, 1990, **249**, 852.

22 J.C. Juskevich and C.G. Guyer, *Science*, 1990, **249**, 875.

23 M. Douglas and A. Wildavsky, 'Risk and Culture', University of California Press, Berkeley, Cal., 1982.

24 US Atomic Energy Commission, 'In the matter of J. Robert Oppenheimer', US General Printing Office, 1954.

25 W. Ascher and R. Healy, 'Natural Resource Policymaking in Developing Countries', Duke University Press, Durham, N.C., 1990, Ch. 3.

26 D. Nelkin and L. Tancredi, 'Dangerous Diagnostics', Basic Books, New York, 1989.

27 *Pierce v. Ortho Pharmaceutical Company*, 84 N.J. 58 (1980).

28 A. Kohn, 'False Prophets: Fraud and Error in Science and Medicine', Basil Blackwell, Oxford, 1986.

29 J.W. James, *Chemistry in Britain*, 1980, **16**, 534.

24
Epilogue

Mervyn L. Richardson*

PRINCIPAL, BIRCH ASSESSMENT SERVICES FOR INFORMATION ON CHEMICALS (BASIC), 6 BIRCH DRIVE, MAPLE CROSS, RICKMANSWORTH, HERTS. WD3 2UL, UK

1 INTRODUCTION

As stated in The Society's sister publications 'Risk Assessment of Chemicals in the Environment' and 'Chemistry, Agriculture and the Environment' - 'We are all exposed to chemicals' and in today's lifestyle we are totally dependent upon chemicals. Our way of life depends on pharmaceutical chemicals, agrochemicals, surfactants, dyestuffs, *etc.*

Hence, risks from chemicals, whether natural or man-made, are inevitable. 'The way we assess these risks is important not only to ourselves, but to other animals and to the environment, to the soil, water, and to air'. Having assessed firstly the hazards and then the risks, the all important matter of management of the risks - the topic of this book - has vital consequences for us all.

His Royal Highness, the Prince Philip, in His Foreword to 'Chemistry, Agriculture and the Environment', drew attention to the dramatic contribution that chemistry has made to the production of food and to the quality of food. Chemistry has also played a vital part in identifying the damaging effects of inappropriate compounds. Our challenge for the future is to combine the interests of humanity and maintain the integrity of the biosphere on which future generations will have to depend.

We are now unable to return to the times of the hunter and gatherer as the next time around there will be nothing to hunt and nothing to gather. It is therefore essential both for ourselves and for future generations that the risks associated with the use, handling, and disposal of chemicals are properly managed. Only by adopting good risk management techniques will it be possible to regard the use of chemicals as acceptable and safe.

*The views expressed are those of the Editor and not necessarily those of The Royal Society of Chemistry.

1.1 A 'Cradle to Grave' Approach

In current times of increased competition, companies will need to ensure that they are demonstrating a clear responsibility of care for the chemicals they synthesize, use, handle, and discharge and equally that customers adhere to a similar attitude of care when using their products for their approved use, *i.e.* 'cradle to grave' management is now a necessity.

A supplying company may well wish to discontinue supplying a chemical to an end user if the originator finds that the end use is not safe or that the risks are inadequately managed. Protection of the environment is obviously paramount - we cannot go to the corner shop and buy a new environment for the one we abused!

Companies who do not adopt such 'cradle to grave' management and ongoing care of their chemical products may well find that if their customers do not desert them, their bankers, insurers, and shareholders probably will.

1.2 Costs can be Recovered

The foregoing will obviously incur costs, significant costs in a few cases, but in others benefits will be significant. These will be gained from improved housekeeping, low disposal costs, and reduced sickness of both employees and of local, national, and international communities.

The question is often raised as to who should pay for the higher costs for an environmentally sound approach to industrialization in developing countries and how this burden is to be allocated between the developed and the developing countries. Failure to act in this vital area of international co-operation will cost us all dearly. Clearly governments in developed countries and the United Nations Agencies must shoulder the major burden. Industry and governments must be prepared to act and to act promptly. The use of good risk management techniques as detailed in the foregoing pages will lead to a greater understanding of these problems, obviously leading to solutions in the safe use, handling, and disposal of chemicals.

1.3 Long Term Effects

The long term effects that chemicals may have on the environment is only recently becoming known: the effects of chlorofluorocarbons on the ozone layer; mercury found in fish in remote lakes; lead being found in Greenland snow; PCBs, dioxins, and pesticides found in the Arctic; all water is or has been reused in recent times (except perhaps that locked as ice in remote mountain ice caves).

The only reasonable solution to global pollution cannot be increased regulations of isolated point sources, but rather an increased emphasis on waste reduction and minimization, and an increase in material recycling.

1.4 A Case Example

In concluding this introduction, attention can be drawn to a recent personal experience – having double glazing installed. In obtaining quotations, no company representative was aware of, or could obtain details of, the chemicals used in adhesives, sealants, or stains. During the installation the fitters were almost totally devoid of any knowledge associated with the risks and safety involved in handling solvents including 1,1,1-trichloroethane (the container even quoted an out of date TLV but they were unaware of its meaning). Cleaning with solvents and staining of wood was carried out late in the day with no instruction to the house owner of the hazards of sleeping in a room with high levels of organic solvents – this could obviously cause severe consequences for the very young, the old, or the infirm, suffering from respiratory conditions. Their attitude to the environment was equally poor, pouring solvents down surface water drains, spilling solvent on lawns causing phytotoxic effects, *etc*. One wonders about the fate of the discarded wood which undoubtedly had been treated with some of the most persistent chlorohydrocarbons.

This is an illustration of perfectly adequate hazard data being available on the substances handled and well known to the initial manufacturers being passed to their immediate customers, but not effectively passed on to the end user and the ultimate customer, the public. This example clearly demonstrates that the chemical industry needs to improve greatly its management of risks – it is of little use compiling detailed data which are clearly neither understood nor passed through to the ultimate customer. It is for reasons such as these that the public image of the chemical industry is poor. Hence, it must be the duty of a primary supplier/manufacturer to ensure that the chemicals entrusted to his customers are used, handled, and disposed of in the recommended manner and not abused. Without such assurance the supplier must refuse to continue supply.

2 HIGHLIGHTS FROM PRECEDING CHAPTERS

The relationship between chemistry, the biological effects generated by chemicals, and their effects on both the workplace and the environmental matrices have been detailed in the preceding chapters by eminent scientists citing risk management case studies and societal consequence on an international basis.

2.2 Introduction

Firstly, it must be stressed that for effects to occur, exposure to a substance is needed. Time and science resources should not be expended on imaginary risks generated from equally imaginary hazards. The risks associated with a potential for harm due to exposure to chemicals need to be identified, assessed, and managed appropriately. It is also important, as discussed in the Preview, that the terminology in the complex area of risk management is clearly appreciated.

The exchange of good quality chemical information on an International basis such as is provided by the International Labour Office Health Information Centre is of paramount importance.

It should be remembered that there must be a compromise between scientific completeness and economic feasibility. It is not an easy matter for manufacturers of a chemical substance to estimate the exposure profile at his customers' plants, or at the consumers' premises since they are unaware of all the purposes for which that substance is likely to be used. In order to help customers, the manufacturer will have to indicate the conditions under which he undertook his evaluation far more clearly than in current practice. This is of importance as exposure for many chemical products occurs to a far greater extent during use than manufacture.

In view of the vast number of existing chemicals on which, in many cases, few relevant or reliable hazard data are available, priority setting schemes, both for the workplace and the environment, will be needed to utilize scarce resources for the testing of those substances which are considered to present the greatest hazards.

2.2 Introduction to the Management of Risk

It is important to recall that long before man-made chemicals were known, mankind utilized many naturally occurring chemicals, the hazard from many of which is far greater than for many xenobiotics.

Whilst there are no safe chemicals, almost all chemicals can be handled safely.

In all cases prevention is better than cure. Promotion of the use of cleaner products and techniques, emission inventories, product labelling, use limitation, economic incentives, and phasing out or banning of chemicals all have their place in modern society.

Everyone having contact with a chemical has to manage the situation in such a manner that possible negative effects of the chemical are minimal. Until the 1960s one of the principal driving forces for reducing the probability of accidents came from learning from mistakes by others – this is now no longer acceptable. In considering whether it is possible to ensure that an accident will never occur, management must be judged on their safety record as well as their production record. It is vital that companies satisfy themselves, their neighbours, their customers, and not least the authorities, that their plants are properly operated and maintained.

In considering new chemicals, competent authorities have drawn up a programme of tests to be undertaken by the notifier in order to enable the competent authority to evaluate the risks of the substance for both man and the environment. This in turn requires a deeper understanding of dose–response relationships, no-effect levels, safety factors, mechanisms of action, metabolic comparability, and the significance of exposure in making a risk assessment. Risk management involves societal factors so that it is no longer the province of the scientist alone.

2.3 Managing Risk in Manufacture

In the 1990s and beyond, increasingly industry and the authorities will be questioned by the public regarding their technical competence. In particular, the State must increase its control function in a credible manner.

Increasingly, every effort has to be made to minimize the possible consequences of industrial environmental pollution, *e.g.* by the use of a 'second barrier', *i.e.* double-walled pipelines. This should assist in preventing human errors leading to catastrophic events. One of the greatest problems is to explain complex chemical technology and techniques to the layman and the media. Fear of the unknown often brings a desire for 'zero-risk' which amounts to a wish to live forever. Companies are therefore advised to use their employees as their ambassadors.

Companies must remember that every accident is a management failure. Improved methods of working explained in a language that people can understand rather than in legal language is of paramount importance. It is essential that such working standards are achieved otherwise benefits from improvements in equipment and even from more stringent legislation are likely to be dissipated by inadequate management control. Total commitment throughout all management levels is vital. Conforming with safety standards must be a way of life within the enterprise and junior staff must see that senior management are completely committed.

Risk management is a multi-faceted activity that often results in decisions on how corporate resources should be spent to reduce risks. Safety problems that appear minor on the surface may have catastrophic potential.

2.4 Risk Management from Waste

Atmospheric emission risks can be assessed by the use of regionalized high level computer system models. Such effective models have time spans of 100 years and must embody dose-effect relationships to calculate potential damage to agricultural production and sensitive materials. It must be remembered that the effects of certain pollutants, including acidification, are transboundary. There is only one sky and we all breath the same air.

In managing risks from chemicals in liquid effluents, there has been a dramatic increase in public and potential awareness of environmental issues. Environmental management is playing a key role within evolving company strategies. In the 1990s and beyond, industry will face increasingly tighter regulatory control of wastewater discharges. This will involve considerable additional costs through application of 'the polluter pays' principle and the adoption of best available techniques not entailing excessive costs (BATNEEC). This will affect all of us as consumers as effluent treatment costs become incorporated in the price of a product.

In the past, attention to wastewater treatment has tended to be superficial. Very few companies have an adequate understanding of the composition of their wastewater; yet this is the essential basis to practical wastewater treatment design and operation. Hence, effluent treatment and disposal have tended to be dealt with as a necessary addition to the production process instead of being an integral part of it. Equally, enforcement of pollution control legislation by regulatory authorities has been slow and superficial. Far too great an emphasis is placed on revenue generation compared with consideration of environmental impact. Secrecy, often under the guise of protecting proprietary information, has kept the public largely uninformed of the risks from effluent discharges and penalties for non-compliance have been (and still are) trivial. In the future the application of BATNEEC must be continuously primed with new technology to reduce economic constraints.

In minimizing the risk from solid waste disposal the ultimate responsibility must rest with public officials. Control of substances entering landfill sites is impossible as inspection of large truck loads of wastes is impracticable. The extent to which the components of wastestreams can supplement demand for raw materials or for energy will affect the survival of sustainable life on our planet.

As water-borne diseases are a leading cause of early death in developing countries and as human excrement plays a key role in these deaths, it can be argued that human excrement is the most hazardous substance in developing countries. Lack of proper sanitation leads in some cases to human excrement being part of the solid waste stream.

There has to be a greater sense of urgency to reduce environmental pollution and to improve industrial production through the implementation of low- and non-waste technologies on a global basis. There will be a direct positive influence of waste minimization action in the social, technical, and economic situation with respect to environmental health criteria.

At all times, prevention is better than cure. It is better to minimize the creation of waste at source, to encourage recycling whenever practical, and to dispose of the resulting wastes in the most environmentally acceptable manner. As we have had a hundred years of non-sustainability, reversing a history of flawed development, we must now have a sense of solidarity as a future for civilization.

2.5 Managing Risk During Chemical Use

Both risk assessment and risk communication are integral parts of risk management as a form of public participation. This is different from 'public relations', which is a one-way process leaving the public uninformed, non-expert, passive, and inactive.

In real public participation, the public must be kept informed completely at a proper level in the early stages of any decision-making process. In these circumstances the public will demand and become a formal factor in decision-making and have the right to object and to appeal. This involves both openness and access to all relevant data. There is no point in getting into a panic concerning the risks of life until one has compared the risks which worry with those which do not.

We all live with risk. Each of us continuously assesses risks in our everyday life – objectivity may not be a part of this process and the so-called 'facts' may only confuse the issue or, if unacceptable, be rejected.

The use of pesticides, particularly in developing countries where control, training, and education as to use are often less advanced, is a cause for concern, especially as economies often direct such countries towards the use of older, less safe, and cheaper products. It is a prerequisite for the scientific community, and the farmer, to ensure that the consumer is aware of the safety of modern pesticides and the benefits that they convey.

Trade Unions can and must play an active role in risk management. Whilst the most blatantly unsafe procedures of the past have disappeared generally in the more responsible companies of the industrialized world, the number of deaths and diseases which continue to occur as a result of using chemicals testifies to the fact that working conditions still fall far short of an acceptable standard in many workplaces around the world. Chronic pesticide poisoning for example is a work-related risk which frequently threatens not only the worker but also family members and others.

In the future, we must aim for fully informed, mutually agreed controls at the workplace, with safety through co-operation rather than confrontation. For those exposed to chemicals, safety must come as a function of education, *i.e.* developing knowledge and skills, or an awareness and understanding of chemicals through experience and study. In the USA, it was found that hazard warning labels were used as a matter of routine in only 25% of the workplaces surveyed.

It is important that management is aware that very often it is the worker who first notices the health problem at the workplace. Management must be aware that the root cause of major accidents is a failure of management rather than a worker related failure. Above all, it is the workers themselves who stand ultimately to benefit, or lose most from workplace policies on risk management of chemicals.

Workplaces that are non–unionized are more likely to receive health and safety inspections, face greater scrutiny during these inspections, and pay high fines for violations, than those where unions are involved.

In many developing countries factory inspectors are virtually non–existent or vastly undertrained. This can lead to inducements such as 'risk allowances' which encourage workers to take greater risks and these practices cannot be supported. It is hence important that multinational corporations should not set up operations or export products which are considered too dangerous for use in their home countries.

All losses and discharges caused by the deeds of any person, through negligence, thoughtlessness, ignorance of what one should know, or shortcomings, however trivial they may be, must be compensated by the person whose fault or negligence has created the deed. In this way, legal liability must be linked to operational responsibility. From experience in the USA, the impact of certain court cases based on 'absolute producer liability' has resulted in enormous insurance costs and availability of cover.

A product (or service) must be deemed defective if it is not in accordance with the material criteria but with reference to the safety which a person is entitled to expect.

One of the most significant aspects of civil liability and current work in the international sphere is the extension of the concept of death, personal injury, damage to property, and ecological damage. Environmental damage concerns the community rather than the individual and is difficult to quantify, but it is essential that we maintain a sustainable ecology.

It is vital to remember the importance of public opinion. When there is a conflict over the safety of an individual substance or an entire class of products (*e.g.* pesticides), the public appears to insist that safety is not obtainable at any price, even though the available science points in the opposite direction.

Researchers interested in the psychology of risk perception have recognized for a long time that lay estimates of the magnitude of risk do not always accord with the corresponding judgements made by technical experts. It is stressed that 'risk' means different things to different people. When experts judge risk, their responses correlate highly with technical estimates of annual fatalities. Lay people can assess annual fatalities if they are asked to. However, their judgements of risks are sensitive to other factors as well, *e.g.* catastrophic potential, threat to future generations, *etc.*

Given adequate data, experts will generally agree with each other in their assessment of risk. However, disagreement among experts is an important common ingredient in consideration of the safety of chemicals. Competing risk assessment methodologies yield estimates varying by several orders of magnitude.

Science needs to project a truer image of its internal divisions and limitations before it can expect the public to arrive at a coherent and realistic accounting of risk. It can be deduced that non-experts sometimes differ from experts, not because they have failed to 'get the science right' but because as a result of knowing the social context of the hazardous activity they have actually 'got the sociology right'.

Chemical safety in a world that is economically and ecologically interdependent will entail increasingly the protection of human values and lifestyles along with the more familiar aspects of health, safety, and the environment.

3 THE FUTURE

By the time the conference on which this book is based has taken place, the United Nations will have held its conference in Brazil on Environment and Development.

National boundaries do not prevent the transfer of pollutants in air or water, and contaminated crops grown on polluted soil can be transported for thousands of kilometres.

In the future, the risk management of chemicals will need to take into account all risks from chemical exposure from small workshops and point emissions, through to the global effects of emissions from diffuse sources.

3.1 Promotion of Human Well-being

The promotion of human well-being and man's development depends on the use of chemicals and chemical processes. Chemicals are used intensively worldwide by all societies independent of their developmental stages. The benefits of chemicals are inestimable. Without adequate risk management, however, chemical usage can result in adverse effects on human health and can have harmful consequences upon the environment.

Exposure to potentially toxic chemicals may occur during their synthesis, formulation, transport, use, or final disposal. Exposure pathways can involve air, water, food, consumer products, and occupational, domestic, and leisure activities. There is a vast number of chemicals and indeed processes that have not been tested adequately for their effects either on health or on the environment.

One aspect of risk management must therefore be priority setting schemes for selecting substances to be so tested.

3.2 The Role of International Organizations

In the management of risk associated with chemicals, society has to be given a much greater awareness of what is involved. The United Nations agencies, WHO, ILO, UNEP, UNIDO, UNESCO, FAO, and others, including the OECD and the EEC, need to work together increasingly to reduce and manage the risks inherent in the use, handling, and disposal of chemicals. The UN London Guidelines for the Exchange of Information on Chemicals in International Trade, and the Basle Convention on the Control of Trans-Boundary Movement of Hazardous Wastes and their Disposal will go a long way to improving safety in international trade.

3.3 Waste Minimization

As part of international risk management, there have to be new concepts and principles for the environmentally sound management of toxic chemicals and wastes, particularly hazardous waste. The use of chemicals that have toxic properties needs to be minimized since the use of certain chemicals as intermediates is vital to the chemical industry and hence their use cannot be eliminated. Risk management, coupled with greater efficiency in use and careful handling, storage, and transportation, can only result in reduced hazardous waste generation.

We have learned that we cannot use the Earth as the ultimate repository of whatever wastes we generate. Thus we are faced with an enormous requirement for effort in waste minimization that must affect ultimately our daily lives in numerous ways. Nevertheless, wastes do have to be disposed of someplace, somehow. Unfortunately, rather than progressing with this difficult task, we are mired by the concept of siting each and every landfill or incinerator any place but in our own back yard.

3.4 Application of Risk Management to Waste Disposal

The application of risk management to waste is of paramount importance. Wastes are by-products from inefficient and/or incomplete reaction/production processes, resulting from the manufacture of finished goods from their constituent raw materials. There is no specific intention to manufacture waste; it is an unintentional on-cost. Such substances, if they cannot be used in some other processes, have no economic use or value. If inappropriately managed they can be become expensive liabilities.

Industry, governments, the international agencies, and the professional bodies all have a role and a need to be involved in prescribing the most effective control measures to ensure that the risks arising from the disposal of chemical waste are properly managed.

3.5 The Complexity of Problems in Waste Handling

The quantity, complexity, and often the unknown nature of hazardous chemical wastes lead to many problems, accidents, and serious situations, due to the improper disposal of these wastes. These problems include:

Deficiencies in the assessment, management, and prediction of health risks;

Difficulty in defining acceptable exposure levels;

Inadequate management locally, nationally, and internationally on the prevention and management of both health and environmental aspects;

Uncoordinated investigations and regulatory approaches at both national and international levels (IUPAC should have a role to play here);

Difficulties with inadequate information sources both in quality and quantity.

In developing countries, the situation is often more serious since there are inadequate facilities to access the information and expertise to assess the hazards of chemicals. Matters are compounded by lack of knowledge on the nature and purity of of imported and domestically synthesized or formulated products. There is increasing extensive use of chemicals by inadequately informed or trained operatives employed by small cottage industries or self–employed, in addition to similar exposure problems from DIY and domestic household exposure.

3.6 Training and Information

One of the most important aspects of the foregoing is the requirement for good and adequate facilities for training and for the dissemination of information. The World Health Organization's International Programme for Chemical Safety, the International Labour Office Health Information System, and in particular, The United Nations Environment Programme International Register for Potentially Toxic Substances, publish many excellent works. For example, the UNEP/IRPTC Reviews of the Soviet Literature on Toxicity and Hazards of Chemicals are of value to both developed and developing countries. Equally, the detailed monographs of the International Agency for Research on Cancer Monographs on The Evaluation of the Carcinogenic Risk of Chemicals to Humans are excellent, peer reviewed works on aspects of the hazardous properties of chemicals, and, as described in more recent monographs, on chemical processes or diffuse activities such as smoking, welding, printing, chlorination of water, and drinking of tea, coffee, and mate.

The London Guidelines will be of particular importance in promoting increased chemical safety through the exchange of information on international trade in chemicals. The general provisions will enhance the sound management of chemicals through exchange of scientific, technical, economic, and legal information, and in particular provide specific guidance on those chemicals which have been banned or severely restricted. Of special importance will be the *prior informed consent* procedure which will enable governments to concentrate management activities on the exchange of information on such chemicals. This will enable greater opportunities for governments to establish their priorities towards safe management of chemicals.

In order to achieve the requirements of risk management there will need to be increased efforts in both monitoring and review. Risk management is a dynamic process and must never be allowed to be static as this will lead to stagnant thinking. It should be remembered that criticism is never inhibited by ignorance. Effective risk management will assist the establishment both of national and international systems in chemical safety and in the longer term lead to harmonization of different existing systems both national and international. This in turn will lead to improved codes of practice for the prevention of major industrial disasters.

Companies, in undertaking environmental audits, will need to specify the likely consequences of a disaster and whether it can be dealt with by their own, local or national facilities, or whether international assistance may be required. The latter could be of particular importance when high technology plants are sited in developing countries which may not yet have acquired the necessary disaster control hierarchy.

3.7 Safe Use of Chemicals

Risk management of chemicals and hence the safe use of chemicals can only be achieved by a systematic approach and the provision of data on both existing and new chemicals. This should include:

i) Internationally agreed guidelines for listing of chemicals;

ii) Application of Good Laboratory Practice;

iii) Minimum pre-marketing sets of data;

iv) Mutual acceptance of data;

v) Recommendations for the exchange of information on chemicals;

vi) Introduction of internationally acceptable qualifications and training schemes;

vii) An internationally recognized scheme for the registration of chemists in health, safety and environmental matters (The Royal Society of Chemistry's Registered Health and Safety Specialists and Registered Professional Water Chemists is a step in the right direction);

viii) Increased international activity in chemical risk management;

3.8 Lack of Awareness

In general, there is a lack of full awareness and application of proper management techniques of health and environmental problems. Overall, there is a lack of trained scientific and management staff and lack of research devoted to these problems and this is being exacerbated by the current recession in developed countries.

Adequate note is not being taken with regard to substitution of newer and less persistent or toxic chemicals, especially when these necessitate changes in knowledge base and/or technologies.

There is a much greater requirement for the provision of adequate scientific information both in presentation and dissemination, and in a form which is readily understandable at international, national, local, and user levels. The use of pictograms without adequate training can lead to these being interpreted in the opposite way to that intended.

Improving the communication of risk information between lay persons, technical experts, and decision makers is a paramount task for research in risk analysis and in policy making. In particular, studies in risk perception must examine the judgements people make when they are invited to characterize and evaluate hazardous activities and techniques. It cannot be assumed necessarily that those who promote and regulate health and safety understand how people respond to, or think about risk.

3.9 Protection of Both Workplace and the External Environment

In order to elaborate a sound strategy to promote both workplace and environmentally sound management of toxic discharges, consideration has to be given to the following:

The right to information for both workers and the public;

Risk reduction or minimization or phasing out of hazardous chemicals and processes, and the development of environmentally safe processes and products.

In particular, for the developing countries, there has to be increased provision of both comprehensive and readily usable information on all the safety, health, and environmental effects of chemicals.

Great improvements are necessary in international co-operation and co-ordination within the UN and other international bodies, not excluding IUPAC, one of the sponsors of the conference on which this book is based.

Protection methods for the environment must not pit man against nature; environment protection must be reinforced concurrently with economic growth. It is difficult for us not to be violent with nature, when we are violent amongst ourselves.

4 CONCLUSIONS

In order to achieve the foregoing the following actions are required:

Improved capability on an international and national level for the management of chemicals through improved and readily enforceable legislative systems, infrastructure, education and training, monitoring, information retrieval, and assessment;

Improvement in the co-ordination of management of both the workplace and health and environmental risks associated with chemicals (governments should consider this as the duty of one and only one department);

Development of improved criteria for decision making and the setting of acceptable limits for chemicals emitted to any environment;

Development of sound principles of risk management for chemicals in production, formulation, transportation, storage, use, and disposal;

Strengthened mechanisms for accident prevention and for emergency response;

Establishment of priority setting schemes, especially for chemicals of global concern and the ongoing evaluation of such schemes;

Improvement in monitoring facilities, bearing in mind that even when using the most sophisticated techniques it is only possible to monitor some 15% of organic pollutants;

International availability of data on the quantities of chemicals synthesized, traded, and their registration. This must include a notification procedure on a global basis for new chemicals;

Harmonized international labelling;

Technology exchange with particular reference to chemical safety, together with the means of the assessment and management techniques employed;

Promotion of scientific resources to improve assessment techniques;

Acceptance by industry of the responsibility in risk management by making adequate data available on safety;

Implementation of product stewardship and duty of care from 'cradle to grave'.

These actions must take into account the critical pathways and mechanisms for global environmental protection and protection of human and national resources, *viz.* land, marine and fresh water, air, climates, the biological diversity, and living and working conditions. The enormous economic and social benefits obtained from chemicals must always be considered in perspective with risk. Risk assessments and risk management must never mislead the public about the real trade-offs necessary between risks, costs, and benefits, and hence distort priorities.

Reading List

The publications in this list are arranged in alphabetical order of the first-named author/editor or, where no author/editor is named, in alphabetical order of title.

Abstracts of the American Industrial Hygiene Conference, Montreal, 1987, American Industrial Hygiene Association, 1987, 500pp.

Abstracts of the American Industrial Hygiene Conference, St. Louis, 1989, American Industrial Hygiene Association, Akron, Ohio, 1989, 210pp.

Abstracts of the Third International Conference on Safety Evaluation and Regulation of Chemicals, Zürich, 1984, Abstracts, Bio-Research Institute Inc., Cambridge, Mass., 1984, 36pp.

'Casarett and Doull's Toxicology - The Basic Science of Poisons', 4th Edn., ed. M.O. Amdur, J. Doull, and C.D. Klaasen, Pergamon Press, New York, 1991.

J.G. Atherton, 'The Bulk Storage and Handling of Flammable Gases and Liquids', Oyez Publishing, 1980, 76pp.

S. Kris Bandall, G.J. Marco, L. Golberg, and M.L. Leng, 'The Pesticide Chemist and Modern Toxicology', ACS Symposium No. 160, American Chemical Society, Washington, DC, 1981.

S.M. Bartell, R.H. Gardner, and R.V. O'Neill, 'Toxicological Risk in Aquatic Ecosystems', Lewis Publishers, Boca Raton, 1990.

V.K. Brown, 'Acute Toxicity in Theory and Practice: With Special Reference to the Toxicity of Pesticides', Wiley, New York, 1980.

G.C. Butler, 'Principles of Ecotoxicology, SCOPE 12', Wiley, New York, 1978.

J. Cairns, 'Biological Monitoring in Water Pollution', Pergamon Press, Oxford, 1982.

E.J. Calabrese, 'Methodological Approaches to Deriving Environmental and Occupational Health Standards', Wiley, New York, 1978.

E.J. Calabrese and E. Kenyon, 'Air Toxics and Risk Assessment', Lewis Publishers, Boca Raton, 1990.

'Carcinogenic Risk Assessment of Pesticides', International Group of National Associations of Manufacturers of Agrochemical Products, Technical Monograph No. 12, Brussels, 1987, 7pp.

A.D. Chambers, 'Internal Auditing: Theory and Practice', Pitman Books, 1981, 368pp.

S.S. Chissick and R. Derricott, 'Occupational Health and Safety Management', Wiley, 1981, 705pp.

'Patty's Industrial Hygiene and Toxicology', 3rd Rev. Edn., Vol. 1, ed. G.D. Clayton and F.E. Clayton, Wiley Interscience, New York, 1978.

'Patty's Industrial Hygiene and Toxicology', 3rd Rev. Edn., Vol. 2 (A,B,C), ed. G.D. Clayton and F.E. Clayton, Wiley Interscience, New York, 1981.

'Patty's Industrial Hygiene and Toxicology', Vol. 3, Theory and Rationale of Industrial Hygiene Practice, 2nd Edn., Parts A (The Work Environment) and B (Biological Responses), ed. G.D. Clayton and F.E. Clayton, Wiley Interscience, New York, 1978.

Confederation of British Industry, 'Developing a Safety Culture. Business for Safety', CBI, London, 1990.

'Environmental Risk Analysis for Chemicals, ed. R.A. Conway, Van Nostrand Reinhold, New York, 1982.

T. Corfield, 'Safety Management', The Supervisors' 'Do-it-Yourself' Series No. 6, Institute of Supervisory Management, Lichfield, 1987, 266pp.

'Costs of Industrial Accidents: A Basic Reading List', 2nd Edn., Royal Society for the Prevention of Accidents, Library Bibliography No. 5, Birmingham, 1981, 6pp.

'Directory of Safety Related Computer Resources', Vol. 1, 'Software and References', American Society of Safety Engineers, Des Plaines, Ill., 1987, 70pp.

'Dow's Fire and Explosion Index Hazard Classification Guide', American Institute of Chemical Engineers, New York, 1987.

J.H. Duffus, 'Environmental Toxicology, Arnold, London, 1980.

Environmental Program of the Netherlands, 1986–1990, Netherlands Ministry of Housing, Physical Planning and Environment, and Netherlands Ministry of Agriculture and Fisheries, Ministry of Housing, Physical Planning and Environment, The Hague, 1985, 169pp.

'EPA Study of Asbestos-Containing Materials in Public Buildings, A Report to Congress', United States Environmental Protection Agency, Washington, DC, 1988.

U. Forster and G.T.W. Wittman, 'Metal Pollution in the Aquatic Environment', Springer-Verlag, Berlin, 1979.

Fourth South East Asian and Pacific Fire Safety Conference (SEAPAC IV) and 15th General Conference of the International Fire Chiefs' Association of Asia (IFCAA), Singapore, 1988, Conference Papers 2 November 1988, Institution of Fire Engineers and International Fire Chiefs' Association of Asia, Institution of Fire Engineers, Singapore, 1988, 3 volumes.

D.S. Gloss and M.G. Wardle, 'Introduction to Safety Engineering', Wiley, 1984, 612pp.

P.D. Goulden, 'Environmental Pollution Analysis', Heyden, London, 1978.

F.E. Guthrie and J.J Perry, 'Introduction to Environmental Toxicology', Blackwell, Oxford, 1980.

'Reliability on the Move: Safety and Reliability in Transportation', ed. G.B. Guy, Proceedings of the Safety and Reliability Society Symposium, Bath, 1989, Elsevier, 1989, 268pp.

'Handbook of Industrial Safety and Health', Trade and Technical Press, Morden, UK, 1980, 576pp.

'Health and Safety Component of Environmental Impact Assessment', Report on a WHO Meeting, Copenhagen, 1986, Environmental Health No. 15, World Health Organization, Copenhagen, 1987, 92pp.

H.W. Heinrich, D. Peterson, *et al*. 'Industrial Accident Prevention: A Safety Management Approach', 5th Edn., McGraw-Hill, 1980, 468pp.

M. Henderson, 'Living with Risk', The British Medical Association Guide, John Wiley, Chichester, 1987.

'Understanding our Environment', ed. R.E. Hester, The Royal Society of Chemistry, London, 1986.

C. Hohenemser and J.X. Kasperson, 'Risk in The Technological Society', AAAS Selected Symposium No. 65, American Association for the Advancement of Science, Westview Press, Boulder, Col., 1982, 339pp.

M.W. Holdgate, 'A Perspective of Environmental Pollution', Cambridge University Press, Cambridge, 1979.

HSE, 'The Tolerability of Risk from Nuclear Power Stations', HMSO, London, 1988.

HSE, 'Quantified Risk Assessment: its Input to Decision Making, HMSO, London, 1989.

HSE, 'Risk Criteria for Land-use Planning in the Vicinity of Major Industrial Hazards', HMSO, London, 1990.

'Progress in Pesticide Biochemistry and Toxicology', multivolume series, ed. D.H. Hutson and T. R. Roberts, John Wiley, Chichester, 1981 onwards.

H.P.A. Illing, 'Toxicology and Disasters', in 'A Textbook of Basic and Applied Toxicology', ed. B. Ballantyne, T.C. Marrs, and P.M. Turner, MacMillan, Basingstoke, 1992.

O.P. Kharbanda and E.A. Stallworthy, 'Safety in the Chemical Industry: Lessons from Major Disasters', Heinemann, 1988, 345pp.

P.R. Kleindorfer and H.C. Kunreuther, 'Insuring and Managing Hazardous Risks: From Seveso to Bhopal and Beyond', Springer-Verlag, Berlin, 1987, 543pp.

J. Locke and K.L.W. Saxton, Draft Outline Syllabus for the New Diploma in Occupational Safety and Health: Appendix to the Consultative Document on Qualifications in Occupational Safety and Health, National Examination Board in Occupational Safety and Health, Leicester, 1987, 51pp.

'Risk Assessment at Hazardous Waste Sites', ed. F.A. Long and G.E. Schweitzer, Based on a symposium sponsored by the ACS Committee on Environmental Improvement at the 183rd meeting of the American Chemical Society, Las Vegas, Nevada, ACS Symposium No. 204, American Chemical Society, Washington, DC, 1982.

J.S. Maini, M. Peltu, *et al.*, 'Regulating Industrial Risks: An Executive Summary of a Workshop', Commission of the European Communities, Joint Research Centre, Executive Report 8 (EUR 10300), Luxembourg, 18pp.

H.F. Mark, D.F. Othmer, C.G. Overberger, and G.T. Seaborg, 'Kirk-Othmer Concise Encyclopaedia of Chemical Technology', Wiley Interscience, New York, 1989.

V.C. Marshall, 'Major Chemical Hazards', Ellis Horwood, Chichester, 1987.

M.H. Martin and P.J. Coughtrey, 'Biological Monitoring of Heavy Metal Pollution', Applied Science Publishers, London, 1982.

F. Moriarty, 'Ecotoxicology – The Study of Pollutants in Ecosystems', 2nd Edn., Academic Press, London, 1988.

'Major Chemical Disasters – Medical Aspects of Management', ed. V. Murray, Proceedings of a meeting arranged by the Section of Occupational Medicine of the Royal Society of Medicine, London, 1989, Royal Society of Medicine, London, 1990.

A.S. Murty, 'Toxicity of Pesticides to Fish', 2 Volumes, CRC Press, Boca Raton, 1986.

National Research Council, 'Risk Assessment in the Federal Government Managing the Process', National Academy Press, Washington, DC, 1983.

NIOSH/OSHA, 'Pocket Guide to Chemical Hazards', US National Institute of Occupational Safety and Health (NIOSH)/Occupational Safety and Health Agency (OSHA), Washington, DC, 1990.

'Occupational Safety and Health: A Basic Reading List', 10th Edn., Royal Society for the Prevention of Accidents, Library Bibliography No. 1, Birmingham, 1987, 80pp.

OECD, 'Guidelines for Testing of Chemicals', Organization for Economic Cooperation and Development (OECD), Paris, 1982 onward.

H. Otway and M. Peltu, 'Regulating Industrial Risks: Science, Hazards and Public Protection', Commission of the European Communities, Joint Research Centre, International Institute for Applied Systems Analysis, Butterworth, 1985, 181pp.

J.R. Parkinson, 'The Role of the Risk Manager in Industry and Commerce', Keith Shipton Developments Special Study No. 6, Woodhead–Faulkner, Cambridge, 1976, 40pp.

D.J. Paustenbach, 'The Risk Assessment of Environmental and Human Health Hazards: A Textbook of Case Studies', Wiley, 1989.

'Ecotoxicology Testing for the Marine Environment', 2 Volumes, ed. G. Persoone, E. Jaspers, and C. Claus, State University of Ghent and Institute for Marine Scientific Research, Ghent and Bredene, 1984.

'Quantified Risk Assessment', Cancawe Report No. 88/56, Conservation of Clean Air and Water – Europe, The Hague, 1988, 24pp.

G.M. Rand and S.R. Petrocelli, 'Fundamentals of Aquatic Toxicology', McGraw–Hill, Washington, DC, 1984.

'A Review of Risk Assessment Methodologies', Report for the Sub-committee on Science, Research and Technology Transmitted to the Committee on Science and Technology, US House of Representatives, Ninety Eighth Congress, First Session, Serial B, Library of Congress, USGPO, Washington, DC, 1983, 75pp.

M.L. Richardson, 'Nitrification Inhibition in the Treatment of Sewage', The Royal Society of Chemistry, London, 1985.

'Toxic Hazard Assessment of Chemicals', ed. M.L. Richardson, The Royal Society of Chemistry, London, 1986.

'Risk Assessment of Chemicals in the Environment', ed. M.L. Richardson, The Royal Society of Chemistry, London, 1988.

'Chemistry, Agriculture and the Environment', ed. M.L. Richardson, The Royal Society of Chemistry, Cambridge, 1991.

J.R. Ridley, 'Safety at Work, 3rd Edn., Butterworth-Heinemann, 1990.

A.J. Rowland and P. Cooper, 'Environment and Health', Arnold, London, 1983.

The Royal Society of Chemistry, 'The Agrochemicals Handbook', 3rd Edn., Cambridge, 1991.

The Royal Society of Chemistry, 'Chemical Safety Data Sheets, Volume 1 - Solvents', The Royal Society of Chemistry, Cambridge, 1989.

The Royal Society of Chemistry, 'Chemical Safety Data Sheets, Volume 2 - Main Group Metals and their Compounds', The Royal Society of Chemistry, Cambridge, 1990.

The Royal Society of Chemistry, 'Chemical Safety Data Sheets, Volume 3 - Corrosives and Irritants', The Royal Society of Chemistry, Cambridge, 1990.

The Royal Society of Chemistry, 'Chemical Safety Data Sheets, Volume 4 - Toxic Chemicals', The Royal Society of Chemistry, Cambridge, 1991.

The Royal Society Study Group, 'Risk Assessment', The Royal Society, London, 1983.

'Safe Products: The Consumer Protection Act and How it Will Work in Practice', Legal Studies and Services Ltd., London, 1987.

'Dangerous Goods Movements', Proceedings of the 1984 Waterloo Workshop, ed. J.H. Shortreed, University of Waterloo, University of Waterloo Press, 1985, 235pp.

'The COSHH Regulations: A Practical Guide', ed. D. Simpson and W.G. Simpson, The Royal Society of Chemistry, Cambridge, 1991, 192pp.

'State of the Environment': An Assessment at Mid-Decade. A Report from the Conservation Foundation', Conservation Foundation, Washington, DC, 1984, 586pp.

A. St.John-Holt, 'Health and Safety: Towards the Millennium', Institution of Occupational Safety and Health, IOSH Publishing, Leicester, 1987, 124pp.

D.L. Stoner, J.B. Smathers, *et al.*, 'Engineering a Safe Hospital Environment', (Biomedical Engineering and Health Systems), Wiley Interscience, 1982, 195pp.

'Successful Procurement Strategies for Large Real Time Systems: A Guide to their Construction', Starts Purchasers' Group, Manchester, 1988, 49pp.

M.J. Suess, 'Examination of Water for Pollution Control', 3 Volumes, Pergamon, Oxford, 1982.

'Toxic Substances and Human Risk; Principles of Data Interpretation', ed. R.G. Tardiff and J.V. Rodricks, Life Science Monographs, Plenum Press, New York, 1987.

'Technologies and Management Strategies for Hazardous Waste Control', United States Office of Technology Assessment, PB83-189241, 1983, 407pp.

P.E. Terry, 'Employers Scrutinizing Health Plans for Value and Impact on Employees', *Occup. Health. Saf.*, 1990, **59**, No.3, 45, 47-51.

L. Tronstad, J.K. Lund, *et al.*, 'Risk Analysis - New Concepts. Experience from Previous Analyses and Presentation of Recommended Method for Conducting Risk Analysis', Veritas Report No. 82-0235, Royal Norwegian Council for Scientific and Industrial Research, Det Norske Veritas, Oslo, 1982, 54pp.

G.W. Underdown, 'Practical Fire Precautions', 2nd Edn., Gower, 1979, 559pp.

A. Wellburn, 'Air Pollution and Acid Rain: the Biological Impact', Longman Scientific and Technical, Harlow, 1988.

P. Wexler, 'Information Resources in Toxicology', Elsevier North Holland, New York, 1982.

D.M. Wilson, 'A Bibliography of Occupational Health and Safety', British Safety Council, London, 1989.

World Health Organization Regional Office for Europe, 'Health Aspects of Chemical Safety Series', WHO Regional Office for Europe, Copenhagen, 1982–1984.

World Health Organization Regional Office for Europe, 'Environmental Health Series', WHO Regional Office for Europe, Copenhagen, 1985 onward.

World Health Organization, 'IPCS Environmental Health Criteria Series', WHO, Geneva, 1976 onward.

'The Pesticide Manual – A World Compendium', 9th Edn., ed. C.R. Worthing and R.J. Hance, British Crop Protection Council, Farnham, 1991.

Subject Index